OPTICAL AND FIBER COMMUNICATIONS REPORTS

For further volumes:
http://www.springer.com/series/4810

OPTICAL AND FIBER COMMUNICATIONS REPORTS

The Optical and Fiber Communications Reports (OFCR) book series provides a survey of selected topics at the forefront of research. Each book is a topical collection of contributions from leading research scientists that gives an up-to-date and broad-spectrum overview of various subjects.

The main topics in this expanding field will cover for example:

- specialty fibers (periodic fibers, holey fibers, erbium-doped fibers)
- broadband lasers
- optical switching (MEMS or others)
- polarization and chromatic mode dispersion and compensation
- long-haul transmission
- optical networks (LAN, MAN, WAN)
- protection and restoration
- further topics of contemporary interest.

Including both general information and a highly technical presentation of the results, this series satisfies the needs of experts as well as graduates and researchers starting in the field. Books in this series establish themselves as comprehensive guides and reference texts following the impressive evolution of this area of science and technology.

The editors encourage prospective authors to correspond with them in advance of submitting a manuscript. Submission of manuscripts should be made to one of the editors. See also http://springeronline.com/series/4810.

Masataka Nakazawa · Kazuro Kikuchi ·
Tetsuya Miyazaki
Editors

High Spectral Density Optical Communication Technologies

Springer

Editors
Distinguished Prof. and Director Masataka Nakazawa
Research Institute of Electrical Communication
Tohoku University
Katahira 2-1-1, Aoba-ku
980-8577 Sendai-shi, Miyagiken
Japan
nakazawa@riec.tohoku.ac.jp

Dr. Tetsuya Miyazaki
National Institute of Information
and Communication Technology
New Generation Network
Research Center
Nukui-Kitamachi, Koganei 4-2-1
184-8795 Tokyo
Japan
tmiyazaki@nict.go.jp

Prof. Kazuro Kikuchi
Department of Electrical Engineering
and Information Systems
University of Tokyo
7-3-1 Hongo, Bunkyo-Ku
Tokyo 113-8656
Japan
kikuchi@ginjo.t.u-tokyo.ac.jp

ISBN 978-3-642-10418-3 e-ISBN 978-3-642-10419-0
DOI 10.1007/978-3-642-10419-0
Springer Heidelberg Dordrecht London New York

Library of Congress Control Number: 2010930425

Cover design: eStudio Calamar S.L.

Printed on acid-free paper

Springer is part of Springer Science+Business Media (www.springer.com)

Preface

The growth of Internet traffic in recent years surpassed the prediction of one decade ago. Data stream in individual countries already reached terabit/s level. To cope with the petabit class demands of traffic in coming years the communication engineers are required to go beyond the incremental improvement of today's technology.

A most promising breakthrough would be the introduction of modulation formats enabling higher spectral efficiency than that of binary on–off keying scheme, virtually the global standard of fiber-optic communication systems. In wireless communication systems, techniques of high spectral density modulation have been well developed, but the required techniques in optical frequency domain are much more complicated because of the heavier fluctuation levels. Therefore the past trials of coherent optical modulation/detection schemes were not successful. However, the addition of high-speed digital signal processing technology is the fundamental difference between now and two decades ago, when trials of optical coherent communication systems were investigated very seriously. This approach of digital coherent technology has attracted keen interest among communication specialists, as indicated by the rapid increase in the pioneering presentations at the post-deadline sessions of major international conferences. For example, 32 terabit/s transmission in a fiber experiment based on this technology was reported in post-deadline session of Optical Fiber Communication Conference (OFC) 2009. The advancement of the digital coherent technologies will inevitably affect the network architecture in terms of the network resource management for the new generation photonic networks, rather than will simply provide with huge transmission capacity.

In order to discuss about the merits and the challenges of various modulation/detection schemes enabling high spectral density transmission, an international workshop was organized under the sponsorship of NICT. "International Workshop on High Density Optical Communication" was held in Tokyo Japan, June 25–26, 2008. Pioneering researchers from all over the world attended the workshop and discussed a wide range of topics from device, modulation format, sub-system, and transmission systems. The presentations and discussions were very successful and we decided to publish a book, in which individual invited speakers contribute to the relevant chapters/sections by extending their views.

The book is aimed at providing with extensive overviews on the theoretical and the experimental aspects of high spectral density optical communication technology, which should be a most promising approach for the new generation large capacity photonic network. Potential readers include researchers in the fields of communication, photonics, microwave, integrated circuits, and network architecture. The book will be useful also to the planners of future info-communication systems either in the industry, public institutions, or universities. The book covers the principles and the historical background of coherent transmission, so it should be valuable for university students at graduate level, who wish to be acquainted with the frontier of optical communication.

The co-editors wish to express their thanks to Dr. Yuichi Matsushima, Vice President of National Institute of Technology (Dr. Matsushima is a Prof. of Waseda University from April 2010), and to Prof. Takeshi Kamiya, Professor Emeritus of the University of Tokyo, for their valuable advice and support.

Sendai-shi, Japan — Masataka Nakazawa
Tokyo, Japan — Kazuro Kikuchi
Tokyo, Japan — Tetsuya Miyazaki
June 2010

Contents

Contributors

C.R. Doerr Bell Laboratories, Alcatel-Lucent, Holmdel, NJ 07748, USA, crdoerr@alcatel-lucent.com

Rene-Jean Essiambre Bell Labs, Alcatel-Lucent, Paris, France; 209 Holmdel-Keyport Rd., Holmdel, NJ 07733, USA, rjessiam@alcatel-lucent.com

Takeshi Hoshida Fujitsu Laboratories Limited, 1-1 Kamikodanaka 4-Chome, Nakahara-ku, Kawasaki 211-8588, hoshida@jp.fujitsu.com

Sander L. Jansen Nokia Siemens Networks, 81541 Munich, Germany, sander.jansen@nsn.com

Takeshi Kamiya National Institute of Information and Communications Technology (NICT), 4-2-1 Nukui-Kitamachi, Koganei, Tokyo 184-8795, Japan, takeshi.kamiya1@mbn.nifty.com

Tetsuya Kawanishi National Institute of Information and Communications Technology (NICT), 4-2-1 Nukui-Kitamachi, Koganei, Tokyo 184-8795, Japan, kawanish@nict.go.jp

Kazuro Kikuchi Department of Electrical Engineering and Information Systems, University of Tokyo, 7-3-1 Hongo, Bunkyo-Ku, Tokyo 113-8656, Japan, kikuchi@ginjo.t.u-tokyo.ac.jp

Nobuhiko Kikuchi Central Research Laboratory, Hitachi Ltd., Higashi-Koigakubo 1-280, Kokubunji, Tokyo 185-8601, Japan, nobuhiko.kikuchi.ca@hitachi.com

Fumito Kubota National Institute of Information and Communications Technology (NICT), 4-2-1 Nukui-Kitamachi, Koganei, Tokyo 184-8795, Japan, kubota@nict.go.jp

Hiroyuki Matsuura Department of Marketing, Communication Equipment Headquarters, Yokogawa Electric Corporation, Tokyo, Japan, hiroyuki.matsuura@jp.yokogawa.com

Yutaka Miyamoto NTT Network Innovation Laboratories, NTT Corporation, 1-1 Hikari-no-oka, Yokosuka, Kanagawa, 239-0847 Japan, miyamoto.yutaka@lab.ntt.co.jp

Tetsuya Miyazaki New Generation Network Research Center, National Institute of Information and Communications Technology (NICT), 4-2-1, Nukui-Kitamachi, Koganei, Tokyo, 184-8795, Japan, tmiyazaki@nict.go.jp

Takashi Mizuochi Mitsubishi Electric Corporation, 5-1-1 Ofuna, Kamakura, 247-8501 Japan, mizuochi.takashi@df.MitsubishiElectric.co.jp

Itsuro Morita KDDI R&D Laboratories, Japan, morita@kddilabs.jp

Masataka Nakazawa Research Institute of Electrical Communication, Tohoku University, Katahira 2-1-1, Aoba-ku, 980-8577 Sendai-shi, Miyagiken, Japan, nakazawa@riec.tohoku.ac.jp

Reinhold Noé University of Paderborn, EIM-E, Optical Communication and High-Frequency Engineering, Warburger Str. 100, D-33098 Paderborn, Germany, noe@upb.de

Timo Pfau Bell Laboratories, Alcatel-Lucent, 600-700 Mountain Ave, Murray Hill, NJ 07974, USA, timo.pfau@alcatel-lucent.com

Jens C. Rasmussen Fujitsu Laboratories Limited, 1-1 Kamikodanaka 4-Chome, Nakahara-ku, Kawasaki 211-8588, jens.rasmussen@jp.fujitsu.com

Shiro Ryu Network Division, SoftBank Telecom Corporation, 1-9-1, Higashi-shimbashi, Minato-ku, Tokyo 105-7316, Japan, shiro.ryu@tm.softbank.co.jp

William Shieh The University of Melbourne, Victoria 3010 Australia, w.shieh@ee.unimelb.edu.au

Yasuyuki Takatori NTT Network Innovation Laboratories, NTT Corporation, 1-1 Hikari-no-oka, Yokosuka, Kanagawa, 239-0847 Japan, takatori.yasushi@lab.ntt.co.jp

Hiroyuki Uenohara Precision and Intelligence Laboratory, Tokyo Institute of Technology, 4259 Nagatsuta, Midori-ku, Yokohama 226-8503, Japan, uenohara.h.aa@m.titech.ac.jp

Peter J. Winzer Bell Labs, Alcatel-Lucent, Paris, France; 209 Holmdel-Keyport Rd., Holmdel, NJ 07733, USA, peter.winzer@alcatel-lucent.com

Xingwen Yi School of Communication and Information Engineering, University of Electronic Science and Technology of China, Chengdu, Sichuan, 611731, China, xwyi@uestc.edu.cn

Part I
Overview and System Technologies

Chapter 1
Social Demand of New Generation Information Network: Introduction to High Spectral Density Optical Communication Technology

Takeshi Kamiya, Tetsuya Miyazaki, and Fumito Kubota

Abstract In this section, first, current situation of traffic growth and penetration of broadband services are described. Then social demand, technical issues, and research trend for future information network in the United States, Europe, and Japan are described. Finally, a detailed construction of this book is introduced.

1.1 Achievements and Challenges of Fiber-Optic Communication Technology

During the latter half of twentieth century, several technological innovations influenced fundamentally the structure of our society. Historians may list such big names as atomic bombs, computers, transistors, plastics, and antibiotics. Emerging later in the fourth quarter century, we have also semiconductor integrated circuits, robotics, and fiber-optic communication.

The fundamental influence of the fiber-optic network is visualized by comparing our daily life of two decades ago and today. Daily communications in the past depended on letters, telephones, and facsimiles, in contrast to the present heavy dependence on e-mails with the attached electronic files, which can migrate to anywhere in the world instantaneously. The global interconnection naturally changed the fundamental structure of world economy into real unity. The enterprises of different countries can be either our collaborators or our competitors very easily. Good news as well as bad news at one place propagate all over the world and influence our daily life seriously.

This revolutionary change of telecommunication system has been brought about by the impetus of technological breakthroughs. The first phase of the development was initiated in around 1970 by the realization of ultralow loss quartz fiber and the continuous operation of semiconductor lasers. It took about one decade for the

---

T. Kamiya (✉)
National Institute of Information and Communications Technology (NICT),
4-2-1 Nukui-Kitamachi, Koganei, Tokyo 184-8795, Japan
e-mail: takeshi.kamiya1@mbn.nifty.com

M. Nakazawa et al. (eds.), *High Spectral Density Optical Communication Technologies*, Optical and Fiber Communications Reports 6, DOI 10.1007/978-3-642-10419-0_1,

fiber-optic transmission systems to replace and upgrade the microwave transmission system. In Japan the nationwide network was installed in 1985, whose traffic capacity was 400 Mb/s using single wavelength and time division multiplexing (TDM) scheme. The second phase of the development was prepared during the latter part of the 1980s by the introduction of erbium-doped fiber amplifiers (EDFAs) and by the precise controllability of oscillation wavelength in distributed-feedback (DFB) semiconductor lasers. With the wavelength division multiplexing (WDM) scheme, one fiber can deliver signals of different wavelength channels in parallel, so that the throughput constraint was greatly relieved. The EDFAs have the advantage of simultaneously amplifying the signals of different channels. By cascading the amplifiers 200 times it became possible to transmit broadband signals through submarine optical cables between continents. (The first trans-pacific optical link employing the cascading optical amplifiers was installed in 1996 with a line rate of 5 Gb/s [1]. However, the recent system is designed to have typically more than several terabits per second total capacity [2]. Here 1 terabit $= 10^{12}$ bits.)

These investments of fiber-optic network infrastructure matched very well with the Internet. The rapid growth of domestic and international communication traffic owes to the establishment of Internet systems. However, because of the increasingly heavier dependence of all kinds of social activities on Internet, the shortage of traffic capacity is the major concern of the telecommunication professionals. After having reached the ultimate performance of TDM/WDM schemes, the new frontier of high spectral density communication technology is attracting keen interest.

1.2 Social Demands Requiring Advanced Photonic Network

According to OECD broadband statistics as of 2008, the number of broadband subscribers in the OECD reached 251 million, an increase of 14% from one year before. This growth increased broadband penetration rates to 21.3 subscriptions per 100 inhabitants. The total subscribers by country are shown in Fig. 1.1 [3].

Taking an example of the Internet traffic in Japan, it has been monotonically increasing since its start. In 2008, the traffic in the core optical network of Japan is reaching 1 Tb/s (Fig. 1.2) [4].

Such rapid growth of traffic reflects the shift of usage from narrow band services (mainly composed of numerical and text data) to broadband services (mainly composed of visual data, including still images and video data). Compared to other countries where broadband penetration level is high, Japan is leading in the service of fiber-to-the-home (FTTH). The number of FTTH subscribers already exceeded 14 million [5]. In addition to the Internet through wired network, fiber networks are also supporting back-haul for wireless access networks including cell phone and digital TV broadcast services. All of these hyper-broadband services are now driving the growth of the communication traffic. Because of the global nature of the Internet, the international traffic from Japan to overseas countries is also growing very fast. In anticipation of heavier demand, there are several Tb/s class trans-oceanic submarine

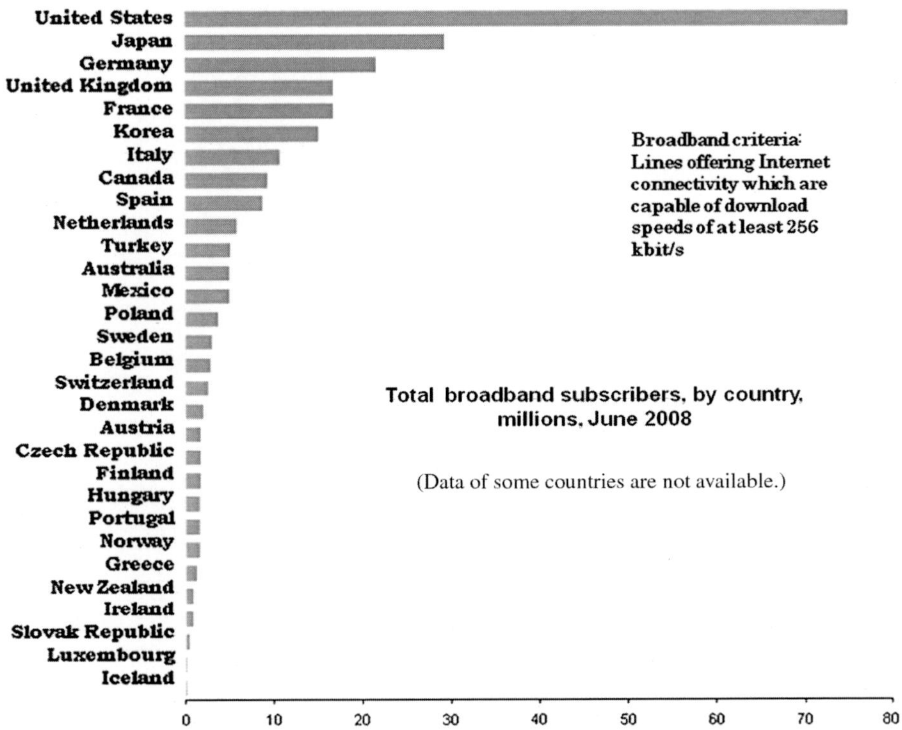

Fig. 1.1 OECD broadband statistics: total broadband subscribers by country

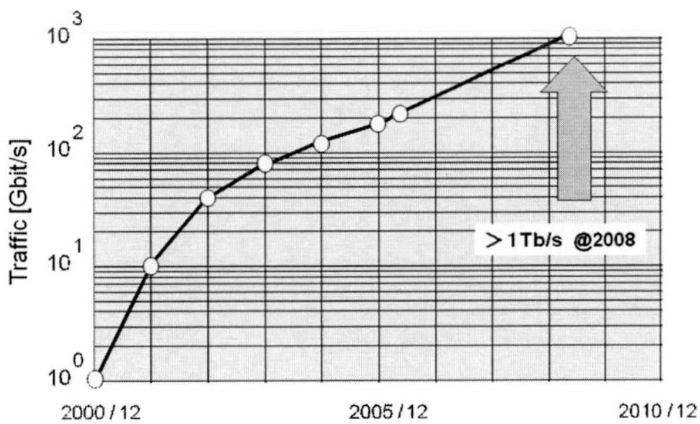

Fig. 1.2 Internet traffic growth in Japan

cable construction projects in the same trans-pacific region [2]. If we can extrapolate the traffic growth rate remaining unchanged, the demand of traffic in 2030 will reach an enormous value of 1 Pb/s (1 petabit = 1000 Tbit).

Figure 1.3 illustrates a sketch of the present and the future photonic networks.

The cell phone services will be upgraded from less than 10 Mb/s level to 1 Gb/s level. The broadcast contents will be expanded to digital cinema of 8 Gb/s. If the

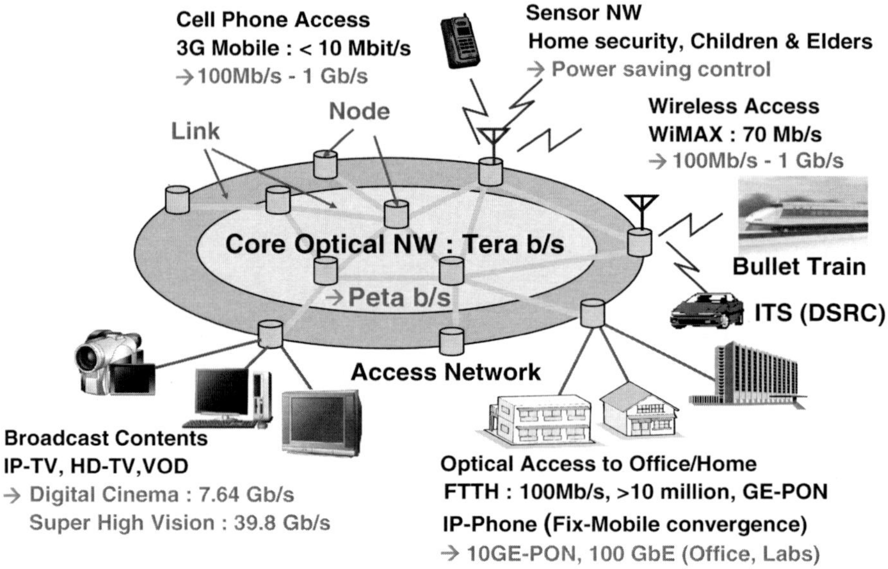

Fig. 1.3 Present and future photonic networks

super high vision trial of 4000 × 4000 pixels will be launched, the bit rate of 40 Gb/s will be required. The FTTH broadband access to offices/home of 100 Mb/s is now used by over 10 million households. The service level will be upgraded to 10–100 Gb/s. The fix-mobile convergence will be further developed. More and more mobile stations (cars, trains, and aircrafts) will be connected to the network. Intelligent transportation systems (ITSs) will incorporate more network technology such as DSRC (dedicated short range communication). The wireless access will be upgraded to 1 Gb/s. The remote surveillance/inspection for offices, factories, and homes will be realized by sensor networks. If applied to electrical power saving systems, the unused electrical appliances are turned to waiting mode of operation by remote control.

However, a simple extrapolation of the present trend will not solve the whole problem. For example, the increasing electrical power consumption in network nodes and the rising possibility of network congestion and failure are becoming serious issues of society. The social demand for future network technology may be summarized by the following to-do-list:

1. Compatibility of high level of service and reliability;
2. Low cost of construction and maintenance;
3. Penetration of services to rural districts and to handicapped users (universal access);
4. Environment-friendly network (minimized pollution and low energy consumption).

1.3 Technical Issues for New Generation Network

In this section we discuss briefly the near future target and the mid-future target. In majority of developed countries including Japan, the information systems are being updated in order to meet the demands of diversified information services such as assurance of quality of services, cost, and broadband channels. This movement is widely called "Next Generation Network (NGN)" [6]. In Japan the "NGN" system is now being deployed, standardized. Some of the initial services have just started. However, it is also widely accepted that NGN cannot afford to solve the whole problems of the existing network management, because not only the modes of services but also the system configurations as well as the management protocols have become formidably complicated after the cumulative evolution of Internet for three decades.

In the United Sates of America "The Global Environment for Network Innovation (GENI)" project has been established under the sponsorship of the National Science Foundation [7]. GENI is defined as a virtual laboratory for exploring future internets at scale, creates major opportunities to understand, innovate, and transform global networks and their interactions with society. A related funding program FIND is stimulating the researchers in creating new ideas for the future networks. In Europe, EU has been promoting science and technology activities through the FP7 framework [8]. One of the major projects of FP7 is the Future Internet. A set of test-bed initiatives have been organized under the FIRE Program [9].

We have been discussing on the further reconstruction of the network with the targeted year of around 2020. In order to discriminate the issue from the currently running "Next Generation Network," this emergent research target is named "New Generation Network (NWGN)" [10]. Its network architecture should be studied based on requirements for the future ubiquitous network society. We have to accelerate R&D on "NWGN" in order to provide the community with advanced and secure ICT infrastructure, which also satisfies the environmental requirements.

Optical technology is one of the most important key technologies for fulfilling these requirements. In the NICT's NWGN Program (AKARI project), a revised version of the conceptual design report was published in 2008 [11]. In this report, Chap. 5 is dedicated to "Basic Configuration of a New Architecture." One of the important concepts in dealing with broadband signals is the "optical packet switching" [12, 13]. The signals on the transmission channel are not converted to electrical signal at the nodes. Conversion processing is performed in the optical level as packets, which are placed on the next transmission channel (O $\rightarrow$ O $\rightarrow$ O). Another important concept is the "lightpath network" [14]. The lightpath from the sender to the receiver is established before the data are transmitted. The path is chosen based on control information of the network itself. The data need not be converted into electrical signal at each node. Some discussions are described concerning the characteristics of either the circuit-switched nodes, the optical packet nodes, or the nodes for lightpath network. Challenges for ideal designing are listed. In short, the advantages of the optical technology in the future network will be best exploited only when the innovations of both the transmission links and switching nodes are fully developed.

1.4 Fundamental Problems of High Spectral Density Modulation Technology

To establish novel photonic networks in the twenty-first century, it will be important not only to employ existing WDM transmission technologies but also to develop new technologies capable of effectively increasing the information transmission capacity per wavelength and also per optical pulse. In transmission links, an increased number of wavelength channels will result in a lack of vacant optical bandwidth in optical repeater amplifiers and will make wavelength management much more complicated. Current optical repeater amplifiers using erbium-doped fiber amplifiers (EDFAs) can amplify the lightwave within the wavelength bandwidth of 30 nm ranging between 1535 and 1565 nm (C-band). Therefore, assuming an attainable spectral efficiency of 0.8 bit/s/Hz, a transmission capacity of only 3 Tb/s per fiber is available [15].

Even if using shorter pulse width ΔT is advantageous for increasing the bit rate in TDM, the corresponding spectral width is broadened proportionally to $1/\Delta T$, so that the number of WDM channels is decreased (Fourier transform principle [16]). Therefore, shortening pulse is of no use in increasing spectral efficiency as far as DWDM and binary modulation scheme are assumed. Within the condition of employing binary codes, the only way to increase the throughput is to enlarge the available spectral width. To enhance the transmission capacity per fiber, a parallel optical repeater configuration involving S band (1460–1530 nm), C band, and L band (1565–1625 nm) was demonstrated in 2001, achieving 11 Tb/s transmission [17]. However, this parallel optical repeater configuration inevitably suffers from a number of constraints such as gain equalization optical loss to obtain flat gain between S-, C-, and L-band amplifiers, and excessive optical loss for dividing/combining these three amplifiers in each optical repeater amplifier.

A possible solution is to introduce multilevel modulation format with or without coherent detection scheme. Although there are still challenging issues, we believe increasing spectral efficiency by introducing multilevel modulation formats is an inevitable choice for the future. This book is dedicated to discuss the fundamental problems of high spectral density optical communication (HSDOC) technology [18]. Major problems are categorized in Table 1.1 together with the corresponding chapters/sections and the authors.

Fundamental problems of HSDOC technology are discussed in Chaps. 2–5, starting from the history of coherent optical communication and its update using advanced digital technology in Chap. 2, followed by the high spectral density achievements in quadrature amplitude modulation (QAM) format in Chaps. 3 and 4, and the record making fiber transmission demonstration in Chap. 5. Various spectral efficient modulation format technologies are reported in Chap. 6. Technical issues of quadrature-phase-shift-keying (QPSK)-based transmission system are discussed in Chap. 7. This book also covers important device technologies such as light sources, integrated modulators, and driver electronics in Chap. 8. Finally, forward error correction techniques for high spectral density optical communications are also covered as an indispensable technology ensuring the overall performance in terms of bit-error-rate.

Table 1.1 Topics of high spectral density optical communication (HSDOC) technology

Category	Topic	Chapter	Author(s)
Fundamentals	Social background	1	T. Kamiya, T. Miyazaki, and F. Kubota
	Coherent detection: history and future	2	K. Kikuchi
	Ultimate HSDOC	3	M. Nakazawa
	Quasi-ultimate HSDOC	4	T. Miyazaki
	High-capacity transmission	5	P.J. Winzer and Rene-Jean Essiambre
Modulation formats	Incoherent multilevel modulation	6	N. Kikuchi
	Optical FDM	7	W. Shieh and Xingwen Yi
	Optical MIMO	8	I. Morita and S.L. Jansen
	No guard interval OFDM	9	Y. Miyamoto and Yasuyuki Takatori
QPSK-based transmission system	Trade-offs between linear and nonlinear impairments	10	T. Hoshida and J.C. Rasmussen
	Real-time digital QPSK	11	T. Pfau and R. Noe
	Control of polarization	12	S. Ryu
Opt-electronic devices	Semiconductor lasers	13	H. Uenohara
	InP integrated devices	14	C.R. Doerr
	LiNbO3 Integrated devices	15	T. Kawanishi
	Digital devices and circuits	16	H. Matsuura
Error correction	Forward eError correction	17	T. Mizuochi

The readers are recommended to refer to the earlier publications on this topic listed in References [19–22].

References

1. W.Ch. Barnet, H. Takahira, J.C. Baroni, Y. Ogi, The TPC-5 Cable Network. IEEE Commun. Mag. **34**(2), 36–40, (1996). http://www.kddi.com/english/corporate/news_release/archive/kdd/press-e96/027.html
2. For example: http://www.dailywireless.org/2008/09/29/trans-pacific-express-completed/ or http://en.wikipedia.org/wiki/Unity_(submarine_cable)
3. http://www.oecd.org/sti/ict/broadband
4. http://www.soumu.go.jp/menu_news/s-news/090227_3.html (In Japanese). This report shows total averaged internet traffic is estimated to be 990 Gb/s, with the annual growth rate of 21.6% at November 2008
5. http://www.soumu.go.jp/main_sosiki/joho_tsusin/eng/Releases/Telecommunications/news090318_1.html
6. "Next Generation Networks Global Standard Initiative" has been established in ITU-T as part of Study Group 13 "Future Network". Information is available from the following website: http://www.itu.int/ITU-T/ngn/
7. For example: http://www.nsf.gov/funding/pgm_summ.jsp?pims_id=500077
8. http://cordis.europa.eu/fp7/dc/index.cfm
9. http://cordis.europa.eu/fp7/ict/fire/
10. New Generation Network Promotion Forum (NWGN Forum) was established in Japan in October 2007. The concept is described in the webpage of NWGN Forum: http://forum.nwgn.jp/

11. AKARI Project group sponsored by NICT issued a document. New Generation Network Architecture AKARI (Conceptual design ver.1.1 in June 2008). The full text is down-loadable from the following website: http://akari-project.nict.go.jp/eng/index2.htm. See particularly Chapt. 5 Basic Configuration of a New Architecture, 5.1. Optical packet Switching and Optical Paths
12. N. Wada, H. Harai, F. Kubota, Optical packet switching network based on ultrafast optical code label process. IEICE Trans. Electron. **E87-C**, 1090–1096 (2004)
13. H. Harai, M. Murata. High speed buffer management for 40 Gb/s based photonic packet Switches. IEEE/ACM Trans. Networking **14**, 191–204 (2006)
14. T. Tachibana, H. Harai, End-to-end lightpath establishment based on rank accounting in multi-domain wdm networks. IEICE Trans. Commun. **E89-B**, 2448–2456 (2006)
15. P.J. Winzer, Modulation and multiplexing in optical communication systems, IEEE LEOS **23**(1), 4–10 (2009)
16. T. Kamiya, F. Saito, O. Wada, H. Yajima (eds.), *Femtosecond Technology*, 2nd edn. Chapt. 1, (Springer, Heidelberg, 1999)
17. K. Fukuchi, T. Kasamatsu, M. Morie, R. Ohira, T. Ito, K. Sekiya, D. Ogasahara, T. Ono, 10.92 = Tb/s (273 × 40 Gb/s) triple-band ultra-dense WDM optical-repeatered transmission experiment. OFC, **4**, PD24 (2001)
18. Technical Digest, Workshop on High Spectral Density Optical Communication for New Generation Network (HDOC-WS 2008), http://www2.nict.go.jp/w/w112/WS080523.pdf
19. T. Okoshi, K. Kikuchi, *Coherent Optical Communication*. (Kluwer/KTK, Dordrecht, 1988)
20. S. Betti, G. De Marchis, E. Iannone, Coherent optical communication systems. (Wiley, New York, 1995)
21. Kean P.H, *Phase modulated optical communication*. (Springer, US, 2004)
22. I.P. Kaminow, T. Li, A.E. Willner, Optical fiber communications VB: systems and networks. (Academic Press/Elsevier, New York, 2008)

Chapter 2
Coherent Optical Communications: Historical Perspectives and Future Directions

Kazuro Kikuchi

Abstract Coherent optical fiber communications were studied extensively in the 1980s mainly because high sensitivity of coherent receivers could elongate the unrepeated transmission distance; however, their research and development have been interrupted for nearly 20 years behind the rapid progress in high-capacity wavelength-division multiplexed (WDM) systems using erbium-doped fiber amplifiers (EDFAs). In 2005, the demonstration of digital carrier phase estimation in coherent receivers has stimulated a widespread interest in coherent optical communications again. This is due to the fact that the digital coherent receiver enables us to employ a variety of spectrally efficient modulation formats such as M-ary phase-shift keying (PSK) and quadrature amplitude modulation (QAM) without relying upon a rather complicated optical phase-locked loop. In addition, since the phase information is preserved after detection, we can realize electrical post-processing functions such as compensation for chromatic dispersion and polarization-mode dispersion in the digital domain. These advantages of the born-again coherent receiver have enormous potential for innovating existing optical communication systems. In this chapter, after reviewing the 20-year history of coherent optical communication systems, we describe the principle of operation of coherent detection, the concept of the digital coherent receiver, and its performance evaluation. Finally, challenges for the future are summarized.

2.1 History of Coherent Optical Communications

This section describes the history of coherent optical communications. Technical details will be summarized in Sect. 2.2, and readers should refer to them appropriately.

K. Kikuchi (✉)
Department of Electrical Engineering and Information Systems, University of Tokyo, 7-3-1 Hongo, Bunkyo-Ku, Tokyo 113-8656, Japan
e-mail: kikuchi@ginjo.t.u-tokyo.ac.jp

M. Nakazawa et al. (eds.), *High Spectral Density Optical Communication Technologies*, Optical and Fiber Communications Reports 6, DOI 10.1007/978-3-642-10419-0_2,

2.1.1 Coherent Optical Communication Systems 20 Years Ago

The research and development in optical fiber communication systems started in the first half of the 1970s. Such systems used intensity modulation of semiconductor lasers, and the optical signal intensity transmitted through an optical fiber was detected by a photodiode, which acted as a square-law detector. This combination of the transmitter and the receiver is called the intensity modulation and direct detection (IMDD) scheme, which has been commonly employed in optical communication systems up to the present date. Such IMDD scheme has a great advantage that the receiver sensitivity is independent of the carrier phase and the state of polarization (SOP) of the incoming signal, which are randomly fluctuating in real systems.

Although the first proposal of coherent optical communications using heterodyne detection was done by DeLange in 1970 [1], it did not attract any attention because the IMDD scheme became mainstream in optical fiber communication systems during the 1970s. On the other hand, in 1980, Okoshi and Kikuchi [2] and Fabre and LeGuen [3] independently demonstrated precise frequency stabilization of semiconductor lasers, which aimed at optical heterodyne detection for optical fiber communications. Figure 2.1 shows the heterodyne receiver proposed in [2] for frequency-division multiplexed (FDM) optical communication systems. Each FDM channel was selected by heterodyne detection with multiple local oscillators (LOs) prepared in the receiver.

Table 2.1 shows the comparison between coherent and IMDD schemes. With coherent receivers, we can restore full information on optical carriers, namely inphase and quadrature components (or amplitude and phase) of the complex amplitude of the optical electric field and the state of polarization (SOP) of the signal. In exchange for such advantages, coherent receivers are sensitive to the phase and SOP of the incoming signal. To cope with this problem, the configuration of coherent systems becomes much more complicated than that of IMDD systems.

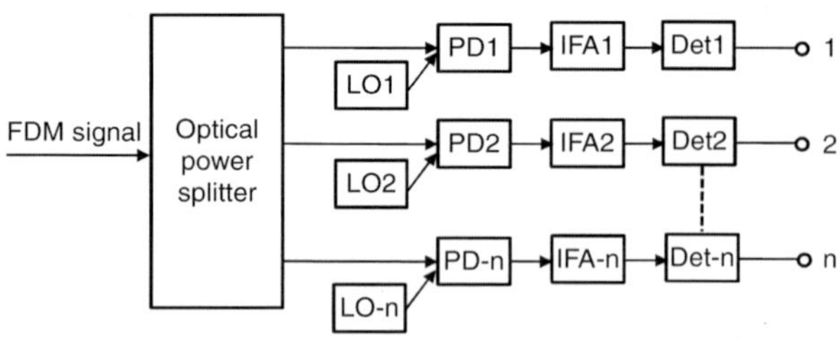

Fig. 2.1 Heterodyne receiver for FDM optical communication systems [2]. LO: local oscillator, PD: photodiode, IFA: intermediate-frequency amplifier, and Det: detector

Table 2.1 Comparison between coherent and IMDD schemes

	Coherent	IMDD
Modulation parameters	I and Q or Amplitude and Phase	Intensity
Detection method	Heterodyne or homodyne detection	Direct detection
Adaptive control	Necessary (carrier phase and SOP)	Not necessary

The center frequency drift of semiconductor lasers for a transmitter and a local oscillator could be maintained below 10 MHz as shown in [2, 3]. Even when the frequency drift of the transmitter laser was suppressed, the carrier phase was fluctuating randomly due to large phase noise of semiconductor lasers. Therefore, narrowing spectral linewidths of semiconductor lasers was the crucial issue for realizing stable heterodyne detection. The method of measuring laser linewidths, called the delayed self-heterodyne method, was invented [4], and the spectral property of semiconductor lasers was studied extensively [5–7]. It was found that the linewidth of GaAlAs lasers was typically in the range of 10 MHz. Such a narrow linewidth together with the precisely controlled center frequency accelerated researches of coherent optical communications based on semiconductor lasers.

The SOP dependence of the receiver sensitivity was overcome by the polarization diversity technique [8]. Each polarization component was detected by orthogonally polarized LOs, and post-processing of the outputs could achieve the SOP-independent receiver sensitivity as shown in Fig. 2.2.

Analyses of the receiver sensitivity of various modulation/demodulation schemes were performed [9, 10]. It was found that the shot-noise-limited receiver sensitivity could be achieved by injecting a sufficient LO power into the receiver to combat against circuit noise. In addition, the use of phase information further improved the receiver sensitivity because the symbol distance could be extended on the IQ plane. For example, let us compare the binary phase-shift keying (PSK) modulation with the binary intensity modulation. When the signal peak power is maintained, the symbol distance on the IQ plane in the PSK modulation format is twice as long as that in the intensity modulation scheme, which improves the receiver sensitivity of the PSK modulation format by 6 dB compared with the intensity modulation scheme. The motivation of R&D of coherent optical communications at this stage lay in this high receiver sensitivity that brought forth longer unrepeated transmission distance.

Many demonstrations of heterodyne systems were reported around 1990. In such systems, the frequency-shift keying (FSK) modulation format was most commonly employed because the semiconductor-laser frequency could be easily modulated

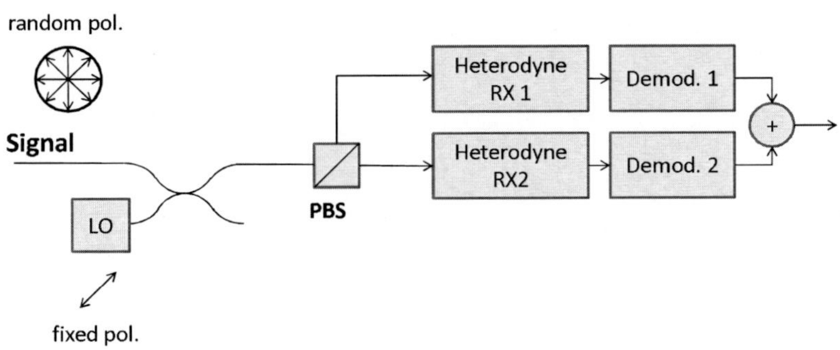

Fig. 2.2 Polarization diversity heterodyne receiver. PBS polarization beam splitter, Demod.: demodulator

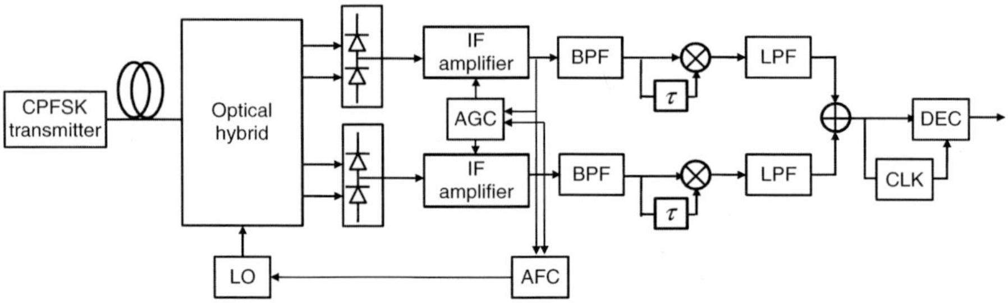

Fig. 2.3 Configuration of optical CPFSK transmitter and receiver [11]. BPF: bandpass filter, LPF: lowpass filter, CLK: clock extraction circuit, AFC: automatic frequency-control circuit, AGC: automatic gain-control circuit, DEC: decision circuit

by direct bias-current modulation. Such frequency modulation was demodulated by differential detection at the IF stage. Field trial of undersea transmission at 2.5 Gbit/s was reported in [11], where advanced heterodyne technologies, such as CPFSK modulation, polarization diversity, automatic frequency control (AFC) of semiconductor lasers, and differential detection, were introduced as shown in Fig. 2.3.

Homodyne receivers were also investigated in the 1980s. The advantage of homodyne receivers was that the baseband signal was directly obtained, in contrast to heterodyne receivers which require a rather high intermediate frequency. Figure 2.4 is an example of the OPLL-type phase-diversity homodyne receiver for BPSK modulation [12]. The most difficult issue for the OPLL-type homodyne receiver is to recover the carrier phase. The optical signal modulated in the PSK format does not have the carrier component; therefore, to recover the carrier phase, the product between the in-phase component $\sin\{\theta_s(t) + \theta_n(t)\}$ and the quadrature component $\cos\{\theta_s(t) + \theta_n(t)\}$ are taken, where $\theta_s(t)$ and $\theta_n(t)$ are phase modulation and phase noise, respectively. Such a product eliminates BPSK modulation, leading to the phase error $\theta_n(t)$ between the transmitter and LO. The phase error is then led to the frequency controlling terminal of LO so that the LO phase tracks the carrier phase. Note that LO acts as a voltage-controlled oscillator (VCO). This type of PLL scheme is well known as the Costas loop. The PLL bandwidth was usually limited below 1 MHz because of large loop delay, and it was difficult to maintain system stability when semiconductor lasers had large phase noise and frequency

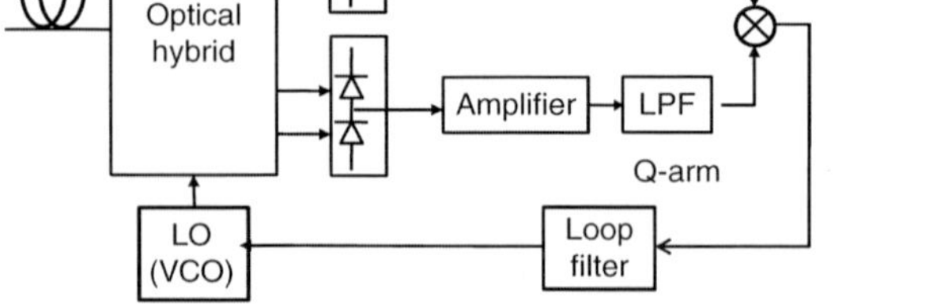

Fig. 2.4 Configuration of Costas OPLL homodyne receiver [12]

drift. This technical difficulty inherent in OPLL has not been solved perfectly even when we use state-of-the-art distributed-feedback (DFB) semiconductor lasers.

In the 1990s, the invention of erbium-doped fiber amplifiers (EDFAs) made the shot-noise-limited receiver sensitivity of the coherent receiver less significant. This is because the carrier-to-noise ratio (CNR) of the signal transmitted through the amplifier chain is determined from the accumulated amplified spontaneous emission (ASE) rather than the shot noise. In addition, even in unrepeated transmission systems, the EDFA used as a low-noise preamplifier eliminated the need for the coherent receiver with superior sensitivity. The coherent receiver actually had advantages other than the high receiver sensitivity. For example, it could cope with multi-level modulation formats and had post-processing functions such as compensation for group-velocity dispersion (GVD) of the fiber. However, such advantages were neither urgent requirements for the system nor cost-effective solutions in the 1990s.

Technical difficulties in coherent receivers had not been solved by that time. The heterodyne receiver required an intermediate frequency (IF), which should be higher than the signal bit rate. The maximum bit rate of the heterodyne receiver was always less than the half of that the square-law detector could achieve. On the other hand, the homodyne receiver was essentially a baseband receiver; however, the complexity in stable locking of the carrier phase drift had prevented its practical applications.

From these reasons, further research and development activities in coherent optical communications have almost been interrupted for nearly 20 years. On the other hand, the EDFA-based IMDD system started to take benefit from wavelength-division multiplexing (WDM) techniques to increase the transmission capacity of a single fiber. The hardware required for WDM networks became widely deployed owing to its simplicity and the relatively low cost associated with optical amplifier repeaters where multiple WDM channels could be amplified simultaneously. The WDM technique marked the beginning of a new era in the history of optical communication systems and brought forth 1,000 times increase in the transmission capacity in the 1990s.

2.1.2 Revival of Coherent Optical Communications

With the transmission-capacity increase in WDM systems, coherent technologies have restarted to attract a large interest over the recent years. The motivation lies in finding methods of meeting the ever-increasing bandwidth demand with multi-level modulation formats based on coherent technologies [13]. The first step of the revival of coherent optical communications research was ignited with the quadrature PSK (QPSK) modulation/demodulation experiment featuring optical in-phase and quadrature (IQ) modulation and optical delay detection [14]. In such a scheme, one symbol carries two bits by using the four-point constellation on the complex plane; therefore, we can double the bit rate, while keeping the symbol rate, or maintain the bit rate even with the halved spectral width.

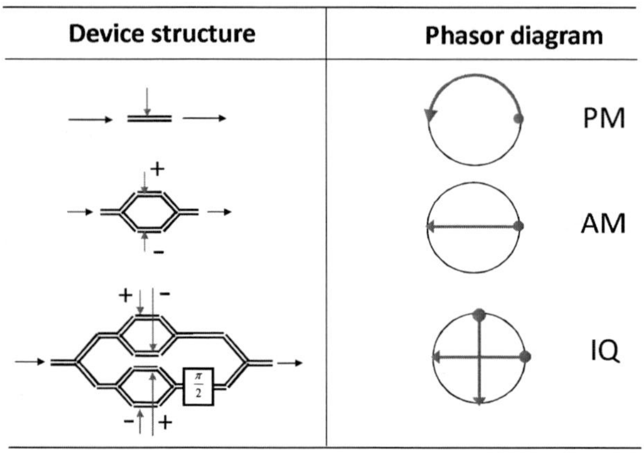

Fig. 2.5 Comparison of the device structure and the phasor diagram among phase modulation, amplitude modulation, and IQ modulation

Figure 2.5 shows the comparison of the device structure and the phasor diagram among phase modulation (PM), amplitude modulation (AM), and IQ modulation. Optical amplitude modulation (AM) can be achieved by using phase modulators in a Mach–Zehnder configuration, which are driven in a push–pull mode of operation [15]. Optical IQ modulation, on the other hand, can be realized with Mach–Zehnder-type push–pull modulators in parallel, between which a $\pi/2$-phase shift is given [16]. The IQ components of the optical carrier is modulated independently with the IQ modulator, enabling any kind of modulation formats. IQ modulators integrated on $LiNbO_3$ substrates are now commercially available.

The optical delay detector is composed of an optical one-bit delay line and a double-balanced photodiode as shown in Fig. 2.6. With such a receiver, we can compare the phase of the transmitted signal with that of the previous symbol and restore the data, which is differentially precoded at the transmitter. Using two such receivers in parallel, we can demodulate the signal IQ components separately without the need for LO. The principle of operation of the DQPSK receiver is shown in Fig. 2.7. Let us assume that the previous symbol constitutes the phase reference. Depending on the phase difference $\Delta\theta$ between the current symbol and the previous symbol, we have the following four cases: (1) $\Delta\theta = 0$, (2) $\Delta\theta = \pi/2$, (3) $\Delta\theta = -\pi/2$, and (4) $\Delta\theta = \pi$. Considering the phase shift of $\pm\pi/4$ in Mach–Zehnder arms in Fig. 2.6, we find that the output from the I port is given as the real part of the phasor

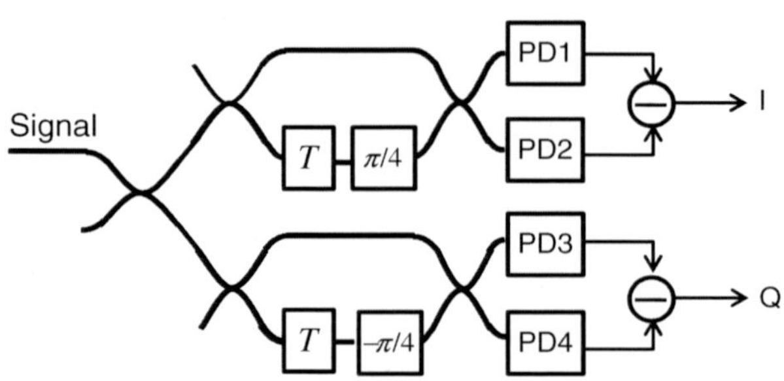

Fig. 2.6 Configuration of differential QPSK receiver. Two differential detectors measure the in-phase and quadrature components of the differential PSK signal

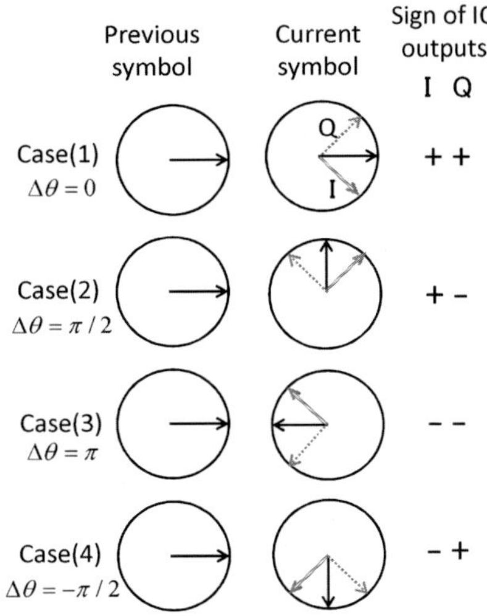

Fig. 2.7 Principle of operation of the DQPSK receiver. From signs of the IQ outputs, we can estimate the phase difference of the QPSK signal

indicated by the double line in Fig. 2.7, whereas that from the Q port as the real part of the phasor indicated by the dotted line. Since signs of these outputs from the IQ ports in Fig. 2.6 have the four different patterns, we can demodulate the DQPSK signal. A number of long-distance WDM QPSK transmission experiments have been reported recently, based on IQ modulation and differential IQ demodulation.

The next stage has opened with high-speed digital signal processing (DSP). In the field of radio communications, digital techniques have been widely introduced into transmitters and receivers. Figure 2.8 shows configurations of (a) DSP-based RF transmitter and (b) receiver. At the transmitter, after appropriate DSP, digital data are converted into two-channel analog data through digital-to-analog converters (DACs), which modulate IQ components of the RF carrier. On the other hand, at the receiver, the transmitted RF signal is mixed with LO, and IQ components are demodulated. Such IQ data are converted to the digital domain by using analog-to-digital converters (ADCs), and symbols are decoded through DSP. Digital signal processing at the transmitter and the receiver enables "software-defined radio communications."

If the RF modulator in Fig. 2.8(a) and the RF mixer in Fig. 2.8(b) are replaced with optical counterparts, which are nothing but the optical IQ modulator and the phase-diversity homodyne receiver, respectively, we can easily imagine the DSP-based optical transmitter and receiver shown in Fig. 2.9. Such an optical transmitter and a receiver were actually reported in [17] and [18–20], respectively.

The recent development of high-speed digital integrated circuits has offered the possibility of treating the electrical signal in a digital signal-processing (DSP) core and retrieving the IQ components of the complex amplitude of the optical carrier from the homodyne-detected signal in a very stable manner. The 20-Gbit/s QPSK signal was demodulated with a phase-diversity homodyne receiver followed by digital carrier phase estimation in [18], although bit-error rate measurements were done still offline. Since the carrier phase is recovered after homodyne detection by means

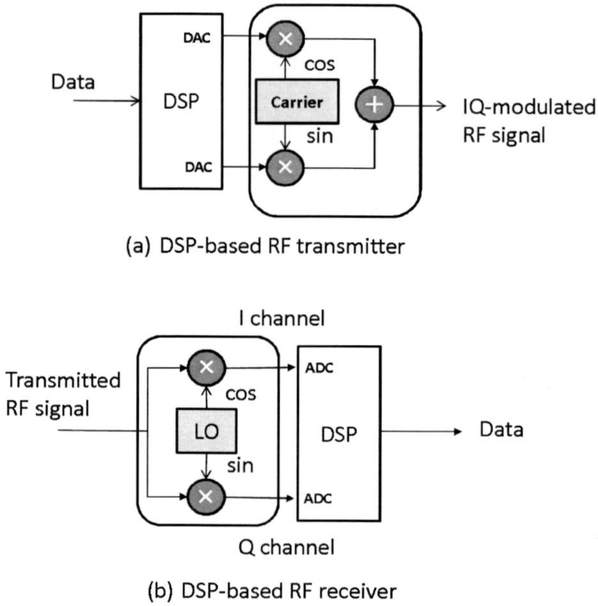

Fig. 2.8 Configurations of (**a**) DSP-based RF transmitter and (**b**) receiver

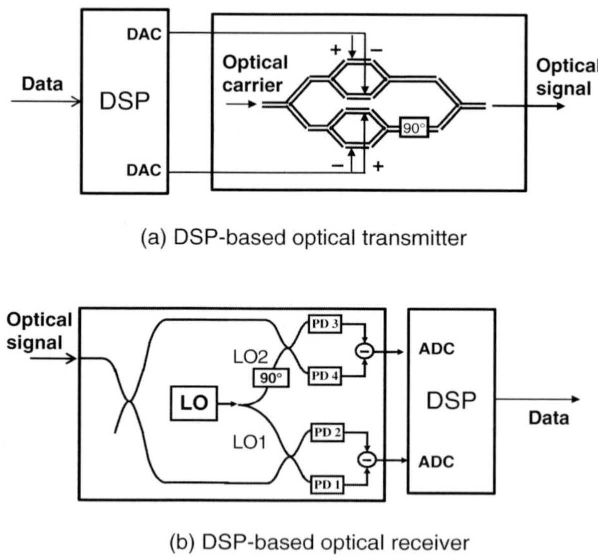

Fig. 2.9 Configurations of (**a**) DSP-based optical transmitter and (**b**) receiver

of DSP, this type of receiver has now commonly been called the "digital coherent receiver." While an optical phase-locked loop (OPLL) that locks the LO phase to the signal phase is still difficult to achieve due to the loop delay problem, DSP circuits are becoming increasingly faster and provide us with simple and efficient means of estimating the carrier phase. Very fast tracking of the carrier phase improves system stability drastically as compared with the OPLL scheme.

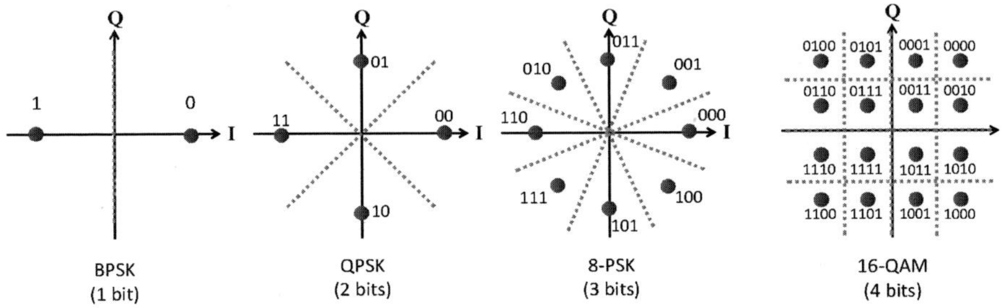

Fig. 2.10 Constellation maps for BPSK, QPSK, 8PSK, and 16QAM modulation formats

Any kind of multi-level modulation formats can be introduced by using the coherent receiver [21]. While the spectral efficiency of binary modulation formats is limited to 1 bit/s/Hz/polarization, which is called the Nyquist limit, modulation formats with M bits of information per symbol can achieve up to the spectral efficiency of M bit/s/Hz/polarization. Although optical delay detection has been employed to demodulate the quadrature phase-shift keying (QPSK) signal ($M = 2$), further increase in multiplicity is hardly achieved with such a scheme. Figure 2.10 shows constellation maps for BPSK, QPSK, 8PSK, and 16QAM formats. These modulation formats can transmit 1 bit, 2 bits, 3 bits, and 4 bits per symbol, respectively. Each symbol is encoded by using the Gray code scheme.

Another and probably more important advantage of the digital coherent receiver is the post-signal-processing function [22]. The IQ demodulation by the digital coherent receiver is the entirely linear process; therefore, all the information on the complex amplitude of the transmitted optical signal is preserved even after detection, and signal-processing functions acting on the optical carrier, such as optical filtering and dispersion compensation, can be performed at the electrical stage after detection.

The polarization alignment is also made possible after detection by introducing the polarization diversity scheme into the homodyne receiver [23]. The complex amplitude of the horizontal polarization and that of the vertical polarization are simultaneously measured and processed by DSP. Polarization demultiplexing and compensation for polarization-mode dispersion (PMD) have also been demonstrated with the digital coherent receiver [24], where bulky and slow optical-polarization controllers as well as optical delay lines are removed.

Since the digital coherent receiver requires high-speed ADC and DSP, most of the experiments have been done still offline. It means that after transmitted data are stored in a computer, bit errors are analyzed offline, However, very recently, an application-specific integrated circuit (ASIC) designed for the 11.5-Gsymbol/s polarization-multiplexed QPSK signal has been developed [25], and the real-time operation of the digital coherent receiver at the bit rate of 46 Gbit/s has been demonstrated by using such an ASIC [26]. This achievement is really a milestone in the history of the modern coherent optical communications. The combination of coherent detection and DSP is thus expected to become a part of the next generation of optical communication systems and provide new capabilities that were not possible without the detection of the phase of the optical signal.

2.2 Principle of Coherent Optical Detection

This section describes the basic operation principle of coherent optical detection. We show how the coherent receiver measures the complex amplitude of the optical signal with the shot-noise-limited sensitivity and how information on the state of polarization can be extracted by the use of polarization diversity.

2.2.1 *Coherent Detection*

Figure 2.11 shows the configuration of the coherent optical receiver. The fundamental concept behind coherent detection is to take the product of electric fields of the modulated signal light and the continuous-wave (CW) local oscillator (LO). Let the optical signal incoming from the transmitter be

$$E_s(t) = A_s(t)\exp(j\omega_s t), \tag{2.1}$$

where $A_s(t)$ is the complex amplitude and ω_s the angular frequency. Similarly, the electric field of LO prepared at the receiver can be written as

$$E_{\mathrm{LO}}(t) = A_{\mathrm{LO}}\exp(j\omega_{\mathrm{LO}} t), \tag{2.2}$$

where A_{LO} is the constant complex amplitude and ω_{LO} the angular frequency of LO. We note here that the complex amplitudes A_s and A_{LO} are related to the signal power P_s and the LO power P_{LO} by $P_s = |A_s|^2/2$ and $P_{\mathrm{LO}} = |A_{\mathrm{LO}}|^2/2$, respectively.

Balanced detection is usually introduced into the coherent receiver as a means to suppress the dc component and maximize the signal photocurrent. The concept resides in using a 3-dB optical coupler that adds a 180° phase shift to either the signal field or the LO field between the two output ports. When the signal and LO

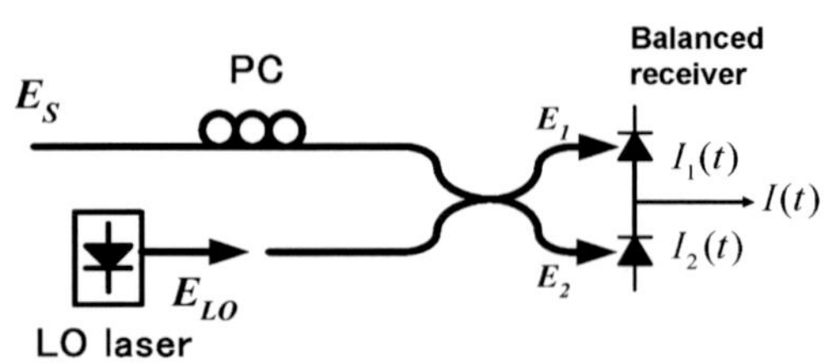

Fig. 2.11 Configuration of the coherent receiver that measures the beat between the signal and LO

are co-polarized, the electric fields incident on the upper and lower photodiodes are given as

$$E_1 = \frac{1}{\sqrt{2}}(E_s + E_{\mathrm{LO}}), \tag{2.3}$$

$$E_2 = \frac{1}{\sqrt{2}}(E_s - E_{\mathrm{LO}}), \tag{2.4}$$

and the output photocurrents are written as

$$\begin{aligned} I_1(t) &= R\left[\mathrm{Re}\left\{\frac{A_s(t)\exp(j\omega_s t) + A_{\mathrm{LO}}\exp(j\omega_{\mathrm{LO}} t)}{\sqrt{2}}\right\}\right]^{\mathrm{ms}} \\ &= \frac{R}{2}\left[\, P_s(t) + P_{\mathrm{LO}} \right. \\ &\quad \left. +2\sqrt{P_s(t)P_{\mathrm{LO}}}\cos\left\{\omega_{\mathrm{IF}} t + \theta_{sig}(t) - \theta_{\mathrm{LO}}(t)\right\}\,\right], \end{aligned} \tag{2.5}$$

$$\begin{aligned} I_2(t) &= R\left[\mathrm{Re}\left\{\frac{A_s(t)\exp(j\omega_s t) - A_{\mathrm{LO}}\exp(j\omega_{\mathrm{LO}} t)}{\sqrt{2}}\right\}\right]^{\mathrm{ms}} \\ &= \frac{R}{2}\left[\, P_s(t) + P_{\mathrm{LO}} \right. \\ &\quad \left. -2\sqrt{P_s(t)P_{\mathrm{LO}}}\cos\left\{\omega_{\mathrm{IF}} t + \theta_{\mathrm{sig}}(t) - \theta_{\mathrm{LO}}(t)\right\}\,\right], \end{aligned} \tag{2.6}$$

where "ms" means the mean square with respect to the optical frequencies, "Re" means to take the real part, ω_{IF} is known as the intermediate frequency (IF) given by $\omega_{\mathrm{IF}} = |\omega_s - \omega_{\mathrm{LO}}|$, and $\theta_{\mathrm{sig}}(t)$ and $\theta_{\mathrm{LO}}(t)$ are phases of the transmitted signal and LO, respectively. R is the responsivity of the photodiode given as

$$R = \frac{e\eta}{\hbar\omega_s}, \tag{2.7}$$

where $\hbar$ stands for the Planck's constant, e the electron charge, and η the quantum efficiency of the photodiode. The balanced detector output is then given as

$$I(t) = I_1(t) - I_2(t) = 2R\sqrt{P_s(t)P_{\mathrm{LO}}}\cos\left\{\omega_{\mathrm{IF}} t + \theta_{\mathrm{sig}}(t) - \theta_{\mathrm{LO}}(t)\right\}. \tag{2.8}$$

P_{LO} is always constant and $\theta_{\mathrm{LO}}(t)$ includes only the phase noise that varies in time.

2.2.2 Heterodyne Receivers

Heterodyne detection refers to the case that $|\omega_{\mathrm{IF}}| >> \omega_b/2$, where ω_b is the modulation bandwidth of the optical carrier determined by the symbol rate. In such a case,

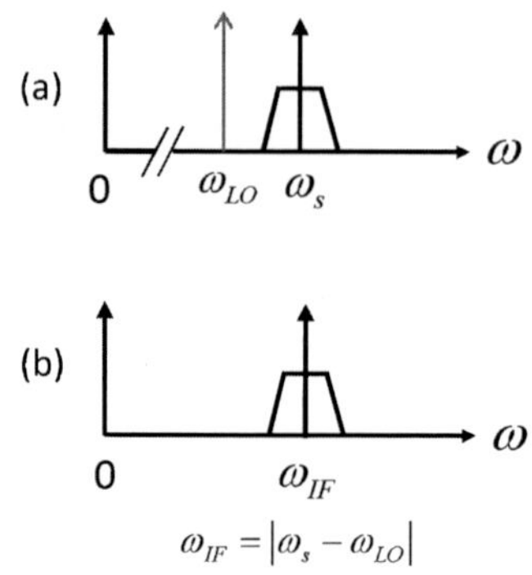

Fig. 2.12 Spectra of (**a**) the optical signal and (**b**) the down-converted IF signal

Eq. (2.8) shows that the electric field of the signal light is down-converted to the IF signal including the amplitude information and the phase information, as shown in Fig. 2.12.

The signal phase is given as $\theta_{\text{sig}}(t) = \theta_s(t) + \theta_{sn}(t)$, where $\theta_s(t)$ is the phase modulation and $\theta_{sn}(t)$ the phase noise. The receiver output is given as

$$I(t) = 2R\sqrt{P_s(t)\,P_{\text{LO}}}\cos\{\omega_{\text{IF}}t + \theta_s(t) + \theta_n(t)\}, \tag{2.9}$$

and we can determine the complex amplitude on $\exp(j\omega_{\text{IF}}t)$ from Eq. (2.9) as

$$I_c(t) = 2R\sqrt{P_s(t)\,P_{\text{LO}}}\exp j\{\theta_s(t) + \theta_n(t)\}, \tag{2.10}$$

where $\theta_n(t)$ is the total phase noise given as

$$\theta_n(t) = \theta_{sn}(t) - \theta_{\text{LO}}(t). \tag{2.11}$$

Note that Eq. (2.10) is equivalent to the complex amplitude $A_s(t)$ of the optical signal except for the phase noise increase stemming from LO.

We have three methods of demodulating $I_c(t)$, which are envelope (noncoherent) detection, differential (delay) detection, and synchronous (coherent) detection as shown in Fig. 2.13. In envelope detection, we measure $|I_c(t)|^2$ from Eq. (2.10), which gives us only the information on $P_s(t)$. Differential detection is effective for constant-envelope modulation formats such as M-ary PSK. In this scheme, we determine the phase difference between the current symbol and the previous one. In the synchronous detection scheme, although the total phase noise $\theta_n(t) = \theta_{sn}(t) - \theta_{\text{LO}}(t)$ might vary in time, the electrical phase-locked loop (PLL) can be used to estimate the phase noise and decode the symbol.

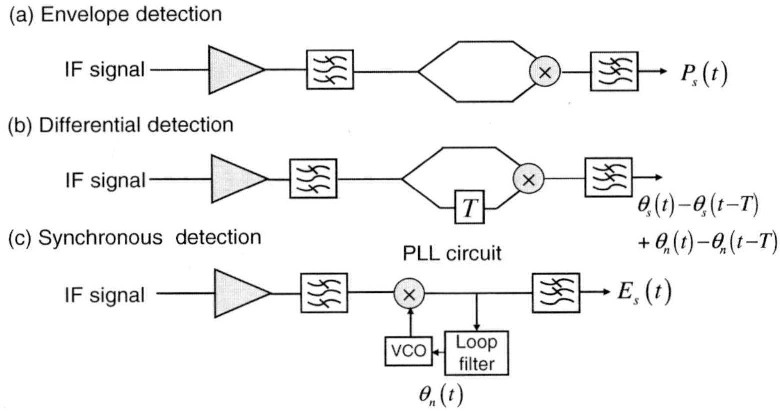

Fig. 2.13 Methods of demodulating the IF signal

2.2.3 Homodyne Receivers

Homodyne detection refers to the case that $\omega_{\text{IF}} = 0$. The photodiode current from the homodyne receiver becomes

$$I(t) = 2R\sqrt{P_s(t)\,P_{\text{LO}}}\cos\left\{\theta_{sig}(t) - \theta_{\text{LO}}(t)\right\}. \tag{2.12}$$

Equation (2.12) means that the homodyne receiver measures the inner product between the signal phasor and the LO phasor as shown in Fig. 2.14. In order to decode the symbol correctly, the LO phase $\theta_{\text{LO}}(t)$ must track the transmitter phase noise $\theta_{sn}(t)$ such that $\theta_n(t) = 0$. This function is realized by the optical phase-locked loop (OPLL); however, in practice, the implementation of such a loop is not simple and adds to homodyne detection the complexity of the configuration. In addition, Eq. (2.12) only gives the cosine component (in other words, the in-phase component with respect to the LO phase), and the sine component (the quadrature component) cannot be detected. Therefore, this type of homodyne receivers is not able to extract the full information on the signal complex amplitude.

Preparing another LO, whose phase is shifted by 90°, in the homodyne receiver, we can detect both in-phase and quadrature components of the signal light as shown in Fig. 2.15. This function is achieved by a 90° optical hybrid shown in Fig. 2.16. Using such 90° optical hybrid, we can obtain four outputs E_1, E_2, E_3, and E_4 from

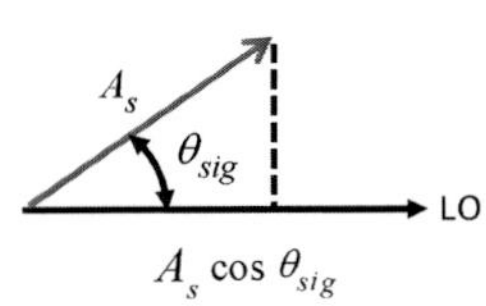

Fig. 2.14 Phasor diagram of the signal and LO for homodyne detection, which measures the inner product between the signal and LO phasors

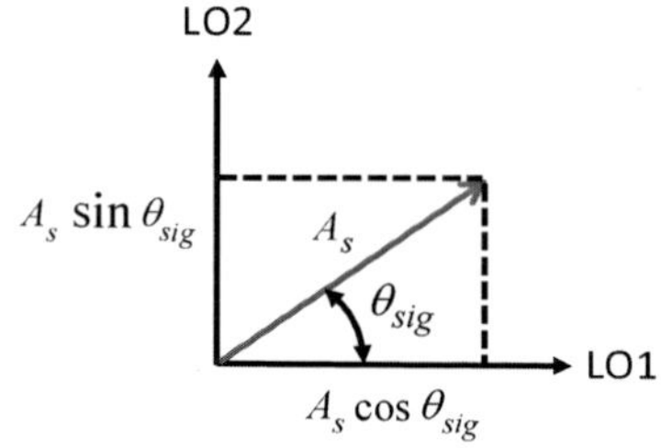

Fig. 2.15 Phasor diagram of the signal and LO for phase-diversity homodyne detection. Both of the in-phase and quadrature components are measured at the same time

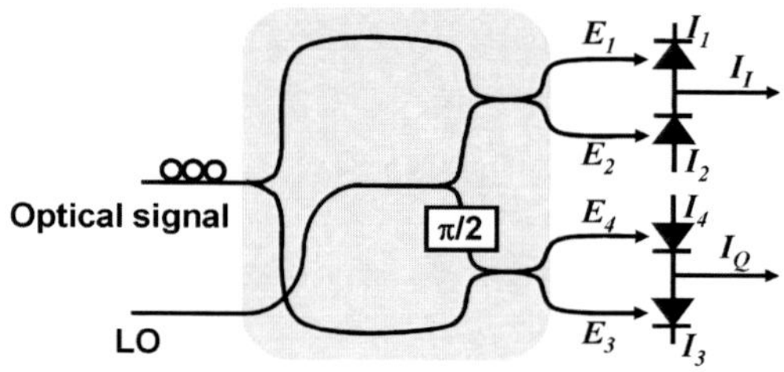

Fig. 2.16 Configuration of the phase-diversity homodyne receiver using a 90° optical hybrid

the two inputs E_s and E_{LO} as

$$E_1 = \frac{1}{2}(E_s + E_{\text{LO}}), \tag{2.13}$$

$$E_2 = \frac{1}{2}(E_s - E_{\text{LO}}), \tag{2.14}$$

$$E_3 = \frac{1}{2}(E_s + jE_{\text{LO}}), \tag{2.15}$$

$$E_4 = \frac{1}{2}(E_s - jE_{\text{LO}}). \tag{2.16}$$

Output photocurrents from balanced photodetectors are then given as

$$I_I(t) = I_{I1}(t) - I_{I2}(t) = R\sqrt{P_s P_{\text{LO}}}\cos\{\theta_{\text{sig}}(t) - \theta_{\text{LO}}(t)\}, \tag{2.17}$$

$$I_Q(t) = I_{Q1}(t) - I_{Q2}(t) = R\sqrt{P_s P_{\text{LO}}}\sin\{\theta_{\text{sig}}(t) - \theta_{\text{LO}}(t)\}. \tag{2.18}$$

Using Eqs. (2.17) and (2.18), we can restore the complex amplitude as

$$I_c(t) = I_I(t) + jI_Q(t) = R\sqrt{P_s(t)\,P_{\text{LO}}}\exp\{j(\theta_s(t) + \theta_n(t))\}, \tag{2.19}$$

which is equivalent to the complex amplitude of the optical signal except for the phase noise increase. Note that the complex amplitude is obtained in the baseband in contrast to heterodyne detection.

Equation (2.19) shows that the electric field of the signal light is down-converted to the baseband. As shown in the down-converted spectrum of Fig. 2.17, we need to allow the negative frequency to express the complex amplitude at the baseband, which contains both the in-phase (or cos) and quadrature (or sin) components. In

Fig. 2.17 Spectra of (a) the optical signal and (b) homodyne-detected baseband signal. The conventional homodyne receiver only measures the real part of the optical complex amplitude, whereas the phase-diversity receiver does the complex amplitude itself whose spectrum exists on both the positive and negative frequency sides

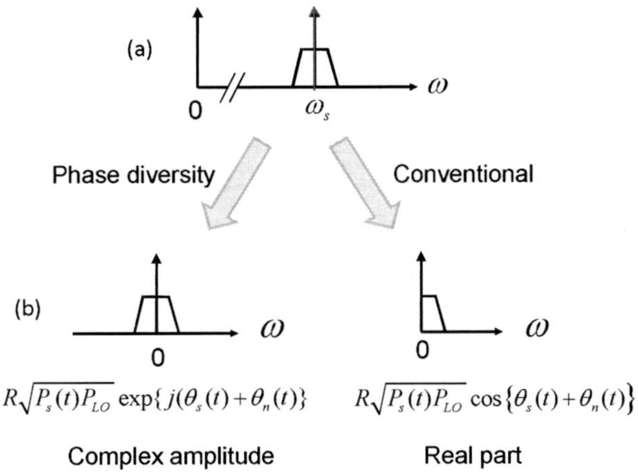

contrast, since the conventional homodyne receiver only measures the in-phase (or cos) component, the baseband signal exists only in the positive frequency side.

This type of receiver is commonly called the "phase-diversity homodyne receiver" [27] or "intradyne receiver" [28]. Eventually, the phase-diversity homodyne receiver and the heterodyne receiver can similarly restore the full information on the optical complex amplitude, as shown by Eqs. (2.10) and (2.19), respectively. However, since the phase-diversity homodyne receiver generates the baseband signal directly, it is more advantageous over the heterodyne receiver which must deal with a rather high intermediate frequency.

Figure 2.18 shows methods of demodulating $I_C(t)$. Similarly to heterodyne detection, we can demodulate $I_C(t)$ by (a) envelope (non-coherent) detection and (b) differential (delay) detection at the baseband. As for synchronous (coherent)

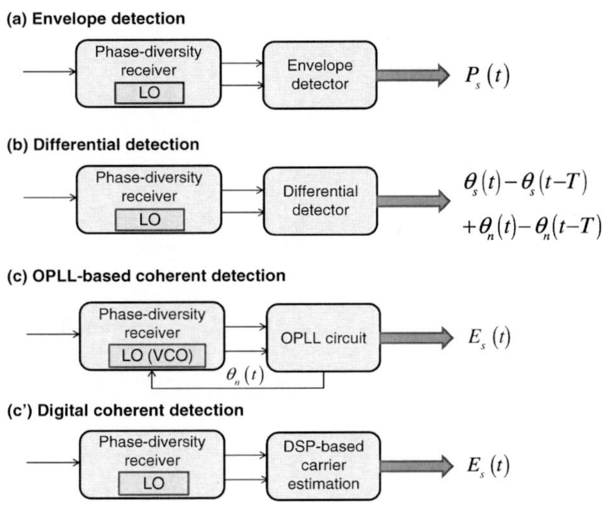

Fig. 2.18 Methods of demodulating the homodyne-detected baseband signal

detection, in addition to relying upon (c) conventional OPLL, where the LO phase tracks the signal phase noise, it is possible to estimate the phase noise $\theta_n(t)$ and restore the signal complex amplitude through digital signal processing on the homodyne-detected signal given by Eq. (2.19) (see (c')). This is the basic idea of the "digital coherent receiver," which has recently been investigated extensively.

2.2.4 Homodyne Receiver Employing Phase and Polarization Diversities

It was assumed up to now that the polarization of the incoming signal was always aligned to that of LO. However, in practical systems, the polarization of the incoming signal is unlikely to remain aligned to the state of polarization (SOP) of LO because of random changes on the birefringence of the transmission fiber. One of the most serious problems of the coherent receiver is that the receiver sensitivity is dependent on SOP of the incoming signal. In this subsection, we show that the polarization diversity receiver can cope with the polarization dependence of the receiver sensitivity.

The receiver employing polarization diversity is shown in Fig. 2.19, where two phase-diversity homodyne receivers are combined with the polarization diversity configuration [23]. The incoming signal having an arbitrary SOP is separated into two linear polarization components with a polarization beam splitter (PBS).

Assuming that a single-polarization component of the optical carrier is modulated at the transmitter, let the x- and y-polarization components after PBS at the receiver be written as

$$\begin{bmatrix} E_{sx} \\ E_{sy} \end{bmatrix} = \begin{bmatrix} \sqrt{\alpha} A_s e^{j\delta} \\ \sqrt{1-\alpha} A_s \end{bmatrix} \exp(j\omega_s t), \tag{2.20}$$

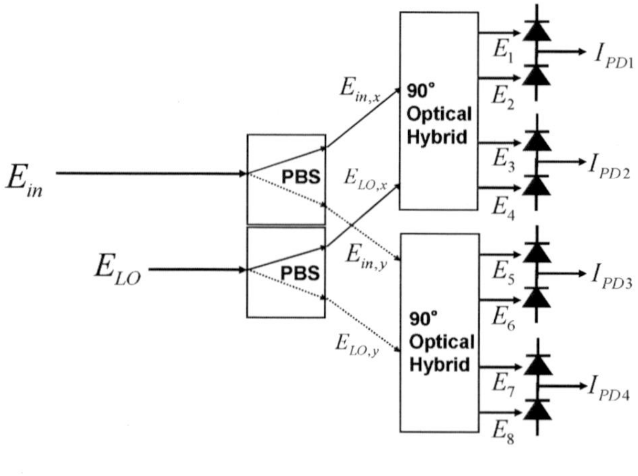

Fig. 2.19 Configuration of the homodyne receiver employing phase and polarization diversities

where α denotes the power ratio of the two polarization components, and δ the phase difference between them. These parameters are dependent on the birefringence of the transmission fiber and time-varying. On the other hand, the x- and y-polarization components equally separated from the linearly polarized LO are written as

$$\begin{bmatrix} E_{\mathrm{LO},x} \\ E_{\mathrm{LO},y} \end{bmatrix} = \frac{1}{\sqrt{2}} \begin{bmatrix} A_{\mathrm{LO}} \\ A_{\mathrm{LO}} \end{bmatrix} \exp(j\omega_{\mathrm{LO}} t). \tag{2.21}$$

Two 90° optical hybrids in Fig. 2.19 generate electric fields $E_{1,\cdots,8}$ at double-balanced photodiodes PD1–PD4:

$$E_{1,2} = \frac{1}{2}\left(E_{sx} \pm E_{\mathrm{LO},x}\right), \tag{2.22}$$

$$E_{3,4} = \frac{1}{2}\left(E_{sx} \pm jE_{\mathrm{LO},x}\right), \tag{2.23}$$

$$E_{5,6} = \frac{1}{2}\left(E_{sy} \pm E_{\mathrm{LO},y}\right), \tag{2.24}$$

$$E_{7,8} = \frac{1}{2}\left(E_{sy} \pm jE_{\mathrm{LO},y}\right). \tag{2.25}$$

Photocurrents from PD1 to PD4 are then given as

$$I_{\mathrm{PD1}} = R\sqrt{\frac{\alpha P_s(t) P_{\mathrm{LO}}}{2}} \cos\{\theta_s(t) - \theta_{\mathrm{LO}}(t) + \delta\}, \tag{2.26}$$

$$I_{\mathrm{PD2}} = R\sqrt{\frac{\alpha P_s(t) P_{\mathrm{LO}}}{2}} \sin\{\theta_s(t) - \theta_{\mathrm{LO}}(t) + \delta\}, \tag{2.27}$$

$$I_{\mathrm{PD3}} = R\sqrt{\frac{(1-\alpha) P_s(t) P_{\mathrm{LO}}}{2}} \cos\{\theta_s(t) - \theta_{\mathrm{LO}}(t)\}, \tag{2.28}$$

$$I_{\mathrm{PD4}} = R\sqrt{\frac{(1-\alpha) P_s(t) P_{\mathrm{LO}}}{2}} \sin\{\theta_s(t) - \theta_{\mathrm{LO}}(t)\}. \tag{2.29}$$

From Eqs. (2.26), (2.27), (2.28), and (2.29), we find that the polarization diversity receiver can separately measure complex amplitudes of the two polarization components as

$$
\begin{aligned}
I_{xc}(t) &= I_{\mathrm{PD1}}(t) + jI_{\mathrm{PD2}}(t) \\
&= R\sqrt{\frac{\alpha P_s(t)P_{\mathrm{LO}}}{2}}\exp j\{\theta_s(t)+\theta_n(t)+\delta\},
\end{aligned}
\tag{2.30}
$$

$$
\begin{aligned}
I_{yc}(t) &= I_{\mathrm{PD3}}(t) + jI_{\mathrm{PD4}}(t) \\
&= R\sqrt{\frac{(1-\alpha)P_s(t)P_{\mathrm{LO}}}{2}}\exp j\{\theta_s(t)+\theta_n(t)+\delta\},
\end{aligned}
\tag{2.31}
$$

from which we can reconstruct the signal complex amplitude $A_s(t)$ in a polarization-independent manner.

2.2.5 Carrier-to-Noise Ratio

From the complex amplitude at the IF stage given by Eq. (2.10), the carrier-to-noise ratio (CNR) of the heterodyne signal is given as

$$
\gamma_s = \frac{\overline{|I_c(t)|^2}/2}{2eRP_{\mathrm{LO}}B} = \frac{\eta P_s}{hfB},
\tag{2.32}
$$

where we assume that the LO shot noise overwhelms the circuit noise with sufficient LO power, and B is the receiver bandwidth at the IF stage. Noting that the minimum bandwidth at the IF stage is given as

$$
B = \frac{1}{T},
\tag{2.33}
$$

where T is the symbol duration, we find

$$
\gamma_s = \frac{\eta P_s T}{hf} = \eta N_s,
\tag{2.34}
$$

where $N_s = P_sT/hf$ means the number of photons per symbol [29]. This is the shot-noise-limited CNR.

On the other hand, the homodyne phase-diversity receiver can generate the complex amplitude given by Eq. (2.19) at the baseband. Therefore, the mean square of the signal photocurrent is given as $\overline{|I_c(t)|^2} = R^2P_sP_{\mathrm{LO}}$. When reconstructing the complex amplitude, we need to add shot noises due to LO powers of $P_{\mathrm{LO}}/2$ from the two ports. The receiver bandwidth at the baseband is $B/2 = 1/2T$; therefore, the total shot-noise current is given as $eP_{\mathrm{LO}}B$. We can thus obtain the CNR $\gamma_s = \eta P_s/hfB = \eta N_s$, which is the same as the heterodyne receiver. Even when the polarization diversity is introduced, signal processing called "maximal-ratio combining" can maintain CNR [29].

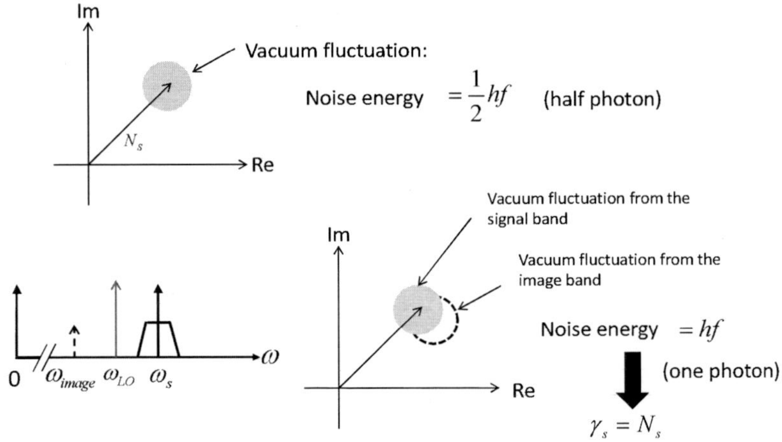

Fig. 2.20 Quantum mechanical picture of the carrier-to-noise ratio of the heterodyne-detected light. Vacuum fluctuations are merged into the IF signal from the image band as well as the signal band. Therefore, the measured CNR is given as $\gamma_s = N_s$

The CNR obtained from coherent detection is the intrinsic CNR of the coherent state of light, which can be understood from the quantum mechanical point of view as follows: Figure 2.20 shows the phasor of light, which is associated with vacuum fluctuations. The average photon energy is hfN_s, where N_s denotes the average number of photons, and the noise energy originating from vacuum fluctuation is $hf/2$, which is the half photon energy. Let us consider the case when we measure the phasor of the light by heterodyne detection. When we measure the beat between the signal at the angular frequency of ω_s and the LO at ω_{LO}, the vacuum fluctuation at ω_{image} is also merged into the IF band at $\omega_s - \omega_{\mathrm{LO}}$. Therefore, the noise energy at IF is hf, resulting in the CNR of the IF signal $\gamma_s = N_s$. In the case of phase-diversity homodyne detection, although the image band noise never merges into the detected signal, branching the signal light into the I and Q ports degrades CNR by 3 dB, which gives us the same CNR as heterodyne detection.

By using this picture, we can understand the impact of EDFA on the receiver sensitivity. Figure 2.21 shows the m-stage EDFA chain, where the amplifier gain G compensates for the fiber loss G^{-1} periodically. Each amplifier generates noise photons of $(G-1)n_{sp}$ due to amplified spontaneous emission, where n_{sp} denotes

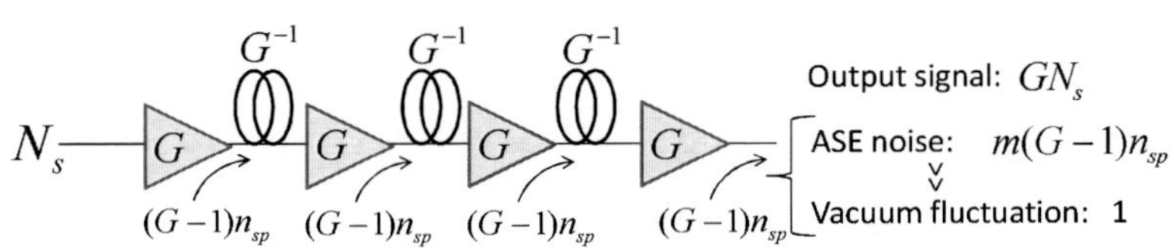

Fig. 2.21 Noise in an optical amplifier chain. Accumulated ASE is usually much larger than vacuum fluctuations, and the shot-noise-limited sensitivity is not necessarily required for the receiver

the spontaneous emission factor. At the output, the accumulated number of noise photons is given as $m(G-1)n_{sp}$, whereas the number of noise photons due to vacuum fluctuations is one.

Since $m(G-1)n_{sp} \gg 1$, CNR of the signal transmitted through the amplifier chain is given as

$$\gamma_s = \frac{GN_s}{m(G-1)n_{sp}} \simeq \frac{N_s}{mn_{sp}}, \tag{2.35}$$

which is much lower than the shot-noise limit given by Eq. (2.34); therefore, the shot-noise-limited receiver sensitivity is not necessary in such an amplifier chain. This is the main reason why R&D in coherent optical communications has been interrupted behind the rapid progress in EDFA technologies as discussed in Sect. 2.1.1.

2.3 Digital Signal Processing in Coherent Receivers

This section describes details of digital signal processing for coherent receivers. Outputs from the homodyne receiver comprising phase and polarization diversities are processed by digital signal-processing circuits, restoring the complex amplitude of the signal in a stable manner despite fluctuations of the carrier phase and the signal SOP. Symbol-by-symbol control of such time-varying parameters in the digital domain can greatly enhance the system stability compared with optical control methods.

2.3.1 Basic Concept of the Digital Coherent Receiver

Figure 2.22 shows the basic concept of the digital coherent receiver. First, the incoming signal is detected linearly with the homodyne receiver comprising phase and polarization diversities. Using this receiver, we can obtain full information on the optical carrier, namely the complex amplitude and the state of polarization. Such complex amplitude measured by the receiver is converted to digital data with ADCs and processed by DSP circuits. The progress in the increased performance, speed, and reliability of integrated circuits now makes digital signal processing an attractive approach to recover the optical complex amplitude from the homodyne-detected baseband signal.

The combination of the optical IQ modulator and the IQ demodulator realizes the linear optical communication system as shown in Fig. 2.23. At the transmitter, we define a vector on the complex plane using two voltages driving the IQ modulator. This vector is mapped on the phasor of the optical carrier through the IQ modulator. Such optical IQ modulation is perfectly restored by IQ demodulation, which is

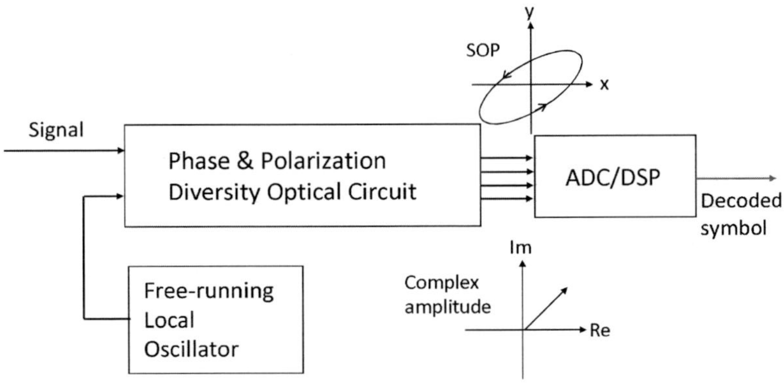

Fig. 2.22 Concept of the digital coherent receiver

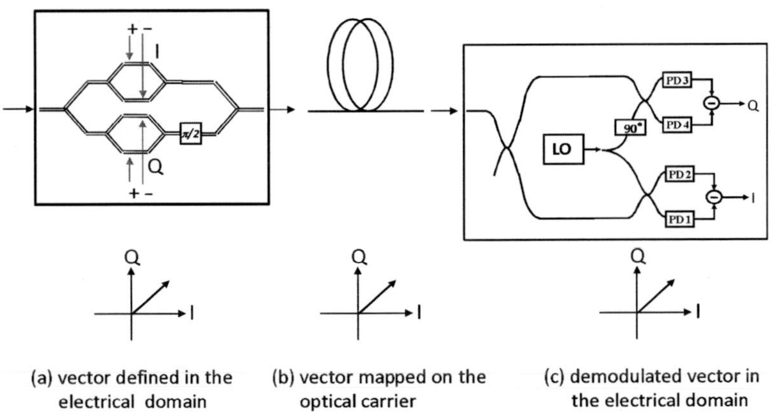

Fig. 2.23 Linear transmission system

performed by the digital coherent receiver. This is exactly the linear system, where IQ information is preserved even with E/O and O/E conversion processes.

The digital signal-processing circuit is typically composed of the sequence of operations shown in Fig. 2.24 to retrieve the information from the received signal. First, the four-channel ADC asynchronously samples the data and restores the complex amplitudes (see Sect. 2.3.2). Such complex amplitudes are equalized by an fixed equalizer which removes inter-symbol interference (ISI) (see Sect. 2.3.5). Next, the dual-polarization signal is demultiplexed and polarization-mode dispersion (PMD) is compensated for, usually by using the constant-modulus algorithm (CMA) (see Sect. 2.3.4). After the clock is extracted from the interpolated data in the time domain, they are resampled to keep one sample within a symbol interval T (see Sect. 2.3.2). Then the carrier phase is estimated, and the remaining ISI is removed by an adaptive equalizer driven by decoded symbols (see Sect. 2.3.5).

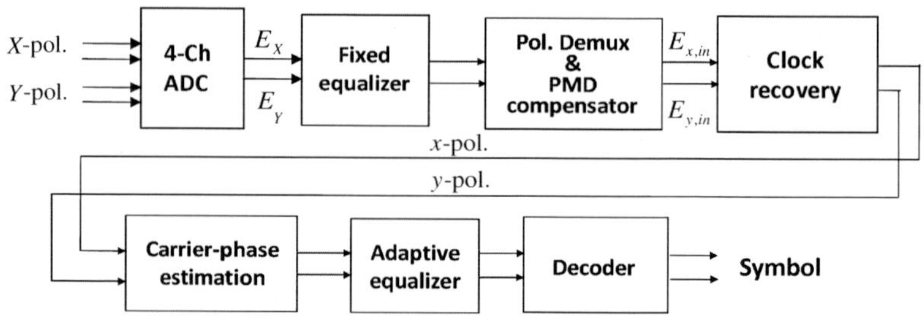

Fig. 2.24 Typical sequence of digital signal processing for decoding the symbol

2.3.2 Sampling of the Signal and Clock Extraction

When the PSK signal is encoded on an RZ pulse train (RZ-PSK), we can realize accurate extraction of the clock from the transmitted signal, detecting the intensity of the received signal and using a standard clock recovery circuit. Photocurrents are then sampled and digitized by ADCs at the timing of the extracted clock. In this case, the digital circuit does not need to resample the data, so that it saves on complexity of the digital circuit.

It is also possible to extract the clock from the asynchronously sampled signal as shown in Fig. 2.24. The necessary condition for asynchronous sampling is that the sampling frequency must be equal to or greater than twice the symbol rate. After the clock is extracted from the interpolated data in the time domain, they are resampled to keep one sample within a symbol interval T.

2.3.3 Phase Estimation

Since the linewidth of semiconductor DFB lasers used as the transmitter and LO typically ranges from 100 kHz to 10 MHz, the phase noise $\theta_n(t)$ varies much more slowly than the phase modulation $\theta_s(t)$. Therefore, by averaging the carrier phase over many symbol intervals, it is possible to obtain an accurate phase estimate. In the following, assuming the M-ary PSK modulation, we explain the phase estimation procedure, but such procedure can easily be extended to quadrature amplitude modulation (QAM).

The phase of the complex amplitude obtained from Eq. (2.21) contains both the phase modulation $\theta_s(i)$ and the phase noise $\theta_n(i)$, where i represents the sample number. The procedure to estimate θ_n is shown in Fig. 2.25, where the case of QPSK is shown for simplicity. We take the Mth power of the measured complex amplitude $I_c(i)$, because the phase modulation is removed from $I_c(i)^M$ in the M-ary PSK modulation format. Subtracting the phase noise thus estimated from the measured phase, we can restore the phase modulation.

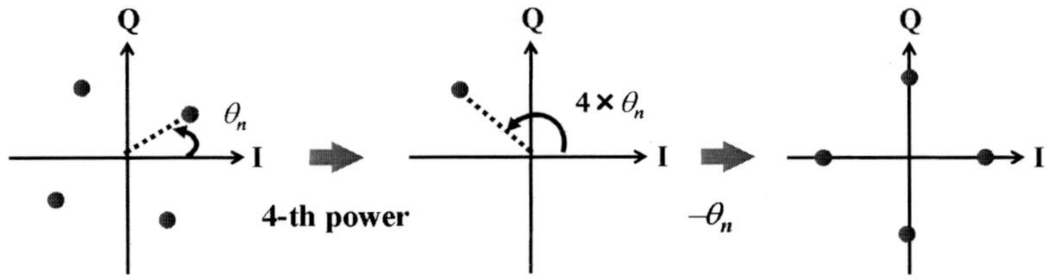

Fig. 2.25 Principle of the Mth-power phase estimation method. For simplicity, the case of QPSK ($M = 4$) is shown. Taking the Mth power of the received complex amplitude, we can eliminate the phase modulation and measure the phase noise

In actual phase estimation, we average $I_c(i)^M$ over $2k + 1$ samples to improve the signal-to-noise ratio of the estimated phase reference. The estimated phase is thus given as

$$\theta_e(i) = \arg\left(\sum_{j=-k}^{k} I_c(i+j)^M\right)/M. \tag{2.36}$$

The phase modulation $\theta_s(i)$ is determined by subtracting $\theta_e(i)$ from the measured phase of $\theta(i)$. The phase modulation is then discriminated among M symbols. Figure 2.26 shows the DPS circuit for such phase estimation.

The symbols thus obtained have the phase ambiguity by $2\pi/M$ because we cannot know the absolute phase. It is important to note that the data should be differentially precoded. Differentially decoding the discriminated symbol after symbol discrimination, we can solve the phase ambiguity problem although the bit-error rate is doubled by error multiplication.

The phase estimate $\theta_e(i)$ ranges between $-\pi/M$ and $+\pi/M$. Therefore, if $|\theta_e(i)|$ exceeds π/M, the phase jump of $2\pi/M$ occurs inevitably as shown in Fig. 2.27. To

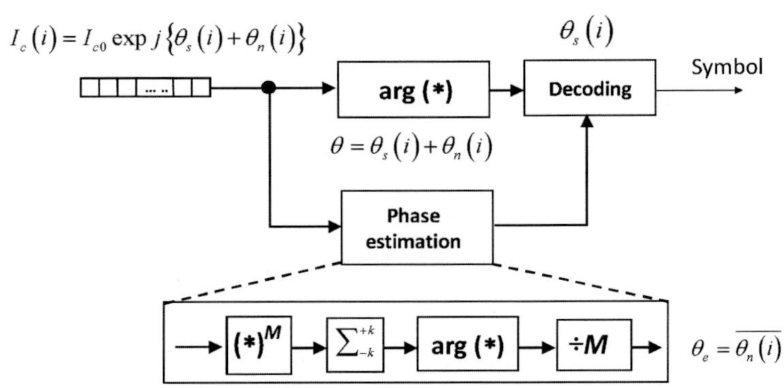

Fig. 2.26 DSP circuit for Mth power phase estimation

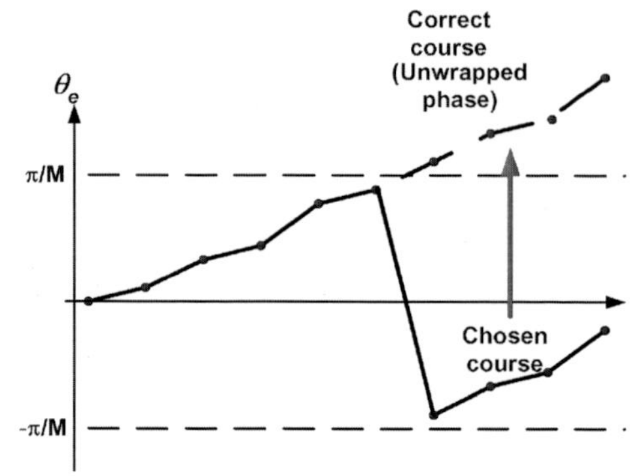

Fig. 2.27 Phase jump and its correction during the phase estimation process. The correct course of the phase drift is determined by removing the phase jump

cope with this problem, the correction for the phase jump is done by obeying the following rule:

$$\theta_e(i) \leftarrow \theta_e(i) + \frac{2\pi}{M} f\left(\theta_e(i) - \theta_e(i-1)\right), \tag{2.37}$$

where $f(x)$ is defined as

$$f(x) = \begin{cases} +1 \text{ for } x < -\frac{\pi}{M} \\ 0 \text{ for } |x| \leq \frac{\pi}{M} \\ -1 \text{ for } x > \frac{\pi}{M} \end{cases}. \tag{2.38}$$

This adjustment ensures that the phase estimate follows the trajectory of the physical phase and cycle slips are avoided [30].

2.3.4 Polarization Alignment

Throughout this subsection, we regard for simplicity that the measured photocurrent is identical to the optical complex amplitude, neglecting conversion factors from the optical domain to the electrical domain. Let the optical complex amplitude in the x-polarization state be E_x and that in the y-polarization state be E_y at the receiver.

When the single-polarization signal $E_{in}(t)$ is transmitted, we can apply the maximal-ratio combining method to align the SOP of the incoming signal [29]. ADCs sample and digitize the four outputs $I_{\mathrm{PD1}}, \ldots, I_{\mathrm{PD4}}$ from the phase/polarization diversity receiver shown in Fig. 2.19. The x- and y-polarization components of the complex amplitude of the signal are given by Eqs. (2.30) and (2.31) as

$$E_x(i) = \sqrt{2\alpha P_s(i)} \exp j\{\theta_s(i) + \theta_n(i) + \delta\}, \tag{2.39}$$

$$E_y(i) = \sqrt{2(1-\alpha)P_s(i)} \exp j\{\theta_s(i) + \theta_n(i)\}. \tag{2.40}$$

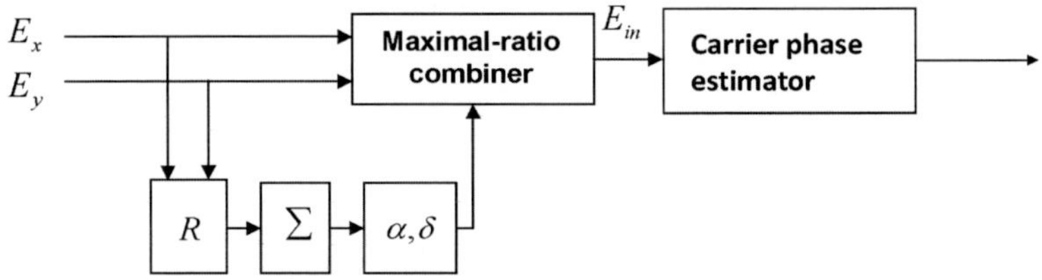

Fig. 2.28 DSP circuit for maximal-ratio polarization combining. The transmitted single polarization is aligned in the digital domain

Figure 2.28 shows the post-processing circuit realizing the maximal-ratio polarization combining process. We define the ratio $r(i)$ as

$$r(i) = E_x(i)/E_y(i). \tag{2.41}$$

Polarization parameters α and δ of the incoming signal vary much more slowly than the phase modulation. Therefore, by averaging $r(i)$ over many symbol intervals, it is possible to obtain accurate values of α and δ : The ratio $r(i)$ averaged over $2\ell+1$ samples is written as

$$\bar{r}(i) = \frac{1}{2\ell+1}\sum_{j=-\ell}^{\ell} r\,(i)\,, \tag{2.42}$$

and α and δ are calculated from

$$|\bar{r}(i)| = \sqrt{\alpha}/\sqrt{\alpha-1}, \tag{2.43}$$

$$\arg(\bar{r}(i)) = \delta. \tag{2.44}$$

The signal complex amplitude, independent of SOP of the incoming signal, is then reconstructed by maximal-ratio combining as

$$E_{in}(i) \propto \bar{r}(i)^* E_x(i) + E_y(i). \tag{2.45}$$

However, the above procedure is not effective when $\alpha \simeq 1$, because δ contains a large error and the term $\bar{r}^* E_x$ is not determined correctly. In such a case, we should use $E_{in}(i) \propto E_x(i) + E_y(i)/\bar{r}(i)^*$ to reconstruct the signal complex amplitude more accurately. The term $\left|E_y/\bar{r}^*\right|$ is much smaller than $|E_x|$, and hence the error in E_{in} is reduced. In summary, the reconstruction formula is given as

$$E_{in}(i) \propto \begin{cases} \bar{r}(i)^* E_x(i) + E_y(i) \; \alpha \le 0.5 \\ E_x(i) + \dfrac{E_y(i)}{\bar{r}(i)^*} \; \alpha > 0.5 \end{cases} . \tag{2.46}$$

On the other hand, when dual polarizations are transmitted, the constant-modulus algorithm (CMA) has been widely applied to polarization demultiplexing [24, 31]. In what follows, we discuss the principle of operation of polarization demultiplexing.

Let the Jones matrix of the fiber for transmission be given as

$$\boldsymbol{T} = \begin{bmatrix} \sqrt{\alpha}e^{i\delta} & -\sqrt{1-\alpha} \\ \sqrt{1-\alpha} & \sqrt{\alpha}e^{-i\delta} \end{bmatrix}, \tag{2.47}$$

where α and δ denote the power splitting ratio and the phase difference between the two polarization modes, respectively. When the polarization-multiplexed signal $\boldsymbol{E}_{in}(t)^T = \left[E_{in,x}(t), E_{in,y}(t)\right]^T$, where "$T$" means to take the transposed matrix, is incident on the fiber, the SOP of the output is given as

$$\begin{bmatrix} E_x(t) \\ E_y(t) \end{bmatrix} = \boldsymbol{T} \begin{bmatrix} E_{in,x}(t) \\ E_{in,y}(t) \end{bmatrix}. \tag{2.48}$$

Assuming M-ary PSK modulation, we normalize the envelope of each input polarization component as

$$\left|E_{in,x}(t)\right|^2 = \left|E_{in,y}(t)\right|^2 = 1. \tag{2.49}$$

Using the received polarization components E_x and E_y, we calculate E_X and E_Y through DSP as

$$\begin{aligned} E_X(t) &= rE_x(t) + kE_y(t) \\ &= \left(r\sqrt{\alpha}e^{j\delta} + k\sqrt{1-\alpha}\right)E_{in,x}(t) \\ &\quad + \left(-r\sqrt{1-\alpha} + k\sqrt{\alpha}e^{-j\delta}\right)E_{in,y}(t), \end{aligned} \tag{2.50}$$

where r and k are complex numbers. Here, we define

$$X \equiv r\sqrt{\alpha}e^{j\delta} + k\sqrt{1-\alpha}, \tag{2.51}$$

$$Y \equiv -r\sqrt{1-\alpha} + k\sqrt{\alpha}e^{-j\delta}. \tag{2.52}$$

Then the power of the X-component is given as

$$\begin{aligned} \left|E_X(t)\right|^2 &= \left|XE_{in,x}(t)\right|^2 + \left|YE_{in,y}(t)\right|^2 \\ &\quad + 2\left|XYE_{in,x}(t)E_{in,y}(t)\right|\cos\theta(t), \end{aligned} \tag{2.53}$$

$$\theta(t) = \arg\left[\frac{YE_{in,y}(t)}{XE_{in,x}(t)}\right]. \tag{2.54}$$

Since the phase difference between initial x and y polarization components are time-varying at the symbol rate due to independent PSK modulations, the third term in Eq. (2.53) changes as a function of time. Here, we consider the condition under which $|E_X|^2$ becomes time-independent. As a matter of fact, such condition is given as $Y = 0$ (hereafter Case (I)) or $X = 0$ (hereafter Case (II)). In each case, (r, k) is expressed as

$$\frac{r_x}{k_x} = \frac{\sqrt{\alpha}}{\sqrt{1-\alpha}} e^{-j\delta}, \tag{2.55}$$

or

$$\frac{r_y}{k_y} = -\frac{\sqrt{1-\alpha}}{\sqrt{\alpha}} e^{-j\delta}. \tag{2.56}$$

In Case (I), Eq. (2.55) leads to

$$E_X = \frac{k_x}{\sqrt{1-\alpha}} E_{in,x}, \tag{2.57}$$

whereas in Case (II), Eq. (2.56) does

$$E_Y = \frac{k_y e^{-j\delta}}{\sqrt{\alpha}} E_{in,y}. \tag{2.58}$$

In Case (I), normalizing E_X such that $|E_X|^2 = 1$, we have

$$r_x = \sqrt{\alpha} e^{-j\delta + j\varphi_x}, \tag{2.59}$$
$$k_x = \sqrt{1-\alpha} e^{j\varphi_x}, \tag{2.60}$$

where φ_x is a real constant. Similarly, in Case (II), we have

$$r_y = -\sqrt{1-\alpha} e^{j\varphi_y}, \tag{2.61}$$
$$k_y = \sqrt{\alpha} e^{j\delta + j\varphi_y}, \tag{2.62}$$

where φ_y is a real constant.

In this way, provided that r and k are controlled so that $|E_X|^2$ does not flucuate at the symbol rate, we can reach either Case (I) or Case (II). In Case (I), we can restore the initial x-polarization as

$$E_X = E_{in,x} e^{j\varphi_x}. \tag{2.63}$$

By using r_x and k_x obtained from Eqs. (2.59) and (2.60), respectively, the y-polarization is derived from the Y port as

$$E_Y = -k_x^* E_x + r_x^* E_y = E_{in,y} e^{-j\varphi_x}. \tag{2.64}$$

On the other hand, in Case (II), the restored x and y polarizations appear at the Y and X ports, respectively, as

$$E_X = E_{in,y} e^{i\varphi_y}, \tag{2.65}$$

$$E_Y = -E_{in,x} e^{-i\varphi_y}. \tag{2.66}$$

The algorithm that controls (r, k) so as to satisfy $|E_X|^2 = 1$ is well known as the constant-modulus algorithm (CMA) or the Godard algorithm [32]. Figure 2.29 shows the DSP circuit for polarization demultiplexing based on CMA. The function of this circuit is expressed as

$$\begin{bmatrix} E_X \\ E_Y \end{bmatrix} = \boldsymbol{p} \begin{bmatrix} E_x \\ E_y \end{bmatrix}. \tag{2.67}$$

The matrix $\boldsymbol{p}$ is written as

$$\boldsymbol{p} = \begin{bmatrix} p_{xx} & p_{xy} \\ p_{yx} & p_{yy} \end{bmatrix}, \tag{2.68}$$

where elements of the matrix must satisfy the unitary condition:

$$p_{xy} = -p_{yx}^*, \tag{2.69}$$

$$p_{yy} = p_{xx}^*, \tag{2.70}$$

$$|p_{xx}|^2 + |p_{xy}|^2 = 1. \tag{2.71}$$

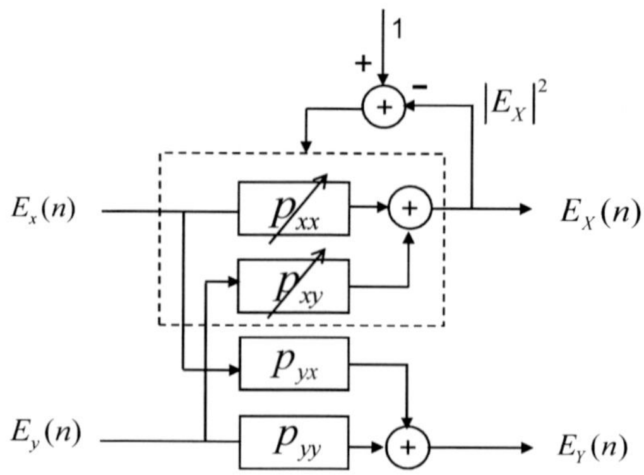

Fig. 2.29 DSP circuit for polarization demultiplexing. $p_{yx} = -p_{xy}^*$ and $p_{yy} = p_{xx}^*$ because of unitarity of the matrix $\boldsymbol{p}$

Following CMA, these matrix elements are updated symbol by symbol as

$$p_{xx}(n+1) = p_{xx}(n) + \mu(1 - |E_X(n)|^2)E_X(n)E_x^*(n), \tag{2.72}$$

$$p_{xy}(n+1) = p_{xy}(n) + \mu(1 - |E_X(n)|^2)E_X(n)E_y^*(n), \tag{2.73}$$

where μ is a step-size parameter and n the number of symbols. By using this algorithm, we can expect that $|E_X(n)|^2 \to 1$ after the symbol-by-symbol iteration process converges. When $|E_X(n)|^2 \to 1$, it has already been shown that $p_{xx}(n) \to r_x$, $p_{xy}(n) \to k_x$, $p_{yx}(n) \to -k_x^*$, and $p_{yy}(n) \to r_x^*$ in Case (I). This result means that the x-polarization component is restored at the X port as shown by Eq. (2.63), and the y-polarization component at the Y port as shown by Eq. (2.64). On the other hand, in Case (II), we find that the x-polarization component appears at the Y port, and the y-polarization at the X port as shown by Eqs. (2.65) and (2.66).

When either polarization-mode dispersion (PMD) or polarization-dependent loss (PDL) is involved, unitarity of the matrix $\boldsymbol{p}$ is not assured. In such a case, the Y polarization should be aligned independently of the X component as shown in Fig. 2.30, where p_{yy} and p_{yx} are updated by using the following equations:

$$p_{yy}(n+1) = p_{yy}(n) + \mu(1 - |E_Y(n)|^2)E_Y(n)E_y^*(n), \tag{2.74}$$

$$p_{yx}(n+1) = p_{yx}(n) + \mu(1 - |E_Y(n)|^2)E_Y(n)E_x^*(n). \tag{2.75}$$

In this iteration process, we need to carefully choose initial matrix elements for $\boldsymbol{p}$ in order to prevent the two outputs converging on the same SOP.

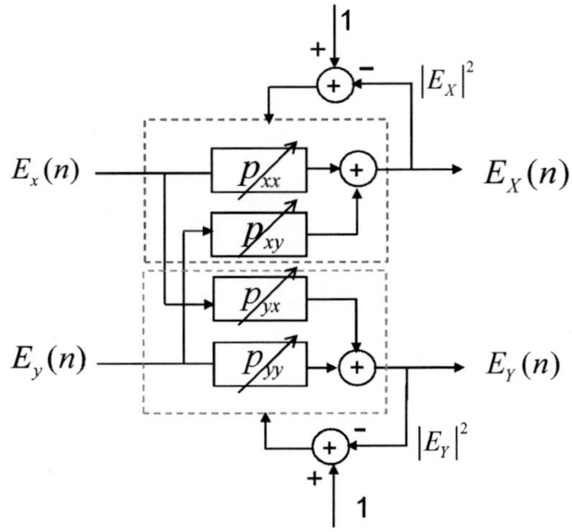

Fig. 2.30 DSP circuit controlling X polarization and Y polarization independently. This circuit should be used when unitarity of the matrix $\boldsymbol{p}$ is not assured

2.3.5 Equalization of Inter-symbol Interference

When the optical signal is transmitted though a fiber link, linear impairments stemming from group-velocity dispersion (GVD) of the fiber, tight optical filtering, and electrical bandwidth limitation induce ISI, which degrades the BER performance. Since the restored complex amplitude contains the information on the amplitude and phase of the optical signal, ISI can be equalized by using the transversal filter shown in Fig. 2.31, where the tap spacing is equal to the symbol interval T. The spacing of $T/2$ is also employed commonly. The transversal filter is also called the finite-impulse response (FIR) filter. When we represent the received complex amplitude as $x(n)$ where n is the sample number, the output from the transversal filter is given as

$$y(n) = \sum_{i=0}^{k} c_i x(n-i), \tag{2.76}$$

where c_i are complex tap weights and k denotes the number of taps. Theoretically, with a sufficient number of taps, this filter can compensate for any kind of linear impairments, because the filter can generate an arbitrary transfer function.

Tap coefficients can be prefixed by using a given transfer function or adaptively controlled. Two kinds of adaptive control methods are explained in the following. We first define the tap-coefficient vector and the input-signal vector as

$$\boldsymbol{c(n)} = \begin{bmatrix} c_0(n) \\ c_1(n) \\ \vdots \\ c_{k-1}(n) \\ c_k(n) \end{bmatrix}, \tag{2.77}$$

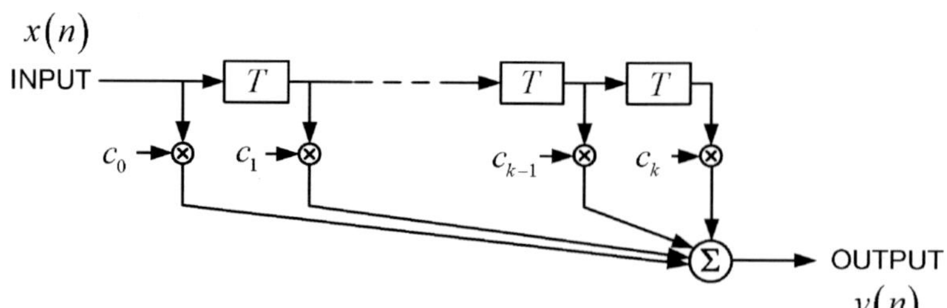

Fig. 2.31 Configuration of the transversal filter for ISI equalization. Delayed replicas are summed up with complex tap weights c_i

$$\boldsymbol{x(n)} = \begin{bmatrix} x_0(n) \\ x_1(n) \\ \vdots \\ x_{k-1}(n) \\ x_k(n) \end{bmatrix}. \tag{2.78}$$

Then the filter output is given as

$$\begin{aligned} y(n) &= \boldsymbol{c}(n)^T \boldsymbol{x}(n) \\ &= \sum_{i=0}^{k} c_i(n) x(n-i). \end{aligned} \tag{2.79}$$

One of the methods of adaptively controlling tap coefficients is the blind equalization based on CMA discussed in Sect. 2.3.4. In this method, the tap coefficients are updated by

$$\boldsymbol{c}\,(n+1) = \boldsymbol{c}\,(n) + \mu \left\{1 - |y\,(n)\,|^2\right\} y\,(n)\,\boldsymbol{x}\,(n)^*, \tag{2.80}$$

where we do not need decoded symbols for updating tap coefficients. The principle is based on the fact that when ISI is suppressed, the envelope of the PSK signal becomes constant. Equation (2.80) controls tap coefficients so that $|y\,(n)\,|^2 \to 1$, and ISI is suppressed in such a case.

Figure 2.32 shows another adaptive-equalization circuit based on the least-mean-square (LMS) algorithm [33]. The tap coefficients are updated by using the LMS algorithm as follows:

$$\boldsymbol{c}\,(n+1) = \boldsymbol{c}\,(n) + \mu\varepsilon\,(n)\,\boldsymbol{x}\,(n)^*. \tag{2.81}$$

In the training mode, we transmit a training signal having a fixed pattern in advance, and the error $\varepsilon\,(n)$ is given as the difference between the training signal and the filter output $y(n)$. The tap coefficients are updated until the error tends to zero. Once the tap coefficients have been fixed in the training mode, the operation mode of the

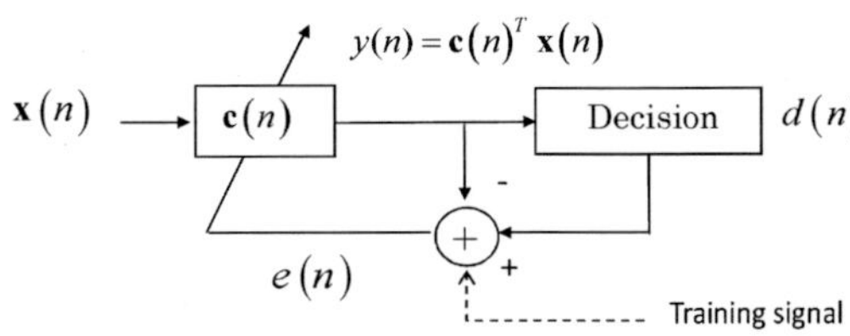

Fig. 2.32 Configuration of the adaptive-equalizer circuit based on LMS algorithm. After tap coefficients are converged in the training mode, the mode of operation is switched to the tracking mode, where the error is given as the difference between the decoded symbol and the filter output

filter is switched to the tracking mode, where the error $\varepsilon(n)$ is given as the difference between the decoded symbol $d(n)$ and the filter output $y(n)$. In the tracking mode, the tap coefficients are adaptively updated so that $|\varepsilon(n)|$ is suppressed even if sources of transmission impairments are time-varying.

If we need to equalize PMD, we should modify the polarization-demultiplexing circuit shown in Fig. 2.29 such that each matrix element becomes a multi-tap transversal filter instead of a one-tap configuration. We define input-signal vectors and tap-coefficient vectors as

$$\boldsymbol{E}_k(n) = \begin{bmatrix} E_k(n) \\ E_k(n-1) \\ \vdots \\ E_k(n-m) \end{bmatrix}, \tag{2.82}$$

$$\boldsymbol{p}_{ij}(n) = \begin{bmatrix} p_{ij,0}(n) \\ p_{ij,1}(n) \\ \vdots \\ p_{ij,m}(n) \end{bmatrix}. \tag{2.83}$$

where i, j, and k are any one of x and y and m is the number of taps. The tap coefficients can be updated by CMA as

$$\boldsymbol{p}_{xx}(n+1) = \boldsymbol{p}_{xx}(n) + \mu(1 - |E_X(n)|^2)E_X(n)\boldsymbol{E}_x^*(n), \tag{2.84}$$

$$\boldsymbol{p}_{xy}(n+1) = \boldsymbol{p}_{xy}(n) + \mu(1 - |E_X(n)|^2)E_X(n)\boldsymbol{E}_y^*(n), \tag{2.85}$$

$$\boldsymbol{p}_{yx}(n+1) = \boldsymbol{p}_{yx}(n) + \mu(1 - |E_Y(n)|^2)E_Y(n)\boldsymbol{E}_x^*(n), \tag{2.86}$$

$$\boldsymbol{p}_{yy}(n+1) = \boldsymbol{p}_{yy}(n) + \mu(1 - |E_Y(n)|^2)E_Y(n)\boldsymbol{E}_y^*(n). \tag{2.87}$$

Updating tap coefficients by Eqs. (2.84), (2.85), (2.86), and (2.87), we can compensate for PMD because the envelope fluctuation is suppressed. At the same time, we can demultiplex dual polarizations as shown in the one-tap case (see Sect. 2.3.4). It is also possible to introduce the least-mean-square (LMS) algorithm instead of CMA.

2.4 Performance of the Digital Coherent Receiver

This section deals with evaluation of basic characteristics of the digital coherent receiver. After showing the optical circuit for the homodyne receiver comprising phase and polarization diversities, we discuss the receiver sensitivity limit, the polarization dependence of the receiver sensitivity, and the phase noise tolerance.

2.4.1 Optical Circuit for the Homodyne Receiver Comprising Phase and Polarization Diversities

The optical 90° hybrid can be implemented into the phase-diversity homodyne receiver by using free-space optical components as shown in Fig. 2.33 [19, 20]. Orthogonal states of polarization for the local oscillator (LO) and the incoming signal create the 90° hybrid necessary for phase diversity. With the $\lambda/4$ waveplate (QWP), the polarization of the LO becomes circular, while the signal remains linearly polarized and its polarization angle is 45° with respect to principal axes of polarization beam splitters (PBSs). After passing through the half mirror (HM), the polarization beam splitters separate the two polarization components of the LO and signal while two balanced photodiodes PD1 and PD2 detect the beat between the LO and signal in each polarization. When the circularly polarized LO is split with a PBS, the phase difference between the split beams is 90°. On the other hand, there is no phase difference between the split signal beams, because the signal is linearly polarized. Photocurrents from PD 1 and 2 are then given by Eqs. (2.17) and (2.18). The optical 90° hybrid using planar lightwave circuits (PLC), where the 90° phase shift is created by an optical path-length difference, has also been demonstrated [34]. This is a promising approach for integrating an LO and PDs on a PLC platform.

The optical circuit for the homodyne phase/polarization diversity receiver composed of free-space optical components is shown in Fig. 2.34. In this receiver, two

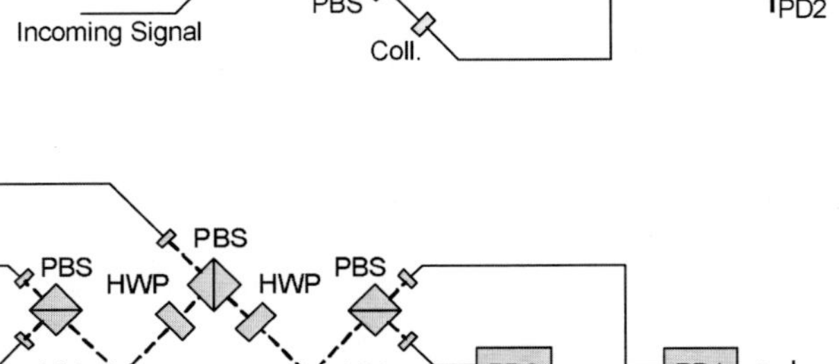

Fig. 2.33 Optical circuit for the homodyne phase-diversity receiver. QWP: quarter-wave plate PBS: polarization beam splitter HM: polarization-independent half mirror, Coll.: collimator

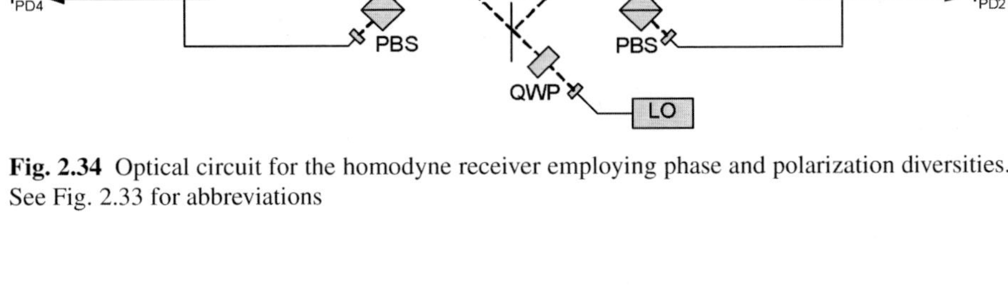

Fig. 2.34 Optical circuit for the homodyne receiver employing phase and polarization diversities. See Fig. 2.33 for abbreviations

homodyne phase-diversity receivers are combined with the polarization diversity configuration. The incoming signal having an arbitrary SOP is separated into two polarization components with a polarization beam splitter (PBS). We refer to the x- and y-polarizations with respect to the principal axes of the PBS. On the other hand, the local oscillator (LO) is split into two paths with a half mirror (HM), after its SOP is made circular by a quarter-wave plate (QWP). Right- and left-hand sides of the receiver then constitute phase-diversity receivers for x- and y-polarizations, respectively. Photocurrents from PD 1, 2, 3, and 4 are then given by Eqs. (2.26), (2.27), (2.28), and (2.29), respectively.

2.4.2 Receiver Sensitivity

The back-to-back BER of the BPSK signal is measured to access the sensitivity of the homodyne-phase-diversity receiver [29]. Data were precoded at the pulse pattern generator (PPG) such that differential decoding of the transmitted data resulted in a pseudo-random binary sequence (PRBS) (2^7–1). Lasers used as a transmitter and an LO were 1.55-μm distributed-feedback (DFB) semiconductor lasers, whose linewidths were about 150 kHz. The laser temperature and bias current were carefully maintained via feedback control to keep frequency drifts of the lasers below 10 MHz. The transmitter laser output was modulated through a $LiNbO_3$ push–pull modulator to generate a BPSK signal at a bit rate of 10 Gbit/s. The received power was adjusted with an attenuator and monitored with a power meter. The received signal was amplified with an erbium-doped fiber amplifier (EDFA) to −10 dBm before it was detected with the coherent receiver. The maximum LO power was 10 dBm, which was determined from the allowable power of the photodiodes. The signals I_{PD1} and I_{PD2} were simultaneously sampled at a rate of 20 Gsample/s with ADCs. The 3-dB bandwidth of ADCs was 8 GHz. The BER measurement was performed offline. The collected samples were resampled to keep only one point per symbol and combined to form a 100-ksymbol-long stream. During phase estimation, we used the averaging span $k = 10$. The SOP of the signal was controlled manually so as to maximize the receiver sensitivity.

Figure 2.35 shows back-to-back BERs measured as a function of the received power with and without optical preamplification. The power was measured before the preamplifier in the case with preamplification. The LO power is also changed when we do not use optical preamplification. The dotted curve is the theoretical one, which is calculated from the formula

$$P_e = \frac{1}{2}\text{erfc}\left(\sqrt{\gamma_s}\right), \tag{2.88}$$

where erfc(*) means the complimentary error function and we assume the ideal CNR given as

$$\gamma_s = N_s = \frac{P_s T}{\hbar\omega_s}. \tag{2.89}$$

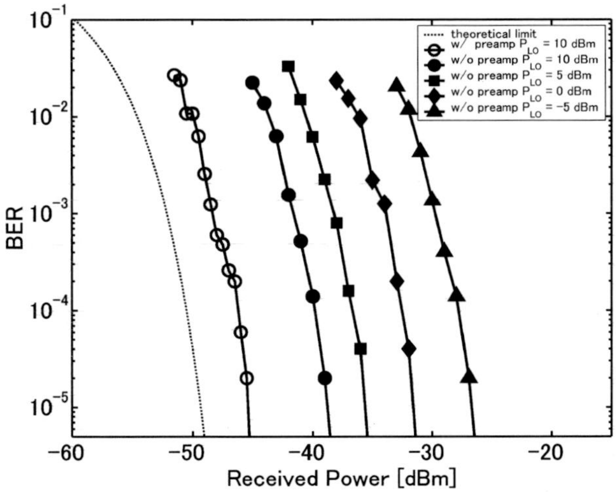

Fig. 2.35 Bit-error rate curves for the BPSK signal with and without optical preamplification. *Broken curve*: the theoretical shot-noise limit, *circles*: with a preamplifier and $P_{LO} = 10$ dBm, dots: without a preamplifier and $P_{LO} = 10$ dBm, *squares*: without a preamplifier and $P_{LO} = 5$ dBm, *diamonds*: without a preamplifier and $P_{LO} = 0$ dBm, and triangles: without a preamplifier and $P_{LO} = -5$ dBm

Without optical preamplification, the increase in the LO power improves the receiver sensitivity; however, 10-dB power penalty still remains from the theoretical limit even at $P_{LO} = 10$ dBm, because we cannot reach the shot-noise limit due to the relatively large circuit noise. On the other hand, with optical preamplification, power penalty is as small as 3 dB, which stems from the spontaneous emission factor larger than 1 and ISI due to bandwidth limitation of the receiver. Since the theoretical power penalty due to preamplification is 0 dB when the spontaneous emission factor of the amplifier $n_{sp} = 1$, the use of the preamplifier is very effective if the LO power is not sufficient. Refer to [29] for sensitivity analyses of coherent receivers with optical preamplification.

2.4.3 Polarization Sensitivity

We evaluate the performance of the homodyne phase/polarization diversity receiver in the single-polarization 10-Gbit/s BPSK system [29]. During digital signal processing for polarization alignment and carrier phase estimation, we used $\ell = 10$ and $k = 10$, respectively.

Figure 2.36 shows back-to-back BERs as a function of the received power before the preamplifier. While we change polarization-parameter values of α(=0, 0.25, 0.5, 0.75, 1), BER curves are almost independent of α. The power penalty is negligible compared with the case where we do not employ polarization diversity but align the signal SOP optimally ($\alpha = 0, 1$). Thus, we find that the receiver is insensitive to the polarization fluctuation of the incoming signal.

2.4.4 Phase Noise Tolerance

The requirement for the laser linewidth in M-ary PSK systems becomes much more stringent as M increases. To evaluate the bit-error rate (BER) performance of the

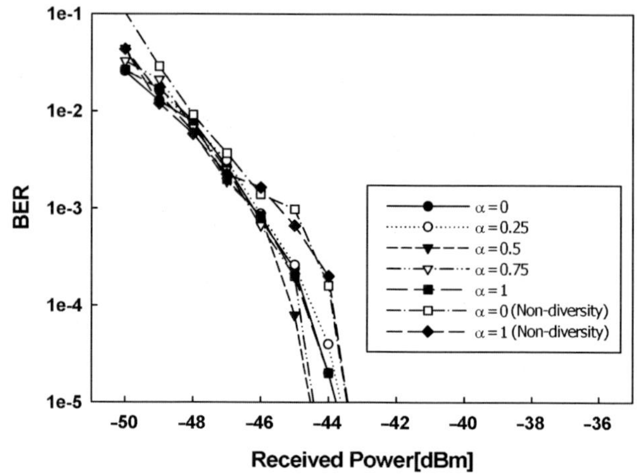

Fig. 2.36 Bit-error rate curves when polarization diversity is introduced. BER curves are almost insensitive to α. The power penalty due to polarization diversity is also negligibly small

homodyne phase-diversity receiver for the 8PSK case, we conduct a computer simulation where the phase noise variance σ_p^2 and the phase estimation span k are used as parameters [21]. We assume that the laser phase noise accumulated during a symbol interval T has a Gaussian distribution with the variance $\sigma_p^2 = 2\pi(\delta f_s + \delta f_{\rm LO})T$, where δf_s and $\delta f_{\rm LO}$ denote 3-dB linewidths of the transmitter and LO, respectively, and that signals are contaminated by additive white Gaussian noise.

Figure 2.37 shows the simulated BER performance of 8PSK systems calculated as a function of the SNR per bit when $\sigma_p = 0$, 1.4×10^{-2} and 4.3×10^{-2}, together with the ideal performance assuming perfect carrier recovery. For comparison, simulated BERs of the differential demodulation scheme are also shown in Fig. 2.37. For 10-Gsymbol/s systems, σ_p of 1.4×10^{-2} gives a linewidth of $\delta f = 150$ kHz for either transmitter or LO laser, and σ_p of 4.3×10^{-2} gives 1.5 MHz. Without phase noise, the BER performance of the phase estimation method approaches the ideal performance when k becomes larger; therefore, our phase estimation method is very

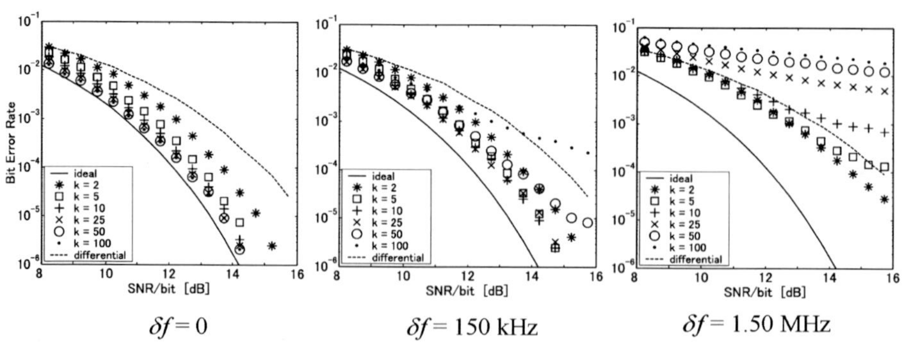

Fig. 2.37 Simulation results on BER performance in 8PSK systems for three cases, i.e., the ideal homodyne scheme without phase noise, the differential detection, and the homodyne detection with phase estimation (k = 2, 5 10, 25, 100), as a function of the SN ratio per bit

effective for obtaining a noiseless accurate phase reference. In contrast, differential demodulation incurs a nearly 3-dB penalty in the SNR, because the reference phase has the same amount of noise as the target phase. As far as the carrier phase is relatively constant over duration of $(2k + 1)\,T$, the phase estimation provides excellent performance. When $\sigma_p = 1.4 \times 10^{-2}$, the phase estimation is still efficient and the SNR penalty is less than 1 dB at BER = 10^{-6} for $k = 5$, 10, and 25. However, the phase noise increase gradually neutralizes the advantage of phase estimation compared to differential demodulation. This is because the phase difference between the first and last samples of $(2k + 1)$ samples becomes significant. For example, in the case of $\sigma_p = 4.3 \times 10^{-2}$, the error floor appears even when $k = 5$, and the merit of phase estimation is small. Also note that independently of σ_p, the BER performance when $k = 0$ is identical to that of differential detection. These results indicate that if the laser linewidth is as low as 150 kHz, the acceptable performance in 8PSK systems is obtained by the optimal choice of $k = 10$.

2.4.5 Coherent Demodulation of Multi-level Encoded Signals

We demonstrate coherent demodulation of optical multi-level coded signals. By using distributed-feedback (DFB) semiconductor lasers with linewidths of 150 kHz as a transmitter and a local oscillator, BPSK, QPSK, 8PSK, and 16QAM signals are successfully demodulated at the symbol rate of 10 Gsymbol/s [35].

Figure 2.38(a) shows back-to-back BERs measured as a function of the received power for 10-Gsymbol/s BPSK, QPSK, 8PSK, and 16QAM. A single polarization component is transmitted. At the receiver, the averaging span of phase estimation is optimized for each modulation format. We introduce an adaptive equalizer for suppressing ISI stemming from the receiver-bandwidth limitation. Figure 2.38(b)

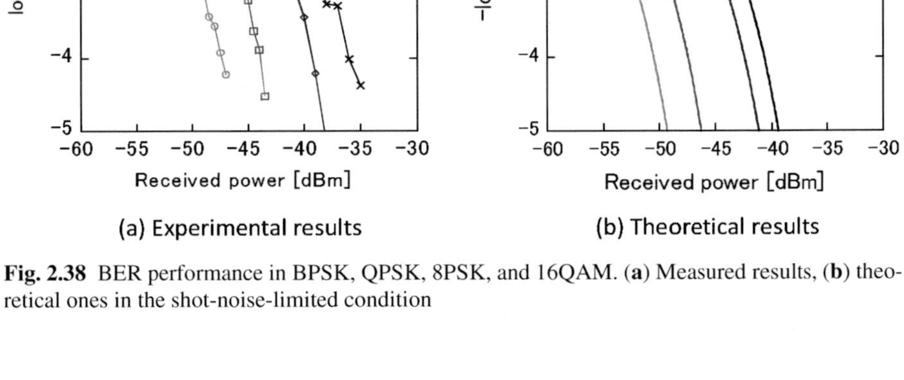

Fig. 2.38 BER performance in BPSK, QPSK, 8PSK, and 16QAM. (**a**) Measured results, (**b**) theoretical ones in the shot-noise-limited condition

shows theoretical BER curves in the shot-noise-limited condition. With increase in the level of modulation, we find that the power penalty from the theoretical sensitivity increases. This is because the BER performance becomes more sensitive against ISI as the level of modulation increases.

2.5 Challenges for the Future

The progress of digital coherent technologies is very rapid. The real-time operation of 46-Gbit/s dual-polarization QPSK receivers has been demonstrated, and the next target is operation at 100 Gbit/s, which aims at transmission of 100-Gbit/s ether signals. As to offline experiments, spectrally efficient transmission has been demonstrated by using higher-level modulation formats such as 8QAM [36] and 16QAM [37]. However, the following is the technical problems that must be tackled and accomplished before the practical coherent optical communication system is realized in the future.

(1) Hybrid integration of planar lightwave circuits (PLCs) for phase and polarization diversities, double-balanced photodiodes, and a local oscillator is an important technical task, which enables cost reduction of the coherent receiver and improves system stability.
(2) A tunable local oscillator with a narrow linewidth is still the key component of the high-performance coherent receiver.
(3) High-speed operation of the coherent receiver relies on the development of high-speed ADC and DSP. The processing speed to cope with $>$ 25-Gsymbol/s systems is desirable for 100-Gbit/s ether applications.
(4) More flexible signal processing such as advanced forward error correction (FEC) should be available in the DSP core.
(5) For long-distance transmission of multi-level coded optical signals, fiber nonlinearity ultimately limits the system performance. Post-compensation for fiber nonlinearity [38] such as self-phase modulation (SPM), cross-phase modulation (XPM), and four-wave mixing (FWM) is an important problem.

The combination of coherent detection and DSP provides new capabilities that were not possible without detection of the phase of the optical signal. We believe that the born-again coherent optical technology will renovate existing optical communication systems in the near future.

References

1. O.E. DeLange, Proc. IEEE **58**, 1683 (1970)
2. T. Okoshi, K. Kikuchi, Electron. Lett. **16**, 179 (1980)
3. F. Favre, D. LeGuen, Electron. Lett. **16**, 709 (1980)
4. T. Okoshi, K. Kikuchi, A. Nakayama, Electron. Lett. **16**, 630 (1980)
5. C. H. Henry, IEEE J. Qunatum Electron. **18**, 159 (1982)

6. Y. Yamamoto, IEEE J. Quantum Electron. **19**, 34 (1983)
7. K. Vahala, A. Yariv, IEEE J. Quantum Electron. **19**, 1096 (1983)
8. B. Glance, J. Lightwave Technol. **LT-5**, 274(1987)
9. Y. Yamamoto, IEEE J. Quantum Electron. **QE-16**, 1251(1980)
10. T. Okoshi, K. Emura, K. Kikuchi, R. Th. Kersten, J. Optical Commun. **2**, 89 (1981)
11. T. Imai, Y. Hayashi, N. Ohkawa, T. Sugie, Y. Ichihashi, T. Ito, Electron. Lett. **26**, 1407 (1990)
12. S. Norimatsu, K. Iwashita, K. Sato, IEEE Photonics Technol. Lett. **2**, 374(1990)
13. J. Kahn, K.-P. Ho, IEEE J. Select. Topics on Quantum Electron. **10**, 259 (2004)
14. R. Griffin, A. Carter, in *Optical Fiber Communication Conference (OFC 2002)*, WX6, Anaheim, CA, USA, 17–22 March 2002
15. F. Koyama, K. Iga, J. Lightwave Technol. **6**, 87 (1988)
16. S. Shimotsu, S. Oikawa, T. Saitou, N. Mitsugi, K. Kubodera, T. Kawanishi, M. Izutsu, IEEE Photonics Technol. Lett. **13**, 364 (2001)
17. D. MacGhan, C. Laperle, A. Savchenko, C. Li, G. Mak, M. O'Sullivan, in *Optical Fiber Communication Conference (OFC 2005)*, PDP27, Anaheim, CA, USA (6–11 March 2005)
18. S. Tsukamoto, D.-S. Ly-Gagnon, K. Katoh, K. Kikuchi, in *Optical Fiber Communication Conference (OFC 2005)*, PDP29, Anaheim, CA, USA (6–11 March 2005)
19. D.-S. Ly-Gagnon, S. Tsukamoto, K. Katoh, K. Kikuchi, J. Lightwave Technol. **24**, 12 (2006)
20. K. Kikuchi, IEEE J. Selected Topics on Quantum. Electron. **12**, 563 (2006)
21. S. Tsukamoto, K. Katoh, K. Kikuchi, IEEE Photoics Technol. Lett. **18**, 1131 (2006)
22. Tsukamoto, K. Katoh, K. Kikuchi, IEEE Photonics Technol. Lett. **18**, 1016 (2006)
23. S. Tsukamoto, Y. Ishikawa, K. Kikuchi, in *European Conference on Optical Communication (ECOC 2006)*, Mo4.2.1, Cannes, France (24–28 Sept. 2006)
24. S. J. Savory, Optics Express **16**, 804 (2008)
25. H. Sun, K.-T. Wu, K. Roberts, Optics Express **16**, 873 (2008)
26. L.E. Nelson, S.L. Woodward, M.D. Feuer, X. Zhou, P.D. Magill, S. Foo, D. Hanson, D. McGhan, H. Sun, M. Moyer, M.O'Sullivan, in *Optical Fiber Communication Conference (OFC 2008)*, PDP9, San Diego, CA (24–28 Feb. 2008)
27. H. Hodgikinson, R.A. Harmon, D.W. Smith, Electron. Lett. **21** 867 (1985)
28. F. Derr, Electron. Lett. **23** 2177 (1991)
29. K. Kikuchi, S. Tsukamoto, J. Lightwave. Technol. **26**, 1817 (2008)
30. R. Noé, in *Opto-Electronics and Communications Conference (OECC 2004)*, 16C2-5, Yokohama, Japan (12–16 July 2004)
31. K. Kikuchi, *LEOS Summer Topicals*, TuC1.1, Acapulco, Mexico (21–23 July 2008)
32. D.N. Godard, IEEE Trans. Commun. **28**, 1867 (1980)
33. S. Haykin, *Adaptive Filter Theory* (Prentice Hall, Englewood, 2001)
34. M. Oguma, Y. Nasu, H. Takahashi, H. Kawakami, E. Yoshida, in *European Conference on Optical Communication (ECOC 2007)*, 10.3.3, Berlin, Germany (16–20 September 2007)
35. Y. Mori, C. Zhang, K. Igarashi, K. Katoh, K. Kikuchi, in *European Conference on Optical Communication (ECOC 2008)*, Tu.1.E.4, Belgium, Brussels (21–25 Sept. 2008)
36. X. Zhou, J. Yu, M-F. Huang, Y. Shao, T. Wang, P. Magill, M. Cvijetic, L. Nelson, M. Birk, G. Zhang, S. Ten, H. B. Matthew, S. K. Mishra, in *Optical Fiber Communication Conference (OFC 2009)*, PDPB4, San Diego, CA, USA (22–26 March 2009)
37. A. H. Gnauck, P. J. Winzer, C. R. Doerr, L. L. Buhl, in *Optical Fiber Communication Conference (OFC 2009)*, PDPB8, San Diego, CA, USA (22–26 March 2009)
38. K. Kikuchi, Opt. Express **16**, 889 (2008)

Chapter 3
Ultrahigh Spectral Density Coherent Optical Transmission Technologies

Masataka Nakazawa

Abstract To meet the increasing demand to expand wavelength division multiplexing (WDM) transmission capacity, ultrahigh spectral density coherent optical transmission employing multi-level modulation formats has attracted a lot of attention. In particular, ultrahigh multi-level quadrature amplitude modulation (QAM) has an enormous advantage as regards expanding the spectral efficiency to 10 bit/s/Hz and even approaching the Shannon limit. This chapter describes fundamental technologies for ultrahigh spectral density coherent QAM transmission and presents experimental results on 128 QAM and 64 QAM-orthogonal frequency division multiplexing (OFDM) transmissions using heterodyne detection with a frequency-stabilized laser and an optical phase-locked loop technique.

3.1 Introduction

Transmission with high spectral efficiency employing multi-level modulation formats has attracted a lot of attention with a view to expanding the capacity of wavelength division multiplexing (WDM) transmission systems, because multi-bit information can be transmitted by one symbol data. Multi-level modulation also enables us to realize a high-speed system with low-speed devices, and therefore helps to enhance tolerance to dispersion and polarization mode dispersion (PMD) as well as to reduce power consumption.

Recently, a number of experimental results have been reported in which multi-level phase-shift keying (PSK) or a combination of PSK and amplitude-shift keying (ASK) has been employed for such a purpose [1–3]. Of these approaches, coherent quadrature amplitude modulation (QAM) [4–8] is one of the most spectrally efficient modulation formats. A 2^N QAM signal processes N bits in a single channel,

M. Nakazawa (✉)
Research Institute of Electrical Communication, Tohoku University, Katahira 2-1-1, Aoba-ku, 980-8577 Sendai-shi, Miyagiken, Japan
e-mail: nakazawa@riec.tohoku.ac.jp

M. Nakazawa et al. (eds.), *High Spectral Density Optical Communication Technologies,* Optical and Fiber Communications Reports 6, DOI 10.1007/978-3-642-10419-0_3,

so it has N times the spectral efficiency of on–off keying (OOK). For example, if we can employ 256∼1024 QAM, which was originally developed for microwaves, we may obtain enormous advantages such as an ultrahigh spectral efficiency exceeding 10 bit/s/Hz.

Orthogonal frequency division multiplexing (OFDM) is another approach that has attracted a lot of attention in relation to transmission with high spectral efficiency. In OFDM transmission, the multi-carrier transmission of low-speed orthogonal subcarriers enables us to improve both spectral efficiency and dispersion tolerance by adopting high-level subcarrier modulation formats and employing coherent detection [9–12].

This chapter describes fundamental technologies and recent progress on coherent QAM transmission using heterodyne detection with a frequency-stabilized fiber laser and an optical PLL circuit. Section 3.2 reviews an analytical description of the bit error rate (BER) and spectral efficiency of QAM signals and discusses the possibility of approaching the Shannon limit. Section 3.3 describes key components for coherent QAM transmission, such as a coherent light source, an optical phase-locked loop (OPLL), an IQ modulator, and a digital demodulator. 1 Gsymbol/s single-channel and frequency division multiplexed (FDM) 128 QAM transmission have been successfully achieved based on these key technologies, demonstrating a spectral efficiency as high as 10 bit/s/Hz. These results are presented in detail in Sects. 3.4 and 3.5. We also present our recent result on optical OFDM coherent transmission with 64 QAM subcarrier modulation in Sect. 3.6.

3.2 Spectral Efficiency of QAM Signal and Shannon Limit

QAM is a modulation format that combines two carriers whose amplitudes are modulated independently with the same optical frequency and whose phases are 90 degrees apart. These carriers are called in-phase carriers (I) and quadrature-phase carriers (Q). The QAM can assign 2^N states by using I and Q, which is called 2^N QAM. Figure 3.1 shows constellation maps for 16 (2^4) QAM and conventional

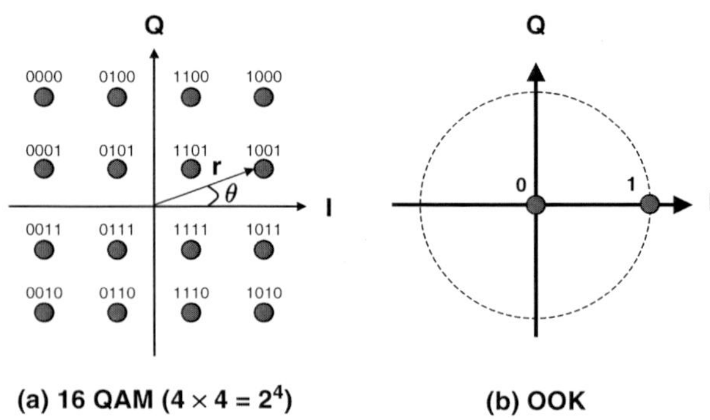

Fig. 3.1 Constellation maps for 16 QAM (**a**) and OOK (**b**)

OOK. As shown in Fig. 3.1, a 2^N QAM signal processes N bits in a single channel, so it can realize N times the spectral efficiency of OOK.

The BER of M-ary QAM signals can be analytically calculated and the result is given by [13]

$$P_b = \frac{3}{8}\text{erfc}\sqrt{\frac{2}{5}E_b/N_0}\ (16\,\text{QAM}) \tag{3.1}$$

$$P_b = \frac{7}{24}\text{erfc}\sqrt{\frac{1}{7}E_b/N_0}\ (64\,\text{QAM}) \tag{3.2}$$

$$P_b = \frac{15}{64}\text{erfc}\sqrt{\frac{4}{85}E_b/N_0}\ (256\ \text{QAM}) \tag{3.3}$$

Here, E_b and N_0 are the energy and noise power per bit, thus E_b/N_0 corresponds to the signal-to-noise ratio (SNR) per bit. Equations (3.1)–(3.3) are plotted in Fig. 3.2 as a function of E_b/N_0. The spectral efficiency of an M-ary QAM signal is shown in Fig. 3.3 as a function of E_b/N_0. Here, the ultimate spectral efficiency is given by the Shannon limit:

$$\frac{C}{W} = \log_2\left(1 + \frac{E_b}{N_0}\frac{C}{W}\right), \tag{3.4}$$

which is known as the Shannon–Hartley theorem [14]. This figure indicates that, as the multiplicity M increases, the spectral efficiency of M-QAM approaches closer to the Shannon limit than other advanced modulation formats such as M-PSK or M-FSK. The increase in M, however, requires a larger E_b/N_0 value under the same BER as shown in Fig. 3.2.

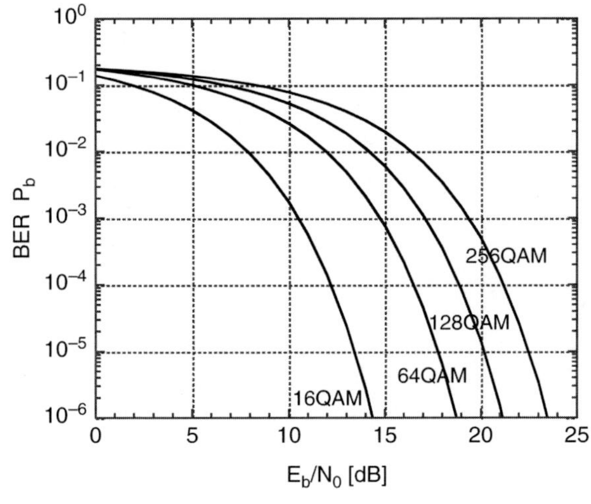

Fig. 3.2 BER of 16, 64, 128, and 256 QAM as a function of E_b/N_0

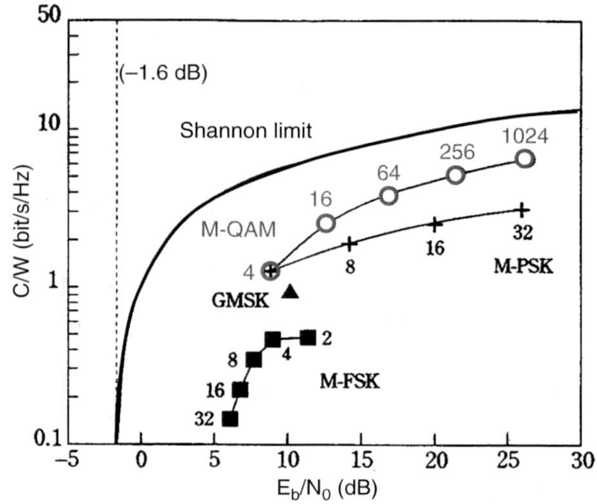

Fig. 3.3 Spectral efficiency of M-ary QAM signal and the Shannon limit. E_b/N_0 at BER $= 10^{-4}$ is shown assuming synchronous detection

The forward error correction (FEC) technique has been developed to realize a better BER performance with a lower E_b/N_0. Figure 3.4 shows the BER after applying FEC vs. Q_{in} value without FEC [15]. Q_{in} is the SNR given in equation (3.5), where I_1 and I_0 are the mean values and σ_1 and σ_0 are the standard deviations of the bits corresponding to 1 and 0, respectively. Here, the Shannon limit describes the lowest Q_{in} value needed to achieve an infinitely low BER by employing FEC under a certain code rate R:

$$\begin{aligned} R &= 1 + \mathrm{BER_{in}} \log_2 \mathrm{BER_{in}} \\ &\quad + (1 - \mathrm{BER_{in}}) \log_2 (1 - \mathrm{BER_{in}}) \\ \mathrm{BER_{in}} &= \frac{1}{2} \mathrm{erfc} \left(\frac{Q_{in}}{\sqrt{2}} \right) \\ Q_{in} &= \frac{I_1 - I_0}{\sigma_1 + \sigma_0}, \end{aligned} \tag{3.5}$$

which is known as the noisy channel coding theorem [14]. These results indicate the distinct possibility of realizing ultrahigh spectral efficiency by combining QAM and FEC.

The advantages of optical QAM transmission are summarized in Fig. 3.5. Although QAM was originally developed for wireless communications, it demonstrates its advantages most significantly in optical communication because of the broad bandwidth and the absence of fading noise in optical fibers. However, fiber nonlinearity is a limiting factor for coherent optical QAM transmission because of the power-dependent phase rotation caused by self-phase modulation (SPM) and cross-phase modulation (XPM) [16, 17]. As described later, SPM can be compensated by applying an appropriate phase rotation using software.

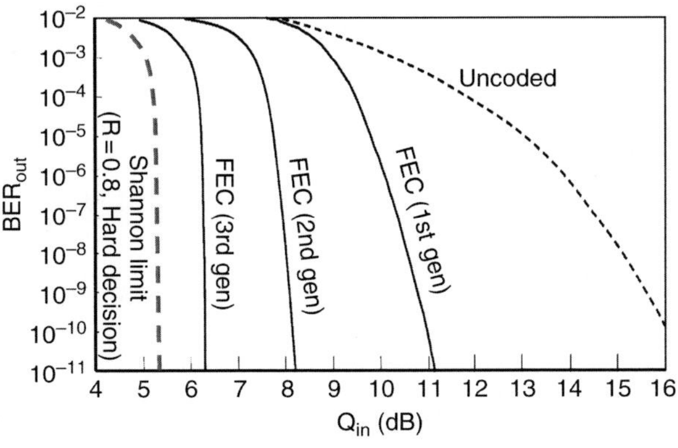

Fig. 3.4 Relationship between BER after FEC and Q value without FEC

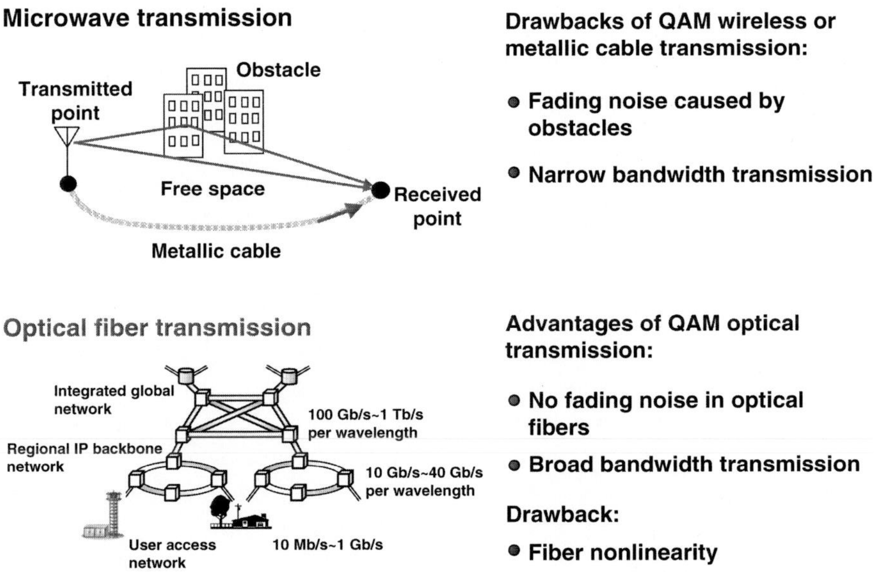

Fig. 3.5 Advantages of optical QAM transmission

3.3 Fundamental Configuration and Key Components of QAM Coherent Optical Transmission

The fundamental configuration of a QAM coherent optical transmission system is shown in Fig. 3.6. A CW, C_2H_2 frequency-stabilized fiber laser is employed as a coherent light source [17]. The optical QAM signal can be easily generated with an IQ modulator [18–21] consisting of two nested MZ modulators and a 90-degree phase shifter driven by QAM signals from an arbitrary waveform generator (AWG). A transmitted QAM signal and an LO signal are heterodyne detected with a photo detector (PD). Then, the optical QAM signal is converted to an IF signal. Here, an

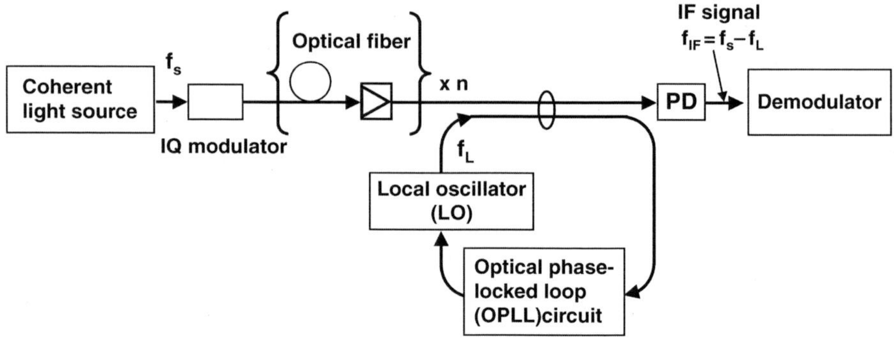

Fig. 3.6 Fundamental configuration for coherent QAM transmission

OPLL technique [22] using a high-speed free-running laser as an LO is also very important as regards the automatic frequency control of the IF carrier. The IF signal is then A/D converted and accumulated in a digital signal processor (DSP). All digital signals are demodulated into I and Q data and finally into a binary sequence in the DSP. Because of the software demodulation, this transmission system operates in an off-line condition.

The fundamental configuration for n-channel FDM-QAM transmission is shown in Fig. 3.7. All coherent sources are frequency locked and only one tone signal is used for the OPLL. After the transmission, a QAM signal is extracted by using a narrowband optical filter such as a fiber Bragg grating (FBG) [23]. At the OPLL, one LO can be used for all carriers by dividing the LO frequency into many frequencies.

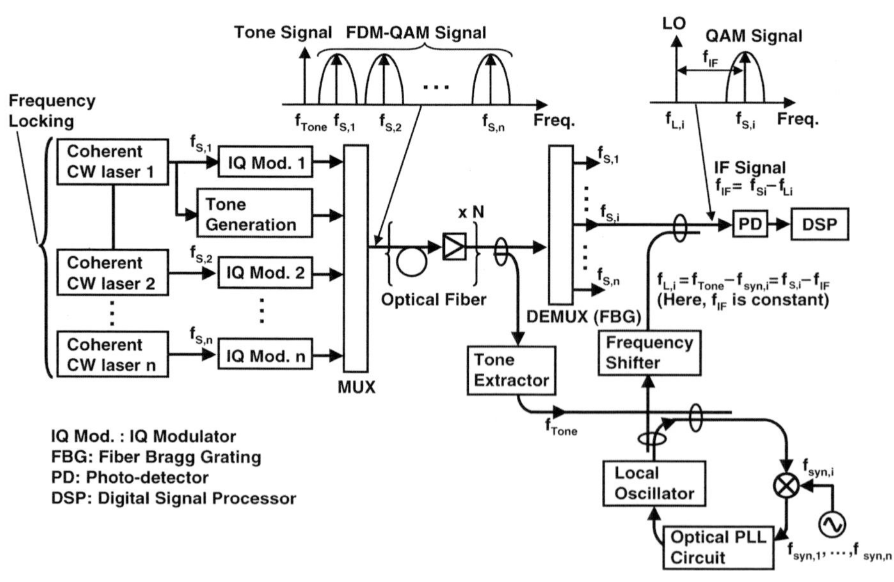

Fig. 3.7 Fundamental configuration for FDM-QAM transmission

The following sections present a detailed description of these key components for QAM coherent transmission, including the coherent light source, OPLL, optical IQ modulator, and digital demodulator.

3.3.1 C_2H_2 Frequency-Stabilized Erbium-Doped Fiber Ring Laser

A stable optical frequency in the 1.5 μm region is indispensable as a light source for QAM coherent optical transmission. C_2H_2 molecules have been utilized as a frequency standard to stabilize the frequency of semiconductor and fiber lasers at 1.55 μm [24]. Fiber lasers are particularly attractive as a coherent light source because of their narrow linewidth. Here we describe a frequency-stabilized, polarization-maintained erbium fiber ring laser that has no frequency modulation at the output beam.

Figure 3.8 shows the configuration of the $^{13}C_2H_2$ frequency-stabilized fiber ring laser [17]. The laser has two main parts: a tunable, polarization-maintained single-frequency fiber ring laser and a laser frequency stabilization circuit. The fiber laser is composed of a 1.48 μm laser diode (LD), polarization-maintained (PM) EDF, a WDM coupler, an output coupler, a PM optical circulator, and a 1.5 GHz ultra-narrow band polarization-maintained fiber Bragg grating (PM-FBG) filter [23]. The FBG filter makes it possible to realize single-frequency operation by selecting only one longitudinal mode from many oscillation modes. The laser has two kinds of frequency controllers. One is a drum-type PZT with EDF wound around it; the other is a multi-layer PZT (MLP) on which an FBG is laid. When these controllers operate synchronously with a phase-sensitive detection circuit, the laser frequency is continuously tuned over 2 GHz without mode hopping.

In the external frequency stabilization unit, we employed a phase-sensitive detection circuit composed of a $LiNbO_3$ (LN) frequency modulator, a $^{13}C_2H_2$ (3 Torr) cell, a photo detector (PD), a double-balanced mixer (DBM), an electrical amplifier,

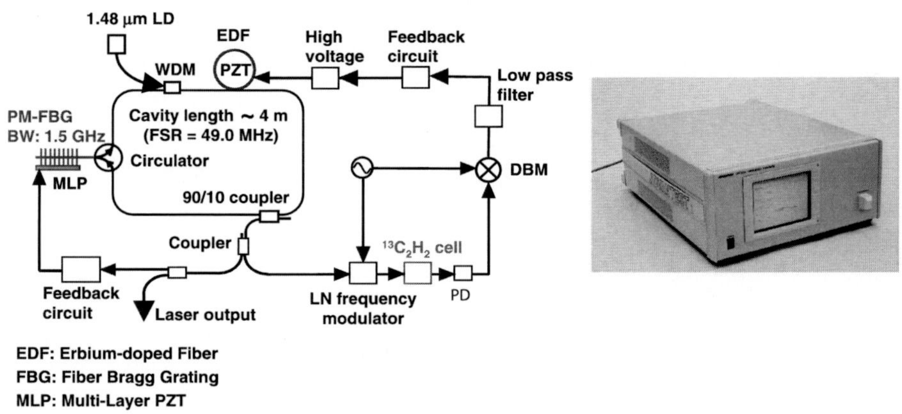

Fig. 3.8 Configuration of a $^{13}C_2H_2$ frequency-stabilized fiber laser and its overview

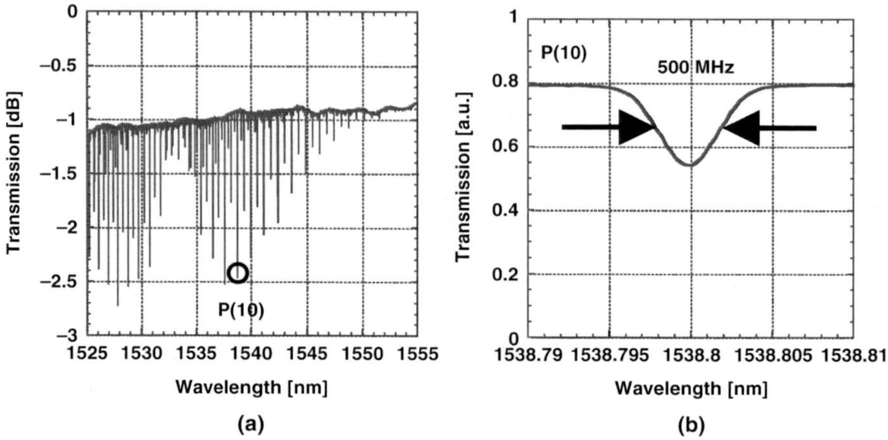

Fig. 3.9 Absorption characteristics of $^{13}C_2H_2$ molecules. **a** Observations of $^{13}C_2H_2$ absorption lines for long span. **b** P(10) linear absorption line

a low-pass filter (LPF), and a feedback circuit providing proportional and integral (PI) control. With this configuration, we can detect the frequency deviation of a fiber laser from the center frequency of a $^{13}C_2H_2$ absorption line. Figure 3.9 shows the absorption characteristics of $^{13}C_2H_2$ molecules. Of these absorption lines, we used the P(10) linear absorption line as a frequency reference as shown in Fig. 3.9b. This absorption line has a center wavelength of 1538.8 nm and a spectral linewidth of 500 MHz. The DBM generates a voltage error signal that is proportional to the frequency deviation from the absorption center. The error signal is fed back to the PZT to control the laser frequency.

We measured the laser linewidth using a delayed self-heterodyne detection method with a 50 km delay fiber [25]. Figure 3.10a shows the electrical spectrum

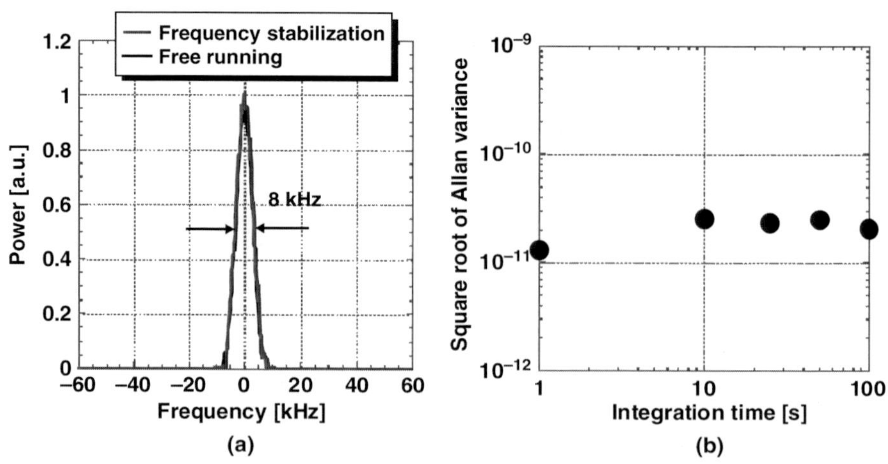

Fig. 3.10 Electrical spectrum of delayed self-heterodyne signal (**a**) and the square root of the Allan variance of the frequency fluctuation of a $^{13}C_2H_2$ frequency-stabilized fiber laser (**b**)

of the heterodyne signal. The linewidth was only 4 kHz. The linewidth of the frequency-stabilized fiber laser was almost the same as that of the free-running fiber laser. We also evaluated the frequency stability from the beat note signals of two lasers. Figure 3.10b shows the square root of the Allan variance estimated from the fluctuation of the beat note signals [26]. For an integration time, τ, of 1 s, the square root of the Allan variance of the fiber laser is 1.3×10^{-11}, which corresponds to a frequency fluctuation of 2.6 kHz. For a τ of 100 s, the Allan variance is 2.0×10^{-11} corresponding to a frequency fluctuation of 4 kHz.

The external frequency modulation scheme has also been applied to the frequency stabilization of an external cavity laser diode. In this case an FM-eliminated output beam with a linewidth of only 4 kHz and a RIN as low as -135 dB/Hz was obtained, and the frequency stability reached as high as 2.3×10^{-11} for $\tau = 1$ s [27]. Such a low-intensity noise is difficult to realize with a fiber laser owing to the small power fluctuation associated with the relaxation oscillation of erbium ions.

3.3.2 Optical PLL for Coherent Transmission Using Heterodyne Detection with Fiber Lasers

The precise optical phase control of light sources is very important for coherent optical transmission with heterodyne detection. The optical frequency difference between a transmitter and a local oscillator (LO) must be kept constant in order to obtain a stable intermediate frequency (IF) signal. In a heterodyne detection system, the use of a high-speed OPLL is a key technique for automatic frequency control. The linewidth of the IF signal is evaluated as

$$\sigma_\phi^2 = \frac{\delta f_\mathrm{T} + \delta f_\mathrm{L}}{2 f_\mathrm{c}}, \tag{3.6}$$

where δf_T and δf_L are the linewidth of the transmitter and LO, and f_c is the bandwidth of the feedback circuit [28]. This indicates that the reduction of the phase noise (linewidth) of the two lasers and the large bandwidth of the feedback circuit are very important factors as regards realizing a precise OPLL. Of the many available lasers, the fiber laser is suitable for an OPLL because of its low phase noise (narrow linewidth), because this allows the laser to be applied directly to an OPLL system.

Figure 3.11 shows the OPLL configuration [22]. This system is composed of an LO, a PD, a DBM, a synthesizer, and two feedback circuits. A $^{13}C_2H_2$ frequency-stabilized fiber ring laser is used as a transmitter, and a free-running fiber laser is used as an LO whose configuration is almost the same as that of the transmitter except that an LN modulator was adopted in the laser cavity for the high-speed tracking of the IF signal. The LO linewidth is also approximately 4 kHz. The signal from the transmitter is heterodyne detected with the LO signal. The phase of the beat signal (IF signal: $f_\mathrm{IF} = |f_\mathrm{S} - f_\mathrm{L}|$) is compared with the phase of the reference signal

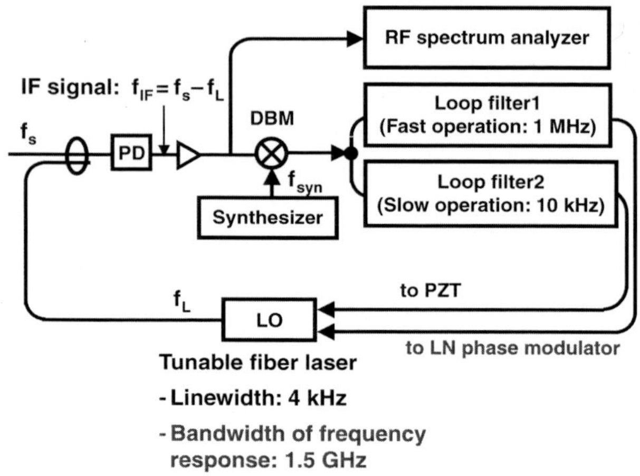

Fig. 3.11 Configuration of OPLL for coherent transmission

from the synthesizer (f_{syn}) by the DBM and the difference between them is fed back to the LO through the feedback circuits. The phase noise of the OPLL is mainly dominated by the loop bandwidth. The bandwidth of the fiber laser shown in Fig. 3.8 was determined by the response characteristic of the PZT, and therefore we have improved the LO response time by using an LN modulator. With this configuration, we could obtain an FM bandwidth for the LO of up to 1 GHz. We also used a PZT tuner to compensate for slow frequency drifts caused by temperature fluctuations. These two feedback circuits have loop filters with different bandwidths. One is a broadband filter (~1 MHz) for fast frequency tuning with the LN modulator, the other is a narrow band filter (~10 kHz) for slow frequency tuning with the PZT. The IF signal is measured with an electrical spectrum analyzer through the PD.

Figure 3.12a and b shows the IF spectrum and the single-sideband (SSB) phase noise spectrum. The linewidth of the spectrum was less than 10 Hz, which was below

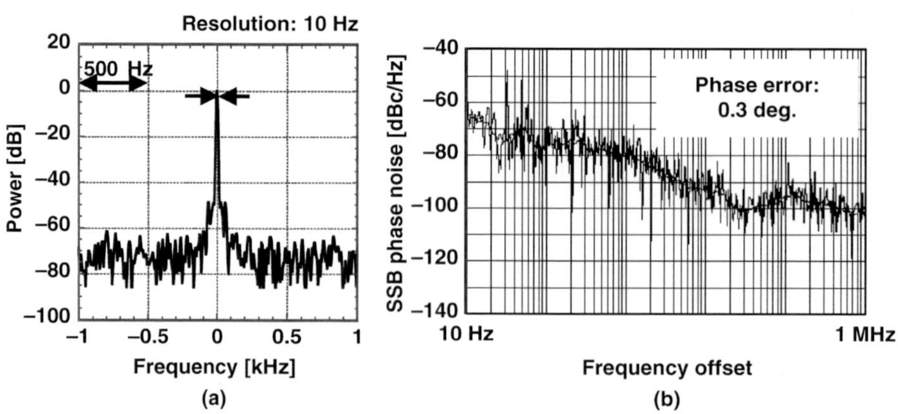

Fig. 3.12 IF signal characteristics. **a)** IF signal spectrum and **b)** SSB phase noise spectrum

the measurement resolution as shown in Fig. 3.12a. The phase noise variance (RMS) of the IF signal estimated by integrating the SSB noise spectrum was as low as 0.3 degree, which indicated that a stable OPLL operation was successfully achieved under a low phase noise condition. Such a low phase noise of the IF signal in spite of the relatively large PLL bandwidth is attributed to the narrow linewidth of the fiber laser.

The tolerance of the phase noise for 64, 128, and 256 QAM signals can be estimated from constellation maps. As shown in Fig. 3.13, the angle between the two closest symbols is $2\delta\phi = 4.7$, 2.7, and 2.0 degree for 64, 128, and 256 QAM, respectively, which correspond to the tolerable phase noise. Therefore, the RMS phase noise of 0.3 degree is sufficiently small for demodulating even a 256 QAM signal.

Recently, a technique using a digital coherent receiver has been intensively studied, in which the phase fluctuation between a signal and an LO is eliminated with a DSP [29]. This technique does not require an OPLL for phase synchronization and therefore enables us to realize coherent transmission using free-running lasers as a light source and an LO. Table 3.1 compares coherent transmission using frequency-stabilized lasers with an OPLL and free-running lasers with a digital coherent receiver. Digital coherent transmission relaxes the requirements as

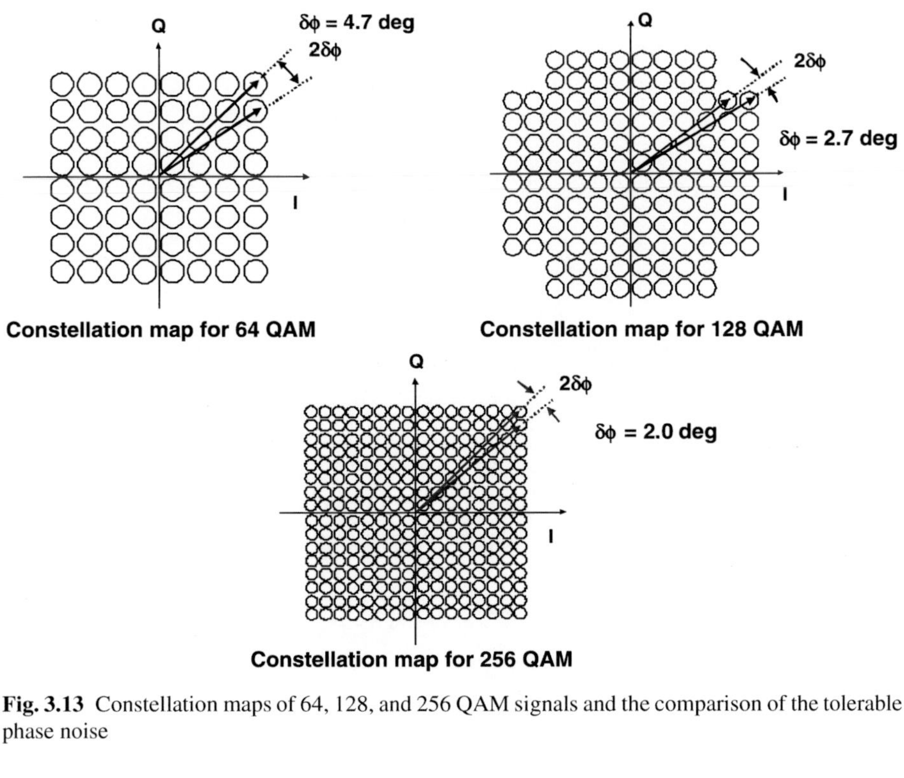

Fig. 3.13 Constellation maps of 64, 128, and 256 QAM signals and the comparison of the tolerable phase noise

Table 3.1 A comparison of coherent transmission using frequency-stabilized lasers with OPLL and free-running lasers with a digital coherent receiver

Lasers	Software processing	PLL	Number of M-ary	FDM	Dispersion compensation, equalization
Frequency-stabilized laser + OPLL	Easy IF processing	OPLL is required	Stable IF →256 QAM is possible	No drift of optical carrier	Software and hardware
Free-running laser	Multiplying the inverse function of IP fluctuation; complicated processing is required	OPLL is not required	Unstable IF → insufficient phase accuracy limited to 16~32 QAM	Drift control of optical carrier is required	Software and hardware

regards the linewidth and frequency stability of a coherent light source, and the IF fluctuation can be reduced in the DSP without employing an OPLL. However, the IF stability thus obtained may not be sufficient for higher QAM multiplicity, and in addition, optical carrier drift control is required for FDM transmission. With frequency-stabilized lasers and an OPLL, an IF signal that is sufficiently stable even for 256 QAM can be obtained as described above, and we can obtain stable optical carriers for FDM transmission. It should be noted that the DSP for dispersion compensation and equalization plays an important role for both transmissions.

3.3.3 Optical IQ Modulator

The configuration of an optical IQ modulator is shown in Fig. 3.14 [19]. It is composed of three Mach–Zehnder interferometers (MZIs), in which the sub-MZIs, MZ_A and MZ_B, are installed in each arm of the main MZI (MZ_C). I and Q optical data are generated individually with MZ_A and MZ_B using the I and Q components of the QAM signal supplied by the AWG. They are combined with MZ_C with a dc bias DC_C so that a 90-degree phase shift is introduced between the two signals. An optical IQ modulator has been realized using an LN modulator [19, 20] and a monolithic InP modulator [21].

In the LN-based IQ modulator, surface acoustic waves are generated by the piezoelectric effect in the LN crystal, which degrades the low-frequency response of the modulator [30]. To suppress the acoustic wave, we tapered the edge of the modulator and reduced its thickness. Figure 3.15a and b shows the E/O characteristics of IQ modulators with the conventional and new structures, respectively. The

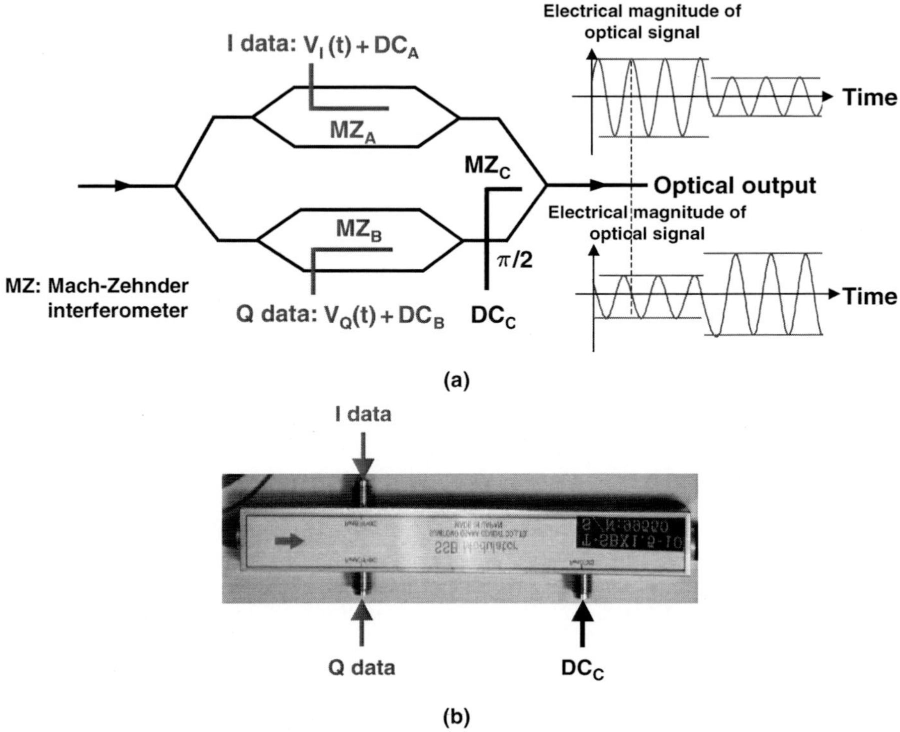

Fig. 3.14 Configuration of IQ modulator (**a**) and overview (**b**)

low-frequency response was successfully improved with the new structure. This improvement plays a very important role in increasing the multiplicity level in QAM transmission.

3.3.4 Digital Demodulator

Figure 3.16 shows a schematic diagram of our digital demodulator. The IF signal data are first A/D-converted and accumulated in a high-speed digital scope, whose sampling frequency, bandwidth, and vertical resolution are 40 Gsample/s, 12 GHz, and 8 bits, respectively. Then I and Q data are demodulated with software by multiplying synchronous cosine and sine functions, respectively, onto $I+Q$ data. Finally the demodulated data are converted into binary data in the software decoder. Here the center frequency of the IF signal is determined by the operating frequency of the synthesizer used in the OPLL circuit. In this off-line system, we send the frequency information to the DSP, and the clock signal used for the IQ demodulation is recovered by software processing.

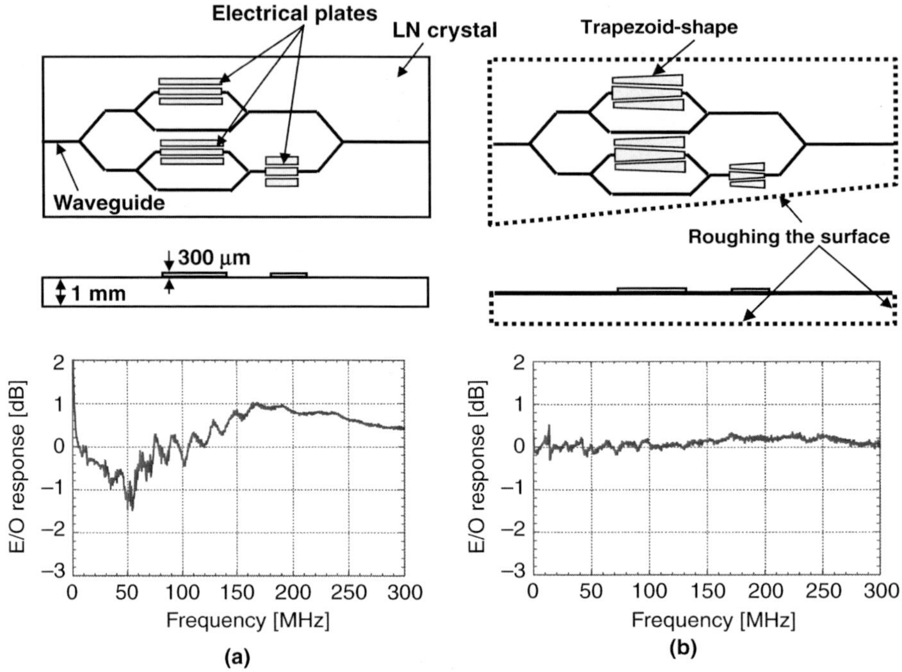

Fig. 3.15 Improvement of IQ modulator. Schematic diagram of the modulator and its E/O characteristics (**a**) before and (**b**) after improvement

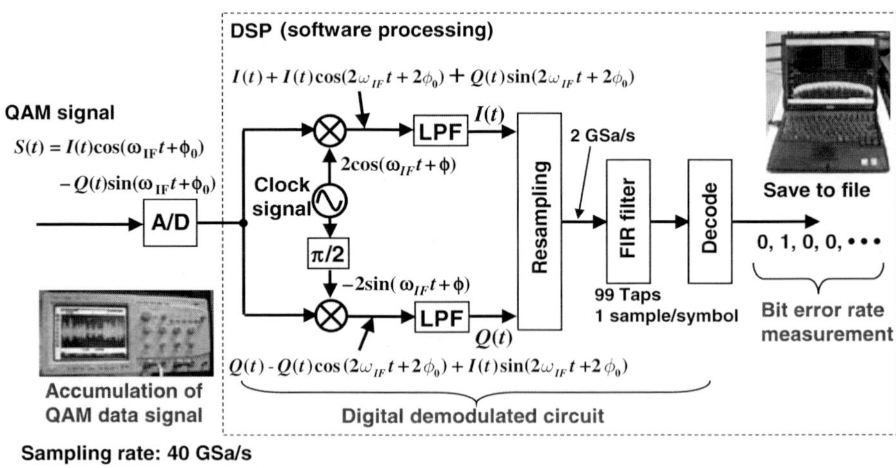

Fig. 3.16 Schematic diagram of digital demodulator

3.4 Single-Channel 1 Gsymbol/s, 128 QAM Transmission

We undertook a single-channel 128 QAM transmission experiment over 160 km based on the configuration described above [31]. By introducing polarization multiplexing in a 1 Gsymbol/s 128 (2^7) QAM transmission, a data speed of 14 Gbit/s was obtained.

3.4.1 1 Gsymbol/s, 128 QAM Transmission Setup

The experimental setup is shown in Fig. 3.17. The laser output is split into two arms via an erbium-doped fiber amplifier (EDFA). One arm is coupled to two IQ modulators. Two AWGs are used to feed independent QAM signals into two orthogonal polarizations. We employed a raised-cosine Nyquist filter at the AWG using a software program to reduce the bandwidth of the QAM signal. It is well known in the microwave communication field that a Nyquist filter is very useful for reducing the bandwidth of a data signal without introducing intersymbol interference [32, 33]. Figure 3.18 shows the transfer function and impulse response of the raised-cosine Nyquist filter. The transfer function is given by

$$H(f) = \begin{cases} \frac{1}{2}\left\{1 - \sin\frac{\pi(f-0.5)}{\alpha}\right\}, & 0.5 - \frac{\alpha}{2} \le |f| < 0.5 + \frac{\alpha}{2} \\ 1, & |f| < 0.5 - \frac{\alpha}{2} \\ 0, & |f| \ge 0.5 + \frac{\alpha}{2} \end{cases} \tag{3.7}$$

where α is called a roll-off factor. As shown in Fig. 3.18b, the impulse response becomes zero at the location of adjacent symbols. This indicates that the bandwidth can be reduced with the Nyquist filter while avoiding the intersymbol interference. We employed a raised-cosine Nyquist filter with $\alpha = 0.35$ so that the bandwidth of the QAM signal is reduced from 2 GHz to less than 1.4 GHz.

The other frequency-stabilized beam is coupled to an optical frequency shifter (OFS), which provides a frequency upshift of 2.5 GHz against the signal. Then the frequency-shifted signal is used as a pilot signal that tracks the optical phase of an LO (tunable tracking laser) under OPLL operation. The polarization of the pilot signal is set so that it is the same as one of the polarization axes of the two QAM signals. After optical amplification with EDFAs, the two orthogonally polarized 128 QAM signals are combined with the pilot signal and these signals are coupled into a 160 km long single-mode fiber (SMF) with a coupled power of −6.5 dBm. After the transmission, the two QAM signals are heterodyne detected with an LO signal whose phase is locked to the pilot signal. Since the LO polarization can be arbitrarily rotated with a polarization controller, two orthogonally polarized QAM signals can be independently detected. Finally an IF data signal is received with a digital demodulator with an A/D converter and a DSP.

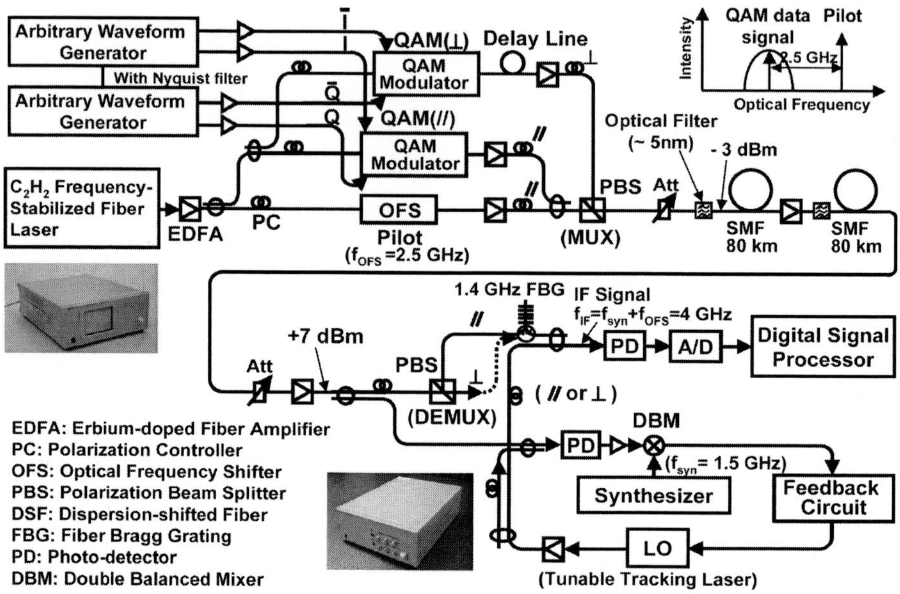

Fig. 3.17 Experimental setup for polarization-multiplexed 128 QAM coherent transmission over 160 km

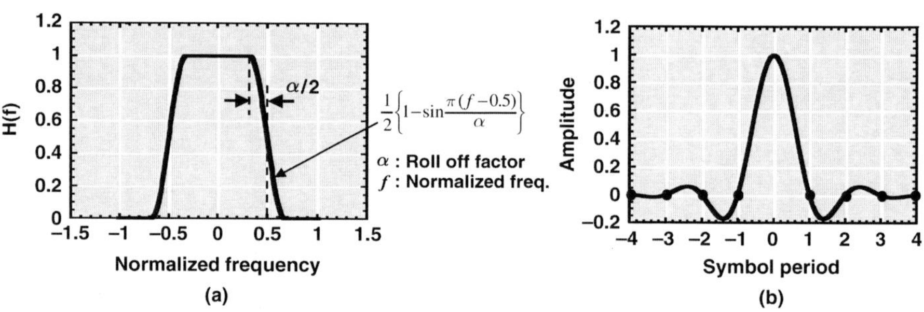

Fig. 3.18 Transfer function (**a**) and impulse response (**b**) of the raised-cosine Nyquist filter

3.4.2 Transmission Results

Figure 3.19 shows the electrical spectrum of the IF data signal when the polarization of the QAM signal is orthogonal to that of the pilot signal. The inset figure shows the relationship between the optical frequencies of the QAM data-modulated signals, the pilot signal, and the LO signal. By using a raised-cosine Nyquist filter, the bandwidth of the QAM data signal was successfully reduced from 2 GHz to less than 1.4 GHz as shown in Fig. 3.19(b). Here the demodulation bandwidth was set at 1.4 GHz in the optical domain by using a narrowband FBG filter [23]. Almost the same spectrum was obtained for the parallel data condition.

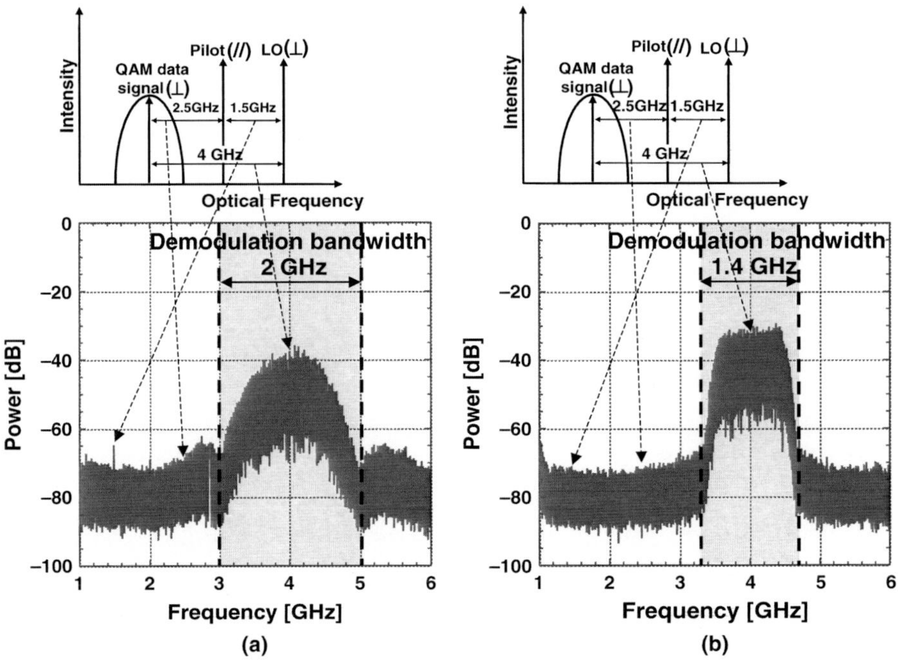

Fig. 3.19 Electrical spectrum of the IF data signal (**a**) without and (**b**) with Nyquist filter

Figure 3.20 shows the BER characteristics of a polarization-multiplexed 128 QAM transmission. The word length of the QAM signal was limited to 4096 symbols by the DSP memory, which corresponds to a BER of up to 3.5×10^{-5}. The power penalty was approximately 3 dB for both polarizations, but no error was observed for a word length of 4096 when the received power level was above −29.5 dBm. Figure 3.21a–c is the constellations and eye patterns for back-to-back and 160 km transmissions with orthogonal and parallel polarizations, respectively. Clear constellations and eye patterns were obtained at received power of −26.5 dBm for both polarization data. As a result, 14 Gbit/s data were successfully transmitted over 160 km with an optical bandwidth of 1.4 GHz.

3.4.3 SPM Compensation

The power penalty is likely to be caused by self-phase modulation (SPM) during the propagation. SPM induces a power-dependent phase shift, which rotates the constellation map and causes bit error [16, 34]. Therefore the transmission power is limited by SPM. It should be noted that the rotation of the constellation map is larger for symbols away from the center. SPM can be compensated using software by

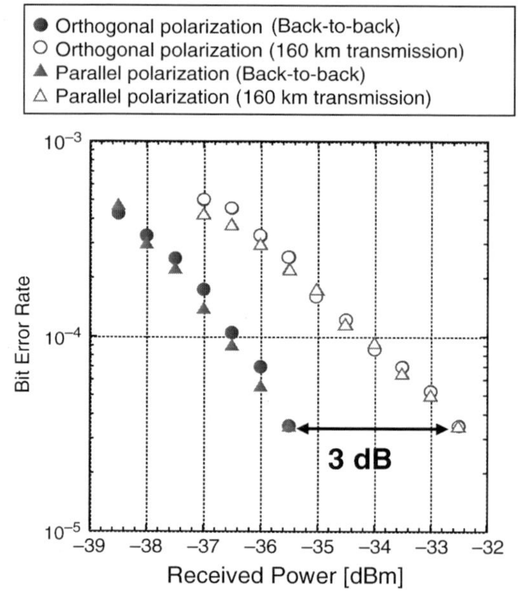

Fig. 3.20 BER characteristics for 1 Gsymbol/s, 128 QAM transmission over 160 km

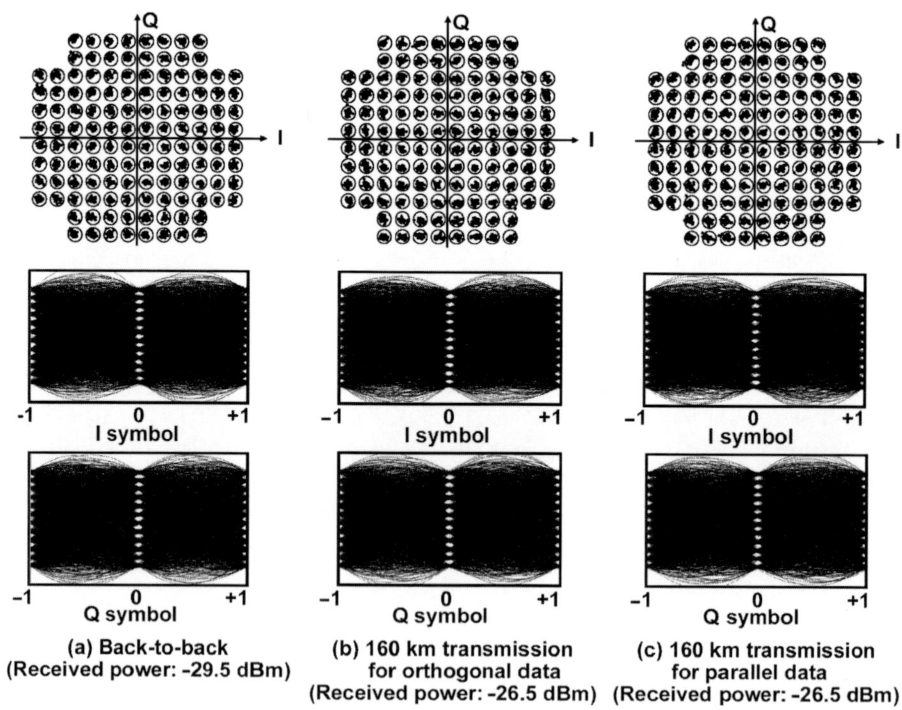

Fig. 3.21 Constellations and eye patterns for back-to-back transmissions (**a**) and 160 km transmissions with (**b**) orthogonal and (**c**) parallel polarization

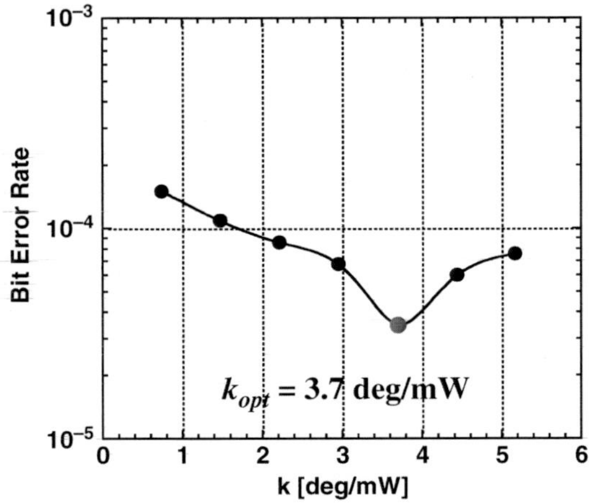

Fig. 3.22 BER vs. SPM compensation factor k for 160 km transmission of 1 Gsymbol/s, 128 QAM signal with a transmission power of 0 dBm

introducing an artificial phase rotation that cancels the SPM-induced phase rotation [35]. The SPM-induced phase rotation is given by

$$\Delta\phi = \gamma P L_{\text{eff}} \times N, \tag{3.8}$$

$$L_{\text{eff}} = \frac{1}{\alpha}\int_0^L e^{-\alpha z} dz, \tag{3.9}$$

where γ is a nonlinear coefficient, P is the transmission power, L is the span length, L_{eff} is the effective span length that takes account of the fiber loss α defined as equation (3.9), and N is the number of spans. From the parameters corresponding to the present experiment, $\gamma = 1.5\,\text{W}^{-1}\text{km}^{-1}$, $\alpha = 0.2$ dB/km ($0.046\,\text{km}^{-1}$), $L = 80$ km, and $N = 2$, we obtain $\Delta\phi/P = 3.6$ deg/mW.

We introduced SPM compensation using software at the AWG by adding a phase shift

$$\Delta\phi_{\text{c}} = -k\left[P_{\text{I}}(t) + P_{\text{Q}}(t)\right] \tag{3.10}$$

to the QAM signal before transmission, where k is a constant and P_{I} and P_{Q} are the optical powers of the I and Q data, respectively. Figure 3.22 shows the relation between k and the BER after transmission. The BER was lowest with $k = 3.7$ deg/mW. This result agrees well with the analytical SPM-induced phase shift of 3.6 deg/mW.

3.4.4 Comparison with Theoretical OSNR Limit

Here we compare the measured BER performance with the theoretical limit shown in Fig. 3.2. To compare the experimental and theoretical results, we need to convert the received optical power P_{rec} to E_{b}/N_0. The received optical power can be converted to OSNR and E_b/N_0 in the following way. The signal and noise power after a preamplifier with gain G and noise figure NF are given by $P_{sig} = GP_{rec}$ and $P_{ASE} = n_{sp}h\nu(G-1)\Delta\nu$, respectively, where $n_{sp} = \mathrm{NF}/2$ is the spontaneous emission factor, h is Planck's constant, ν is the optical frequency, and $\Delta\nu$ is the detection bandwidth of the optical signal. Here, the power is defined in terms of single polarization. Therefore the OSNR is obtained as

$$\begin{aligned} \mathrm{OSNR} &= \frac{P_{sig}}{P_{ASE}} = \frac{P_{rec}G}{n_{sp}h\nu(G-1)\Delta\nu} \cong \frac{P_{rec}}{n_{sp}h\nu\Delta\nu}, \\ &= \frac{P_{rec}}{(\mathrm{NF}/2)\,h\nu\Delta\nu} \end{aligned} \tag{3.11}$$

where we assumed $G >> 1$. In dB unit, Eq. (3.11) is written as

$$\begin{aligned} \mathrm{OSNR[dB]} &= 10\log_{10} P_{rec} - NF[\mathrm{dB}] + 3 - 10\log_{10}(h\nu\Delta\nu) \\ &= P_{rec}[\mathrm{dBm}] - 30 - NF[\mathrm{dB}] + 3 + 88, \\ &= 61 + P_{rec}[\mathrm{dBm}] - NF[\mathrm{dB}] \end{aligned} \tag{3.12}$$

where we used $h = 6.626 \times 10^{-34}$ J s, $\nu = 194$ THz, and $\Delta\nu = 12.5$ GHz [36]. From Eq. (3.11) and (3.12), E_b/N_0 is given by

$$\begin{aligned} E_b/N_0 &= \frac{P_{sig}/R}{N_0} = \frac{\Delta\nu}{R}\frac{P_{sig}}{N_0\Delta\nu}, \\ &= \frac{\Delta\nu}{R}\frac{P_{sig}}{P_{ASE}} = \frac{\Delta\nu}{R}\mathrm{OSNR} \end{aligned} \tag{3.13}$$

$$E_b/N_0[\mathrm{dB}] = 10\log_{10}(\Delta\nu/R) + \mathrm{OSNR[dB]}. \tag{3.14}$$

Here, R is the data rate per polarization. In our experiment, $R = 7$ Gbit/s, $\Delta\nu = 12.5$ GHz, and the received optical power at BER $= 2 \times 10^{-3}$ is $P_{rec} = -41.7$ dBm. From equations (3.12) and (3.14), this corresponds to OSNR $=$ 15.3 dB and $E_b/N_0 = 17.8$ dB. On the other hand, from Fig. 3.2, the theoretical limit of E_b/N_0 for 128 QAM at this BER is 16.4 dB, which is 1.4 dB lower than the experimental value. This difference may be attributed to the imperfect implementation of such hardware as the AWG and the IQ modulator.

3.5 128 QAM-FDM Transmission with a Spectral Efficiency of 10 bit/s/Hz

To meet the increasing demand to expand optical transport capacity, intensive efforts have been made to achieve a high spectral efficiency in a dense WDM (DWDM) transmission system. The highest reported spectral efficiency for a 100 Gbit/s channel is 6.2 bit/s/Hz, which was achieved with polarization-multiplexed 16 QAM [37]. As described in the previous section, a spectral efficiency as high as 10 bit/s/Hz is expected for FDM transmission with 128 QAM. So as a next step, we undertook a six-channel FDM 1 Gsymbol/s, 128 QAM coherent transmission experiment with a 1.4 GHz spacing [38].

Figure 3.23 shows the experimental setup. Six carriers are generated from two coherent light sources, whose frequencies differ by 1.4 GHz, with multi-carrier generators (MCGs) consisting of an LN phase modulator driven at a modulation frequency of 2.8 GHz. These carriers are coupled to IQ modulators and combined to obtain a six-channel 128 QAM-FDM signal. The signal is then polarization multiplexed and combined with the pilot signals. The transmission power is set at −2 dBm. At the receiver, a 1.4 GHz FBG filter is used to extract an FDM channel. In the OPLL circuit, the frequency of the synthesizer, f_{syn}, is varied depending on the channel.

Figure 3.24a shows the electrical spectrum of the beat signal between the QAM signals and a single-frequency fiber laser. There were six FDM channels with the same power levels and residual harmonic signals with a power level of about 10% of the total power of the QAM signals. Figure 3.24b shows the electrical spectrum of the IF signal of the Channel 4 data when the polarization of the signal was orthogonal to that of the pilot tone signal. Although there were some residual components of Channels 3 and 5, they were removed with a digital filter in the DSP. Almost the

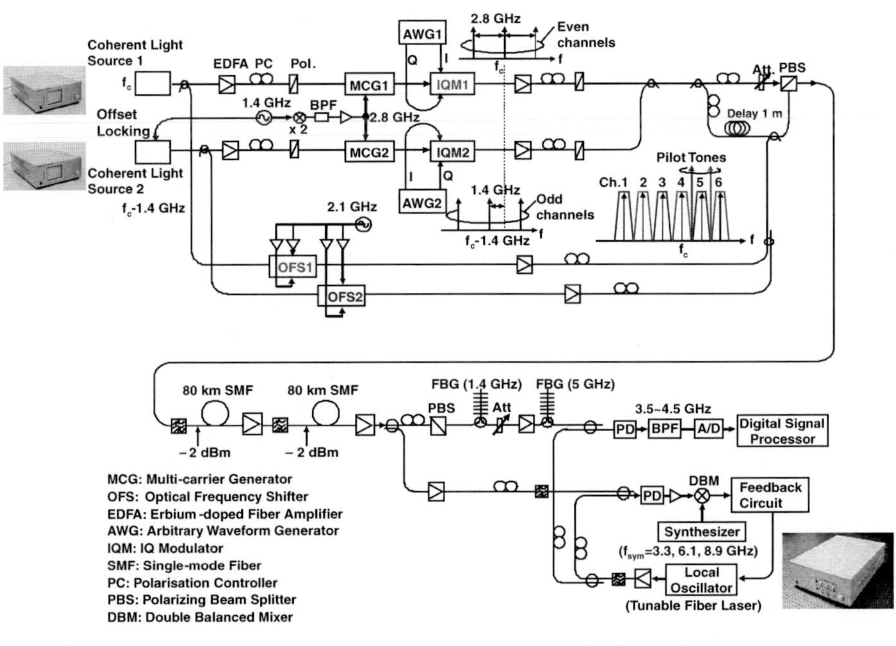

Fig. 3.23 Experimental setup for polarization-multiplexed 6 × 128 QAM-FDM coherent transmission over 160 km

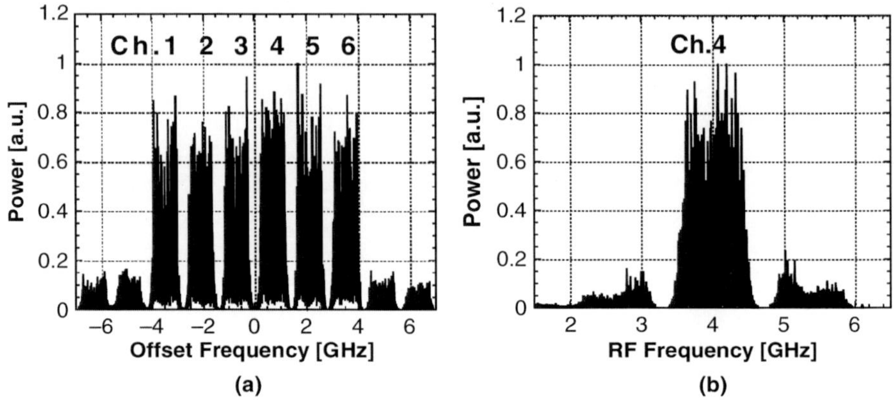

Fig. 3.24 Electrical spectrum of the beat signal between the QAM signals and a single-frequency fiber laser (**a**) and the IF signal of the Channel 4 data (**b**)

same spectrum was obtained for the parallel data condition. A residual pilot tone signal was also removed by the digital filter.

Figure 3.25 shows the BER characteristics. We measured the BER 10 times and plotted the average values in Fig. 3.25. On the horizontal axis, the received power is converted to that of each channel. The BER characteristics were almost the same for all the channels. The power penalty was approximately 5.3 dB at a BER of 3.5×10^{-5} for all the channels, but no error was observed for the received power above −27.5 dBm. These results indicate that polarization-multiplexed 6×1 Gsymbol/s, 128 QAM-FDM data were successfully transmitted over 160 km with a 1.4 GHz spacing, and therefore the spectral efficiency reached as high as 10 bit/s/Hz with a total capacity of 84 Gbit/s. It is important to note that four-wave mixing (FWM) is not detrimental even to FDM coherent transmission over 160 km.

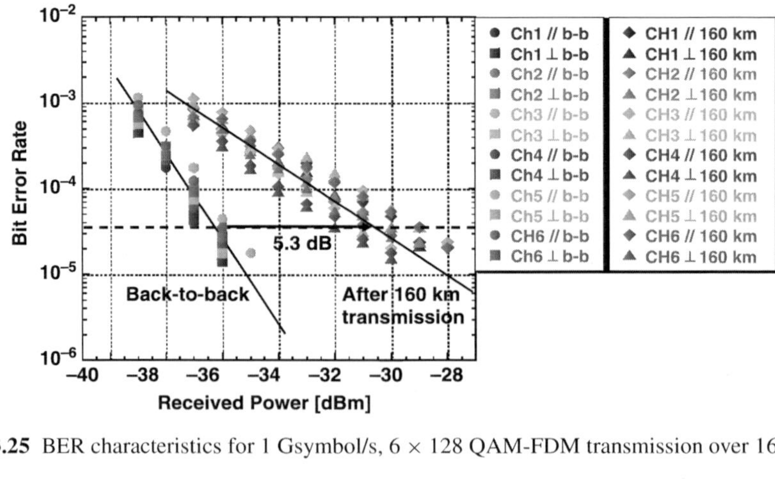

Fig. 3.25 BER characteristics for 1 Gsymbol/s, 6×128 QAM-FDM transmission over 160 km

3.6 64 QAM-OFDM Coherent Transmission

3.6.1 Principle of OFDM Transmission

Another attractive candidate for transmission with high spectral efficiency is orthogonal frequency division multiplexing (OFDM), where the multi-carrier transmission of low-speed orthogonal subcarriers enables us to improve the spectral efficiency [39]. The baseband OFDM signal can be represented as

$$S(t) = \sum_{n=0}^{N-1} \left\{ a_n(t)\cos\left(2\pi n\frac{t}{T}\right) - b_n(t)\sin\left(2\pi n\frac{t}{T}\right) \right\}, \tag{3.15}$$

where $a_n(t)$ and $b_n(t)$ are the symbol data, T is the symbol period, and N is the number of subcarriers. The subcarrier spacing is set so that it is equal to the symbol rate $1/T$, which allows orthogonality between the subcarriers

$$\frac{2}{T}\int_{t_0}^{t_0+T} \cos\left(2\pi n\frac{t}{T}\right)\cos\left(2\pi k\frac{t}{T}\right)dt = \begin{cases} 1, & n = k \\ 0, & n \neq k \end{cases}. \tag{3.16}$$

A typical example of an OFDM signal spectrum and waveform is shown in Fig. 3.26. Although the spectrum of each subcarrier overlaps, the oscillating tail of each subcarrier spectrum intersects at the zero level. The data can be easily demodulated in spite of the spectral overlap by extracting a subcarrier using the orthogonality. For example, the data $a_k(t)$ can be demodulated in the following way:

$$\begin{aligned} &\frac{2}{T}\int_{t_0}^{t_0+T} S(t)\cos\left(2\pi k\frac{t}{T}\right)dt \\ &\quad = \frac{2}{T}\sum_{n=0}^{N-1}\left\{ a_n(t_0)\int_{t_0}^{t_0+T} \cos\left(2\pi n\frac{t}{T}\right)\cos\left(2\pi k\frac{t}{T}\right)dt \right. \\ &\qquad \left. -b_n(t_0)\int_{t_0}^{t_0+T} \sin\left(2\pi n\frac{t}{T}\right)\cos\left(2\pi k\frac{t}{T}\right)dt \right\} \\ &\quad = a_k(t_0)\,, \end{aligned} \tag{3.17}$$

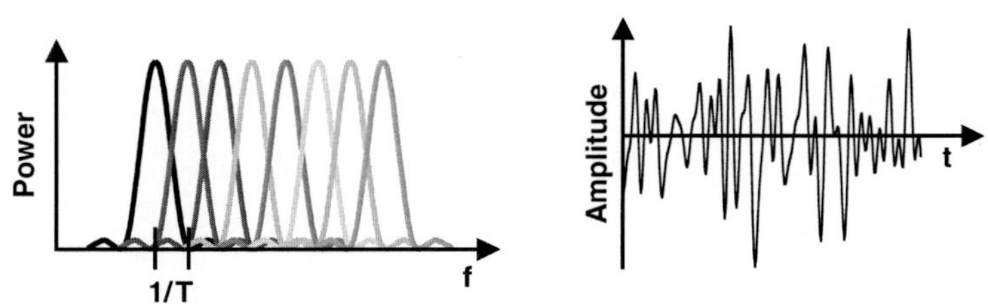

Fig. 3.26 Typical example of an OFDM signal spectrum and waveform

Subcarrier multiplexing with a spectral overlap allows us to improve the spectral efficiency and thus the dispersion tolerance. By virtue of these features, a 1 Tbit/s per channel OFDM transmission has been demonstrated [10]. It should be noted, however, that OFDM is more sensitive to nonlinear effects than a single-carrier transmission owing to the large peak-to-average power ratio (PAPR). This occurs as a result of phase coherence among subcarrier signals.

Figure 3.27 shows a block diagram of baseband OFDM signal generation and demodulation. At the transmitter, as shown in Fig. 3.27a, the binary data are first encoded with a multi-level format such as QPSK or QAM. A training symbol used for amplitude and phase equalization of subcarriers is then added and the data are converted into a parallel sequence and divided into N subcarriers using inverse fast Fourier transformation (IFFT). After adding a guard interval and employing D/A conversion, a baseband OFDM signal is obtained. At the receiver, the IF signal is synchronously detected, and after removing the guard interval and extracting the training symbol, the OFDM signal is demodulated via the fast Fourier transformation (FFT) of the subcarriers. The demodulated signal is finally converted to binary data.

OFDM has been realized with several configurations in optical communication including direct-detection OFDM [9, 40], coherent optical OFDM [10, 11, 41, 42], coherent WDM [43, 44], and no-guard interval OFDM [12]. In the latter two cases, subcarriers are generated in the optical domain. To achieve higher spectral efficiency in OFDM, high-level subcarrier modulation formats and coherent detection play a very important role. A spectral efficiency of 7.0 bit/s/Hz has recently been realized with 32 QAM subcarrier modulation, where an 8 $\times$ 65.1 Gbit/s PDM-OFDM signal was transmitted with a channel spacing of 8 GHz [11]. We can expect the spectral efficiency to be enhanced further by increasing the QAM multiplicity level up to 64 states and beyond.

3.6.2 24 Gbit/s, 64 QAM-OFDM Coherent Transmission Experiment

We realized QAM subcarrier modulation up to 64 levels and performed polarization-multiplexed 2 Gsymbol/s 64 QAM-OFDM transmission over a 160 km standard single-mode fiber (SSMF) [45]. The experimental setup is shown in Fig. 3.28. At the transmitter, the data for transmission are divided into 256 subcarriers and fed into two AWGs after performing the IFFT and adding guard intervals (GI) and training symbols (TS) to the sequence. Here, the OFDM symbol rate per subcarrier is 7.8 Msymbol/s, the FFT size is 512, the GI length is set at 64 samples (64 ns) per OFDM symbol, and 4 out of every 252 symbols in the OFDM signal are assigned as TS. The laser output is split into two arms, and one is coupled to an optical IQ modulator, where the coherent light is modulated with the 2 Gsymbol/s OFDM baseband signal generated at the AWGs. The optical OFDM signal is then polarization multiplexed by splitting the signal with an optical coupler and combining the

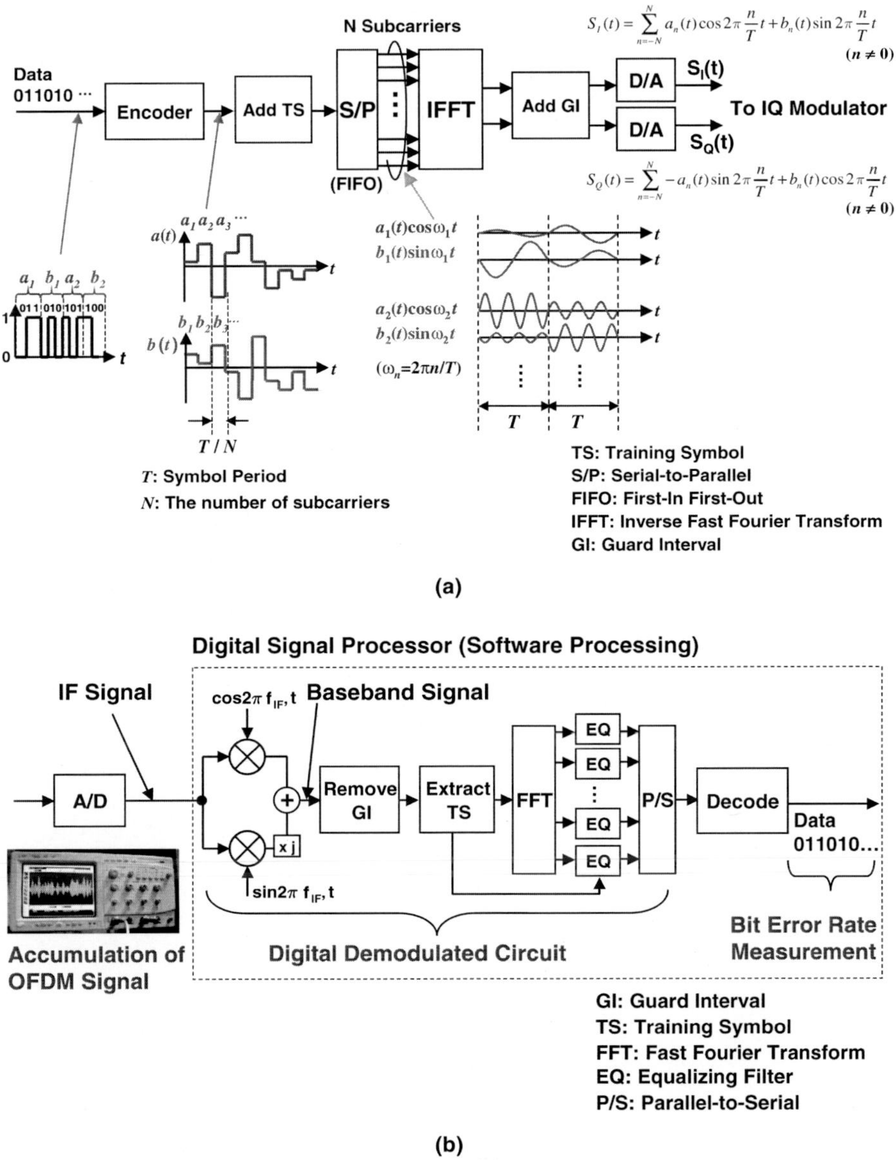

Fig. 3.27 Block diagram of baseband OFDM signal generation (**a**) and demodulation (**b**)

resulting two signals with a polarization beam combiner (PBC) after introducing a path difference with a 40 m delay fiber. The other part of the laser output is frequency upshifted by 2.1 GHz and used as a pilot tone signal. The polarization of the pilot signal is matched to one of the polarization axes of the two OFDM signals. The OFDM signal is combined with the pilot signal and coupled into a 160 km SSMF with a launch power of -6 dBm.

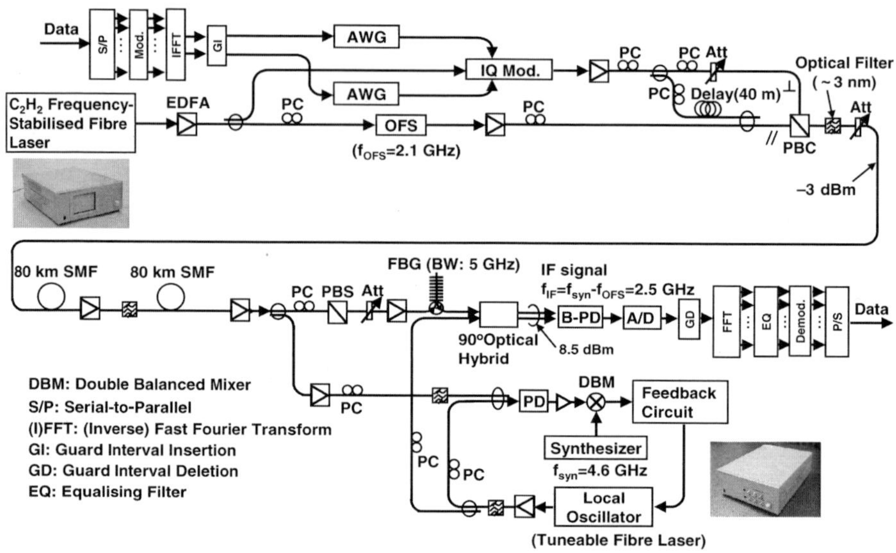

Fig. 3.28 Experimental setup for polarization-multiplexed 64 QAM-OFDM coherent transmission over 160 km

At the receiver, the OFDM signal is divided into two polarizations with a polarization beam splitter (PBS) and heterodyne detected with a 90 degree optical hybrid using an LO signal from a frequency-tunable tracking fiber laser, whose phase is locked to the pilot signal, and then down-converted to an intermediate frequency (IF) signal with $f_{IF} = 2.5$ GHz. After detection with a balanced photodiode (B-PD), the data is A/D converted and post-processed with a DSP in an off-line condition. At the DSP, after demodulating the OFDM signal with an FFT, an equalization (EQ) filter is adopted for each subcarrier to compensate for amplitude and phase distortions with the aid of TS. Finally the BER is evaluated after conversion to a binary data sequence.

Figure 3.29 shows the electrical spectrum of the heterodyne-detected IF signal under a back-to-back condition. The demodulation bandwidth is set at 2.5 GHz by

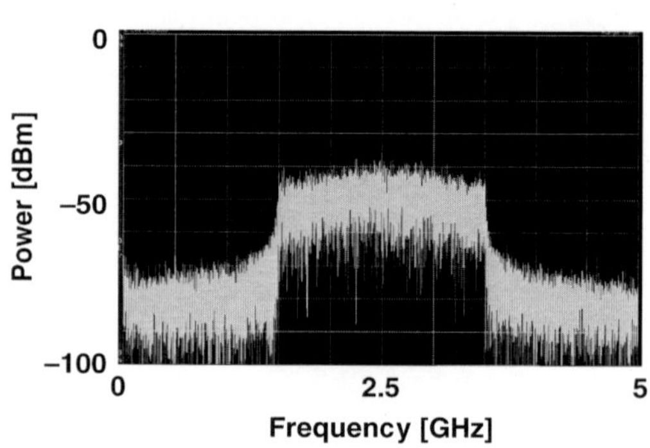

Fig. 3.29 IF spectrum of heterodyne-detected signal

using a digital filter. The BER performance for back-to-back and 160 km transmissions is shown in Fig. 3.30a. Here, the maximum data length for demodulation was limited to 63,488 symbols owing to the DSP memory size, which corresponds to a BER limit of up to 2.6×10^{-6}. The power penalty was approximately 1.2 dB for both polarizations, but error-free performance with a BER of less than the calculation limit was achieved with the received optical power above −25 dBm. Figure 3.30b shows the constellation map after a 160 km transmission for orthogonal polarization at a received power of −25 dBm. A similar clear constellation was also obtained for parallel polarization. As a result, 24 Gbit/s data (corresponding to a raw data rate of 22.2 Gbit/s after excluding 1.6% GI and 5.8% TS overheads) were successfully transmitted over 160 km with a demodulation bandwidth of 2.5 GHz. This indicates that we can expect a spectral efficiency as high as 8.9 bit/s/Hz for FDM transmission.

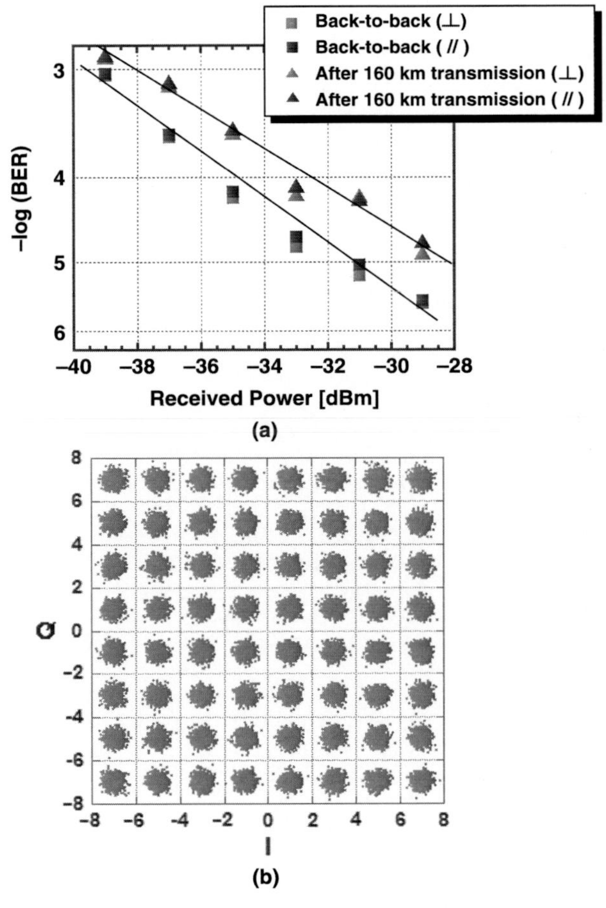

Fig. 3.30 BER characteristics for polarization-multiplexed 2 Gsymbol/s 64 QAM-OFDM transmission over 160 km (**a**) and constellation after 160 km transmission (**b**)

3.7 Conclusion

We have described recent progress on coherent QAM and OFDM transmission using heterodyne detection with a frequency-stabilized fiber laser and an optical PLL circuit. 1 Gsymbol/s single-channel and frequency division multiplexed (FDM) QAM transmissions with the multiplicity up to 128 levels were successfully achieved, demonstrating a spectral efficiency as high as 10 bit/s/Hz. QAM with ultrahigh multiplicity has also been applied to subcarrier modulation in OFDM, where 64 QAM-OFDM coherent transmission was successfully demonstrated with a spectral efficiency of 8.9 bit/s/Hz in a single channel. Such an ultrahigh spectrally efficient transmission system would also play very important roles as regards increasing the total capacity of WDM systems and improving tolerance to chromatic dispersion and PMD as well as in reducing power consumption.

References

1. E. Ip, J.M. Kahn, Carrier synchronization for 3- and 4-bit-per-symbol optical transmission. J. Lightwave Technol. **23**, 4110–4124 (2005).
2. N. Kikuchi, K. Mandai, K. Sekine, S. Sasaki, First experimental demonstration of single-polarization 50-Gbit/s 32-level (QASK and 8-DPSK) incoherent optical multilevel transmission. in *Tech. Digest of the Conference on Optical Fiber Communication*, Postdeadline paper PDP21, 2007.
3. J. Yu, X. Zhou, M.-F. Huang, Y. Shao, D. Qian, T. Wang, M. Cvijetic, P. Magill, L. Nelson, M. Birk, S. Ten, H. B. Mathew, S. K. Mishra, 17 Tb/s (161 × 114 Gb/s) PolMux-RZ-8PSK transmission over 662 km of ultra-low loss fiber using C-band EDFA amplification and digital coherent detection. ECOC Postdeadline paper Th.3.E.5, 2008.
4. M. Nakazawa, M. Yoshida, K. Kasai, J. Hongou, 20 Msymbol/s, 64 and 128 QAM coherent optical transmission over 525 km using heterodyne detection with frequency-stabilized laser. Electron. Lett. **42**, 710–712 (2006).
5. M. Nakamura, Y. Kamio, T. Miyazaki, Linewidth-tolerant 10-Gbit/s 16-QAM transmission using a pilot-carrier based phase-noise canceling technique, Opt. Express. **16**, 10611–10616 (2007).
6. P.J. Winzer, A.H. Gnauck, 112-Gb/s polarization-multiplexed 16-QAM on a 25-GHz WDM grid. ECOC. Postdeadline paper, Th3.E.5, 2008.
7. M. Seimetz, Performance of coherent optical square-16-QAM-systems based on IQ-transmitters and homodyne receivers with digital phase estimation. OFC NWA4 (2006).
8. Y. Mori, C. Zhang, K. Igarashi, K. Katoh, K. Kikuchi, Phase-noise tolerance of optical 16-QAM signals demodulated with decision-directed carrier-phase estimation. OFC. OWG7 (2009).
9. A.J. Lowery, D. Liang, J. Armstrong, Orthogonal frequency division multiplexing for adaptive dispersion compensation in long haul WDM systems. OFC PDP39 (2006).
10. Y. Ma, Q. Yang, Y. Tang, S. Chen, W. Shieh, 1-Tb/s per channel coherent optical OFDM transmission with subwavelength bandwidth access. OFC PDPC1 (2009).
11. H. Takahashi, A. Al Amin, S.L. Jansen, I. Morita, H. Tanaka, DWDM transmission with 7.0-bit/s/Hz spectral efficiency using 8 × 65.1-Gbit/s coherent PDM-OFDM signals. OFC PDPB7, (2009).
12. H. Masuda, E. Yamazaki, A. Sano, T. Yoshimatsu, T. Kobayashi, E. Yoshida, Y. Miyamoto, E. Matsuoka, Y. Takatori, M. Mizoguchi, K. Okada, K. Hagimoto, T. Yamada, S. Kamei, 13.5-Tb/s (135 × 111-Gb/s/ch) no-guard-interval coherent OFDM transmission over 6,248 km using SNR maximized second-order DRA in the extended L-band. OFC PDPB5 (2009).

13. J.G. Proakis, *Digital Communications*, 4th edn. (McGraw Hill, New York, 2000).
14. C.E. Shannon, A mathematical theory of communication. Bell Syst. Tech. J. **27**, 379–423 and 623–656, July and October, (1948).
15. T. Mizuochi, Recent progress in forward error correction for optical communication systems. IEICE Trans. Comm. **E88-B**, 1934–1946, May, (2005).
16. J.P. Gordon, L.F. Mollenauer, Phase noise in photonic communication systems using linear amplifiers. Opt. Lett. **15**, 1351–1353 (1990).
17. H. Kim, Cross-phase-modulation-induced nonlinear phase noise in WDM direct detection DPSK systems. J. Lightwave Technol. **21**, 1770–1774 (2003).
18. K. Kasai, A. Suzuki, M. Yoshida, M. Nakazawa, Performance improvement of an acetylene (C_2H_2) frequency-stabilized fiber laser. IEICE Electron Express **3**, 487–492 (2006).
19. S. Shimotsu, S. Oikawa, T. Saitou, N. Mitsugi, K. Kubodera, T. Kawanishi, M. Izutsu, Single side-band modulation performance of a $LiNbO_3$ integrated modulator consisting of four-phase modulator waveguides. IEEE Photon. Technol. Lett. **13**, 364–366 (2001).
20. T. Sakamoto, A. Chiba, T. Kawanishi, 50-Gb/s 16 QAM by a quad-parallel Mach-Zehnder modulator. presented at the ECOC 2007, Postdeadline Paper 2.8.
21. C.R. Doerr, P.J. Winzer, L. Zhang, L. Buhl, N.J. Sauer, Monolithic InP 16-QAM modulator. OFC PDP20 (2008).
22. K. Kasai, J. Hongo, M. Yoshida, M. Nakazawa, Optical phase-locked loop for coherent transmission over 500 km using heterodyne detection with fiber lasers. IEICE Electron. Express **4**, 77–81 (2007).
23. A. Suzuki, Y. Takahashi, M. Nakazawa, A polarization-maintained, ultranarrow FBG filter with a linewidth of 1.3 GHz. IEICE Electron. Express, **3**, 469–473 (2006).
24. A. Onae, K. Okumura, K. Sugiyama, F.L. Hong, H. Matsumoto, K. Nakagawa, R. Felder, O. Acef, Optical frequency standard at 1.5 μm based on Doppler-free acetylene absorption. Proc. 6th Symposium on Frequency Standards and Metrology, 445 (2002).
25. T. Okoshi, K. Kikuchi, A. Nakayama, Novel method for high resolution measurement of laser output spectrum. Electron. Lett. **16**, 630–631, 1980.
26. D.W. Allan, Statistics of atomic frequency standards. Proc. IEEE **54**, 221–230, (1966).
27. K. Kasai, M. Nakazawa, FM-eliminated C_2H_2 frequency-stabilized laser diode with a RIN of -135 dB/Hz and a linewidth of 4 kHz. Opt. Lett. 34(14), 2225–2227 (2009).
28. K. Kikuchi, T. Okoshi, M. Nagamatsu, N. Henmi, Degradation of bit-error rate in coherent optical communications due to spectral spread of the transmitter and the local oscillator. J. Lightwave Technol. **LT-2**, 1024–1033, (1984).
29. D.-S. Ly-Gagnon, S. Tsukamoto, K. Katoh, K. Kikuchi, Coherent detection of optical quadrature phase-shift keying signals with carrier phase estimation. J. Lightwave Technol. **24**, 12–21 (2006).
30. R.L. Jungerman, C.A. Flory, Low-frequency acoustic anomalies in lithium niobate Mach-Zehnder interferometers. Appl. Phys. Lett. **53** 1477, (1988).
31. H. Goto, K. Kasai, M. Yoshida, M. Nakazawa, Polarization-multiplexed 1 Gsymbol/s, 128 QAM (14 Gbit/s) coherent optical transmission over 160 km using a 1.4 GHz Nyquist filter. Optical Fiber Communication Conference OFC JThA45, February (2008).
32. H. Nyquist, Certain topics in telegraph transmission theory. AIEE Trans. **47**, 617–644 (1928).
33. K. Kasai, J. Hongo, H. Goto, M. Yoshida, M. Nakazawa, The use of a Nyquist filter for reducing an optical signal bandwidth in a coherent QAM optical transmission. IEICE Electron. Express **5**, 6–10 (2008).
34. R.-J. Essiambre, G.J. Foschini, G. Kramer, P.J. Winzer, Capacity limits of information transport in fiber-optic networks. Phys. Rev. Lett. **101**, 163901 (2008).
35. G. Charlet, N. Maaref, J. Renaudier, H. Mardoyan, P. Tran, S. Bigo, Transmission of 40 Gb/s QPSK with coherent detection over ultra long haul distance improved by nonlinearity mitigation. ECOC 2006, Postdeadline paper, Th4.3.4.
36. N.S. Bergano, C.R. Davidson, Circulating loop transmission experiments for the study of long-haul transmission systems using erbium-doped fiber amplifiers. J. Lightwave Technol. **13**, 879–888 (1995).

37. A.H. Gnauck, P.J. Winzer, C.R. Doerr, L.L. Buhl, 10 × 112-Gb/s PDM 16-QAM transmission over 630 km of fiber with 6.2-b/s/Hz spectral efficiency. OFC, PDPB8, (2009).
38. M. Nakazawa, Optical quadrature amplitude modulation (QAM) with coherent detection up to 128 states. OFC OThG1. (2009).
39. S. Hara, R. Prasad, Multicarrier techniques for 4G mobile communications (Artech House, Boston, 2003).
40. B.J.C. Schmidt, Z. Zan, L.B. Du, A.J. Lowery, 100 Gbit/s transmission using single-band direct-detection optical OFDM. OFC Postdeadline paper, PDPC3 (2009).
41. R. Dischler, F. Buchali, Transmission of 1.2 Tbit/s continuous waveband PDM-OFDM-FDM signal with spectral efficiency of 3.3 bit/s/Hz over 400 km of SSMF. OFC Postdeadline paper, PDPC2 (2009).
42. Q. Yang, N. Kaneda, X. Liu, S. Chandrasekhar, W. Shieh, Y.K. Chen, Real-time coherent optical OFDM receiver at 2.5-GS/s for receiving a 54-Gb/s multi-band signal. OFC Postdeadline paper PDPC5 (2009).
43. A.D. Ellis, F.C. Garcia Gunning, Spectral density enhancement using coherent WDM. IEEE Photon. Technol. Lett. **17**, 504–506 (2005).
44. E. Yamazaki, F. Inuzuka, A. Takada, K. Yonenaga, T. Morioka, Inter-channel crosstalk cancellation by encoding with adjacent channels in coherent WDM. OFC JThB6 (2006).
45. T. Omiya, H. Goto, K. Kasai, M. Yoshida, M. Nakazawa, 24 Gbit/s, 64 QAM-OFDM coherent transmission with a bandwidth of 2.5 GHz. to be presented at 1.3.2, ECOC 2009.

Chapter 4
"Quasi Ultimate" Technique

Tetsuya Miyazaki

Abstract In this chapter, a multi-level modulation/demodulation method based on a self-homodyne phase noise canceling scheme using a pilot carrier is introduced as a "Quasi Ultimate" technique. The technique allows us to enjoy simultaneous constellation observation and symbol error rate measurement even with multi-level modulation of high degree, such as 16-QAM or 64-QAM, by relaxing the spectral linewidth requirements of the laser light sources and simplifying the receiver configuration.

In Sect. 4.1, the motivation for and the basic concept of the technique are described. In Sect. 4.2, a pilot symbol insertion technique and its experimental demonstration are described. In Sect. 4.3, a polarization-multiplexed pilot carrier technique and its experimental demonstration in QPSK and 8PSK are described. In Sect. 4.4, an ISI digital pre-equalization technique for M-QAMs and its experimental demonstration are outlined. In Sect. 4.5, simulated results for the 256-QAM case are described.

4.1 Introduction

Multi-level modulation/demodulation formats have been investigated to enhance spectral efficiency in order to cope with the skyrocketing traffic demand toward Peta-bit per second and beyond in future backhaul photonic networks. Differential quadrature phase-shift keying (DQPSK) is the most basic and practical multi-level modulation format with 2 bit/symbol transmission efficiency [1]. In DQPSK transmission systems, no polarization control is required, and the two channels carrying I- and Q-components can be demodulated by using a one-bit delay line interferometer paired with a balanced receiver. Of course, a differential encoder should be properly employed at the transmitter side for decoding [2]. However for higher

T. Miyazaki (✉)
New Generation Network Research Center, National Institute of Information and Communications Technology (NICT), 4-2-1, Nukui-Kitamachi, Koganei, Tokyo, 184-8795, Japan
e-mail: tmiyazaki@nict.go.jp

M. Nakazawa et al. (eds.), *High Spectral Density Optical Communication Technologies*, Optical and Fiber Communications Reports 6, DOI 10.1007/978-3-642-10419-0_4,

order, multi-level modulation formats like 8PSK, 16-QAM, and beyond, absolute optical phase can provide much simpler decoding than a differential encoder and decoder with complex designs.

A synchronous homodyne detection scheme using an optical phase-locked loop (OPLL) is one way to provide an absolute optical phase reference from synchronized local laser light sources. Carrier recovery is indispensable in an OPLL for phase locking and demodulation. A residual carrier technique using incomplete modulation was demonstrated in binary PSK homodyne detection [3], and an optical Costas loop was demonstrated for QPSK homodyne detection [4]. However, these OPLL systems with finite loop delay time require extremely narrow spectral linewidth in both the signal and the local light sources. The requirements are defined using the linewidth ($\Delta\nu$) to symbol rate ratio (inverse of baud interval: T_s) to attain a bit error rate (BER) of less than 10^{-9}. For QPSK, $\Delta\nu \times T_s < 5.2 \times 10^{-5}$, for 8PSK, $\Delta\nu \times T_s < 1.8 \times 10^{-5}$, and for 16-QAM, $\Delta\nu \times T_s < 1.4 \times 10^{-6}$ [5, 6], as shown in Fig. 4.1.

One feature of current digital optical coherent intra-dyne techniques is the absence of "critical" feedback loop control, because the carrier phase can be estimated by digital signal processing [7]. Recently, carrier phase estimation in 16-QAM has been demonstrated [8]. However, it should be noted that even with the digital coherent technique, the stringent spectral linewidth problem still remains unresolved. For example, the required spectral linewidth at 40 Gbit/s is less than 120 kHz for 16-QAM and less than 1.2 kHz for 64-QAM to attain a 2-dB receiver sensitivity penalty for a BER of 10^{-4} [9]. In addition, although a 23-GSample/s 6-bit analog-to-digital converter (ADC) and a digital signal processor for a dual-polarization QPSK transceiver have been developed on a custom CMOS ASIC chip [10], this technology will require further intensive development for higher order multi-level modulation formats such as 16-QAM and beyond. Almost all reported results of digital coherent technologies with multi-level modulation higher than QPSK employ non-real-time offline signal processing. The BER performance obtained in offline mode cannot attain a BER of less than 10^{-9} even in a back-to-back configuration due to the symbol length limitation of the memory storage.

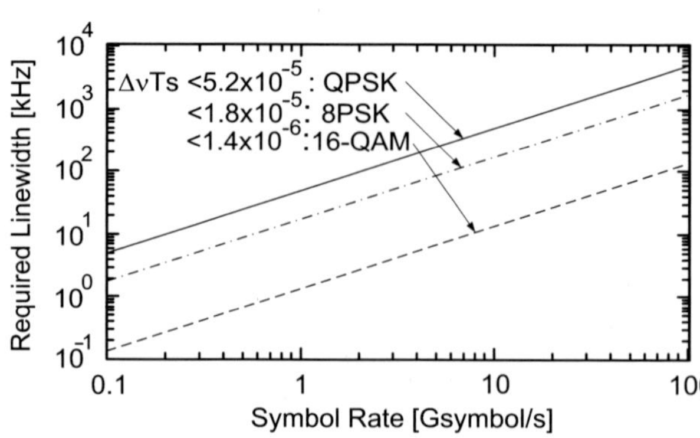

Fig. 4.1 The linewidth versus symbol rate ratio requirement

We have proposed two pilot tone techniques to solve these problems in analogue OPLLs or even current digital coherent techniques. The first one involves periodic insertion of a pilot symbol into the time slots of the multi-level modulated signal, and the second one involves using a pilot carrier that is polarization-multiplexed orthogonal to the multi-level modulated optical signal.

4.2 Pilot Symbol Insertion Technique

4.2.1 8PSK Simulation

In this section, the principle of a pilot symbol-aided self-homodyne scheme is introduced, and the obtained phase noise tolerance is investigated theoretically.

Inserting a synchronization bit in the electrical signal frames at the transmitter side to aide coherent operation in a 565-Mbit/s, 1064-nm PSK homodyne OPLL receiver was reported in 1991 [11]. Although currently available digital signal processing technologies can enhance the bit rate of the scheme to at least multi-Gbit/s, pilot symbol insertion in an optical time domain multiplexing (OTDM) technique is discussed here for bit-rate scalable self-homodyne detection without using a local oscillator in the critical OPLL receiver, or even to neglect rather complex digital signal processing for carrier phase estimation in M-QAM signals.

In the transmitter, an un-modulated pilot symbol serving as an absolute optical phase reference can be periodically inserted by using an OTDM multiplexer (OTDM-MUX) including optical phase modulators (or I–Q vector modulators) in each delay line, as shown in Fig. 4.2. In this case, a pilot symbol is employed for four payload time slots (T_S: slot time) in a frame. It should be noted that all time slots in a frame contain identical optical phase noise because they are generated from a single optical slot by the OTDM-MUX.

Figure 4.3 shows the simulation model. In the receiver, the transmitted optical signal is tapped to extract a pilot symbol in a frame by using an OTDM demultiplexer (Opt. DEMUX) with a synchronous optical time gating function. The extracted optical pilot symbol is supplied to I- and Q-branch-balanced detectors to achieve self-homodyne detection. By tuning the optical delay employed in the main transmission line in the receiver, the detected payload time slot can be selected. By employing simultaneous optical DEMUX [12], all payload time slot signals can be detected in parallel.

Here a light source with Lorentzian spectral line broadening is assumed. We also assume a spectral linewidth $\Delta\nu$ with a Gaussian distribution defined by $\sigma_\varphi^2 = 2\pi\Delta\nu T_S$, where $\Delta\nu << R_S$ and T_S is the reciprocal of the symbol transmission rate. Then, the symbol error rate PE is calculated as follows [13]:

$$\mathrm{PE}(\gamma_s) = \int_{-\infty}^{\infty} \frac{P_{\mathrm{BER}}(\Delta\varphi, \gamma_s)}{\sqrt{2\pi}\sigma_\varphi} \exp\left(-\frac{\Delta\varphi^2}{2\sigma_\varphi^2}\right) d\Delta\varphi.$$

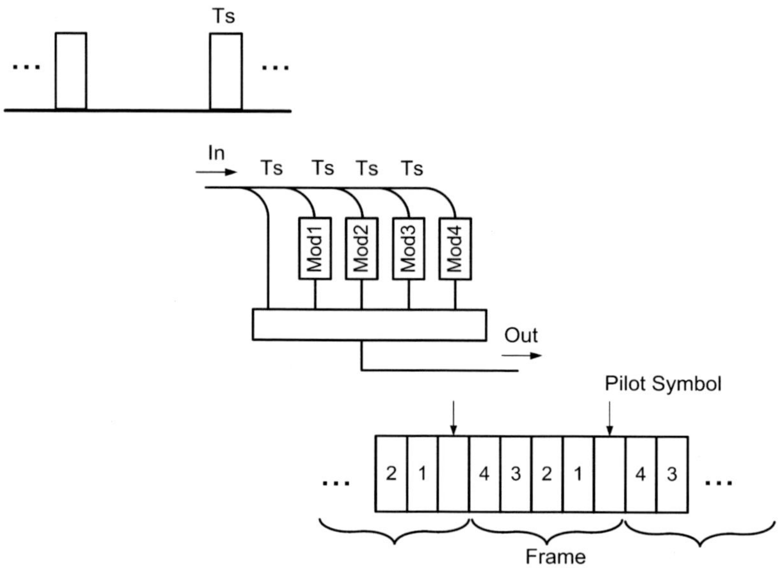

Fig. 4.2 Pilot symbol insertion in the optical time domain

Fig. 4.3 Simulation model for the pilot symbol insertion technique

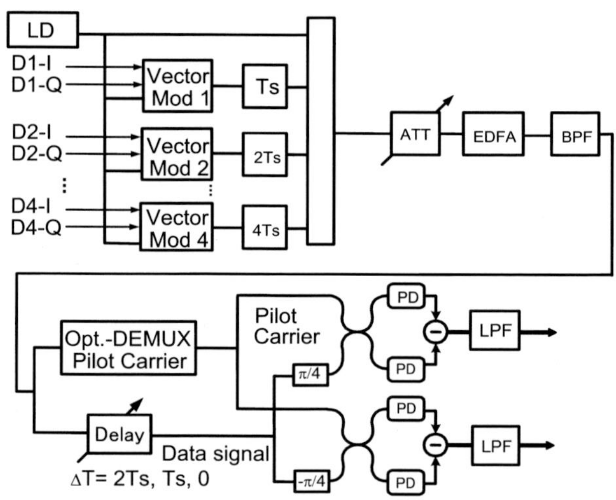

In M-ary PSK, the bit error rate of delayed demodulation becomes [14]

$$P_{\mathrm{BER}M}(\Delta\varphi, \gamma_{\mathrm{s}}) = \frac{1}{\log_2 M}\mathrm{erfc}[1 - 2\int_0^{\frac{\pi}{M}} p(\varphi_n + \Delta\varphi|\gamma_s)]d\varphi_n.$$

The actual BER performance is calculated as an approximation of the above equations. Figure 4.4 shows the relationship between spectral linewidth and power penalty for BER $= 10^{-9}$ in the 8PSK case using the proposed detection scheme,

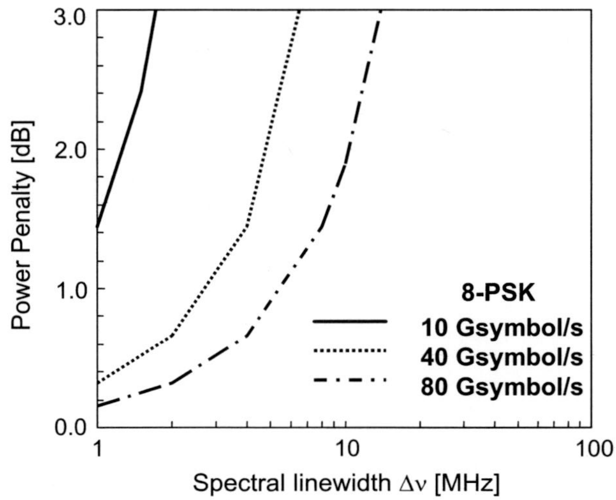

Fig. 4.4 Power penalty (at BER $= 10^{-9}$) versus spectral linewidth using the pilot symbol insertion technique

with symbol rate as a parameter. At a transmission rate of 10 Gsymbol/s, even a linewidth of 1 MHz can achieve a power penalty of less than 2 dB. The linewidth requirement is expected to be drastically relaxed compared to the conventional OPLL case ($\Delta\nu < 180\,\text{kHz}$ ($T_s = 100\,\text{ps}$, $\Delta\nu \times T_s < 1.8 \times 10^{-5}$) for 8PSK in OPLL [6]). To improve the frame utilization efficiency, it is necessary to widen the insertion interval of the pilot symbols and to provide two or more detection modules.

In this proposed technique, multi-level modulation schemes, such as M-ary PSK and M-ary QAM, can be used, and a pre-coding scheme is not necessary because absolute phase mapping is preferable. As a result, the transmitter structure can be simplified compared to conventional differential schemes.

4.2.2 QPSK Homodyne Using Pilot Symbols

In this section, an experimental demonstration based on the periodic pilot symbol insertion in homodyne PSK/QPSK modulation is introduced.

The operating principle of the proposed self-homodyne PSK and QPSK systems is illustrated in Fig. 4.5. For the proposed homodyne PSK signal, in the transmitter, the light from a laser diode is phase modulated by a return-to-zero radio-frequency (RZ-RF) signal with a duty cycle of 50%. Because the RZ-RF format can leave certain time slots un-modulated, pilot symbols are periodically inserted in the data stream, which provides an absolute optical phase reference for the adjacent phase-modulated data slots. The phase information is therefore coded using an absolute phase referenced to the inserted pilot symbols, instead of using a relative phase referenced to the adjacent bits. At the receiver side, an interferometer with only a half-symbol delay performs self-homodyne direct detection. Compared to other existing homodyne or conventional differential PSK/QPSK (DPSK/DQPSK) techniques [1, 7, 8], this scheme offers the following advantages: (1) a high-speed,

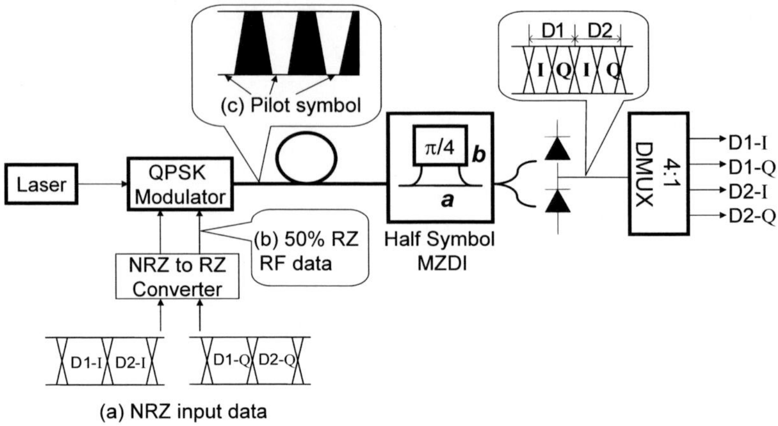

Fig. 4.5 QPSK homodyne using pilot symbol

complicated differential pre-coder is not required in the transmitter, and a high-speed DSP electrical circuit or local oscillator is not required in the receiver to recover the carrier's phase and (2) for the proposed QPSK scheme, simultaneous detection of in-phase (I) and quadrature (Q) components is achieved by using only one delay interferometer and a subsequent balanced photo-detector, which reduces the system complexity and implementation cost. Here it should be noted that the proposed scheme corresponds to the case of one pilot symbol serving one payload slot in the frame, described in previous section, although phase noise components contained in their slots are not identical in this case. However, the linewidth requirements for the laser are relaxed by using an interferometer with only a half-symbol delay, which provides more stable performance under temperature variations and good tolerance against demodulator phase error.

In the PSK case, as the interference demodulation is carried out between the phase-modulated slot and the adjacent two-pilot slots, after the photo-detection, the data is recovered with an NRZ data format, instead of an RZ data format. In the QPSK case, after formatting as 50% RZ-RF signals, I and Q tributaries are employed in the phase modulation to achieve quaternary phase states in the light. At the receiver side, if it is assumed that a $+\pi/4$ phase shift and a half-symbol time delay are introduced for the b-branch in a Mach–Zehnder delay interferometer (MZDI), the relative phase shifts of the data slot relative to the pilot symbol slot are different for the two time slots within one-bit period after the MZDI, as shown in Fig. 4.6. The data slot is phase shifted by $-\pi/4$ relative to the pilot symbol slot at the first time slot, whereas it is phase shifted by $+\pi/4$ at the second time slot. Thus, after the interferometer, the detected data in the two time slots correspond to the I- and Q-components of the QPSK signal. With the proposed inserted pilot symbol technology, simultaneous detection of the I- and Q-components can be achieved. After the MZDI, an optical or electronic demultiplexer can be employed to separate the I- and Q-components for each of the data channels, as shown in Fig. 4.5. After

Fig. 4.6 Time slot timing chart

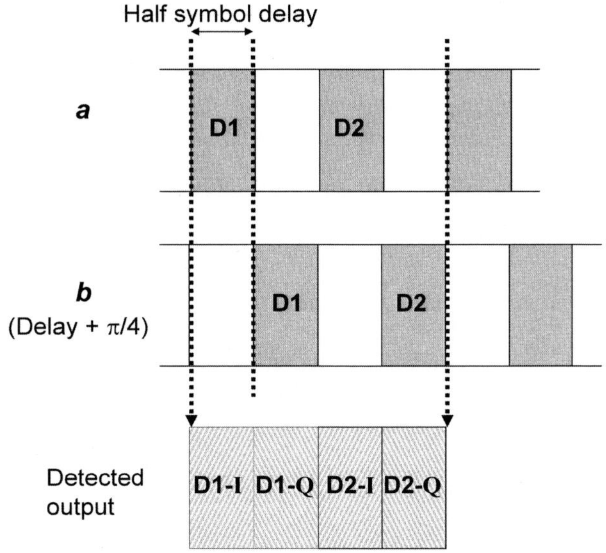

photo-detection, the bit rate of the detected RF data is twice that of the driving RF signals at the transmitter.

To evaluate the performance of the proposed homodyne PSK/QPSK schemes, the performance was experimentally investigated for homodyne PSK and QPSK systems with symbol rates of 20 Gsymbol/s (20 Gb/s for PSK and 40 Gb/s for QPSK). An integrated QPSK modulator was employed where $0-\pi$ and $0-\pi/2$ phase modulations were applied with a dual-drive Mach–Zehnder modulator (MZM) and a subsequent phase modulator, respectively. The details of the experimental setup are shown in [15]. To demonstrate the operating principle of the proposed PSK system, a pattern '1001110010110100001' on CH1 is used as an example. Figure 4.7 shows measured waveforms for the 20-Gbit/s self-homodyne PSK system on (a) input 10-Gbit/s CH1, (b) input 20-Gbit/s CH1+CH2, (c) detected 20-Gbit/s CH1+CH2, and (d) detected 10-Gbit/s CH1.

The measured BER characteristics of two-channel 40-Gbit/s QPSK and 20-Gbit/s PSK signals are shown in Fig. 4.8. Eye diagrams for both cases are also shown in the inset. For both the 40-Gbit/s QPSK and 20-Gb/s PSK signals, a sensitivity difference of less than 0.3 dB was observed for the two de-multiplexed channels, CH1 and CH2. For all cases, error-free operation with a BER of less than 10^{-9} was confirmed. The degraded eye diagrams in the 40-Gbit/s QPSK case are mainly attributed to the limited frequency response of the phase modulator employed for the $0-\pi/2$ phase modulation. Better performance is expected if using an I–Q vector QPSK modulator. The proposed self-homodyne phase modulation scheme with inserted pilot symbols is scalable to higher bit-rate systems. Moreover, it can be applied to other multi-level phase modulation schemes. For example, when applying it to an 8PSK system, it is possible to detect the corresponding three data tributaries using only two MZDIs with half-symbol delay and $+\pi/4$ or $\pi/4$ phase shift.

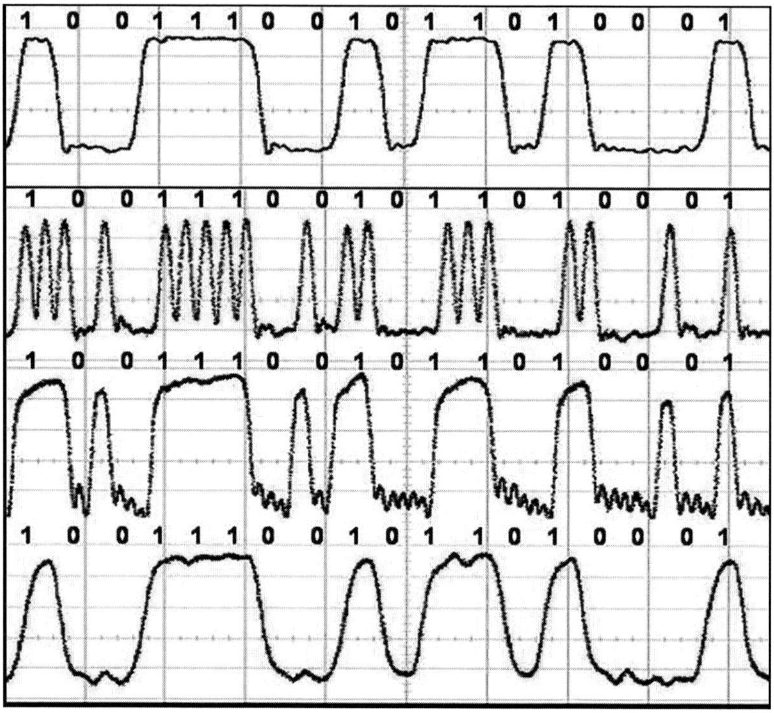

Fig. 4.7 Measured waveforms for the 20-Gb/s self-homodyne PSK system. **a** Input 10-Gb/s CH1, **b** input 20-Gb/s CH1+CH2, **c** detected 20-Gb/s CH1 + CH2, and **d** detected 10-Gb/s CH1

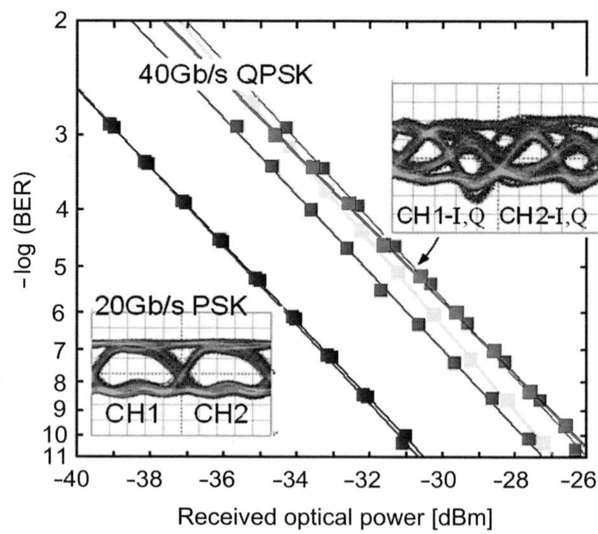

Fig. 4.8 BER characteristics of the two-channel 40-Gb/s QPSK and 20-Gb/s PSK signals

4.3 Polarization-Multiplexed Pilot Carrier Technique

In this section, the operating principle of a polarization-multiplexed pilot carrier-aided multi-level homodyne modulation/demodulation technique is introduced, and a first demonstration in the case of the QPSK format is described. Then, various tolerances to phase noise, polarization-dependent loss (PDL), dispersion, and polarization mode dispersion using this technique in QPSK are discussed.

4.3.1 Principle

The operating principle of the technique for QPSK is schematically shown in Fig. 4.9. A lightwave linearly polarized at 45° is introduced into a $LiNbO_3$ optical phase modulator, which has an effective phase modulation capability only for the TM polarization component of the input lightwave, due to the nature of $LiNbO_3$ waveguide. Then, optical QPSK modulation is encoded by applying electrical DATA1 plus DATA2 to an electrode (for example, DATA1 for $0-\pi$ and DATA2 for $0-\pi/2$). Therefore, the TE polarization component remains un-modulated to act as a pilot carrier to provide an absolute optical phase reference that is multiplexed in the orthogonal polarization state (TE) with respect to the QPSK signal component (TM). At the receiver side, the polarization state of the pilot carrier is rotated by 90° to perform QPSK homodyne detection.

The features of this technique are: first, the phase noise of the light source can be canceled out, thanks to the fact that both polarization components include identical phase noise of the light source; and second, no digital signal processing is required for carrier phase estimation and no local laser source is required. Extension of this technique to 8PSK and M-QAM is discussed in Sects. 4.3.3 and 4.4.

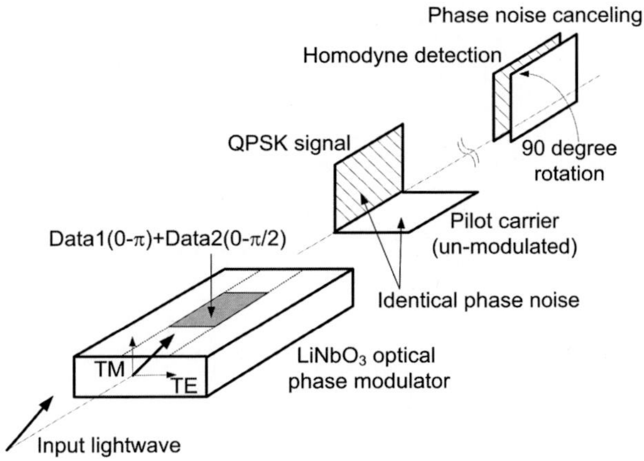

Fig. 4.9 Schematic diagram of polarization-multiplexed pilot carrier technique

4.3.2 QPSK Demonstration

The experimental setup for the polarization-multiplexed QPSK homodyne transmission system is shown in Fig. 4.10 [15–18]. We utilized light sources with various spectral linewidths to evaluate the phase noise tolerance of the proposed technique: three external cavity laser diodes (EC-LDs) with spectral linewidths of 120, 200, and 400 kHz; a DFB-LD with a spectral linewidth of 30 MHz; and an amplified spontaneous emission (ASE) light source with a spectral sliced bandwidth of 380 GHz (3 nm).

A polarization controller (PC1) was used to adjust the polarization state of the CW light for pilot carrier generation. For DGD and dispersion tolerance measurements, a variable DGD generator and a dispersion generator consisting of polarization-maintaining fibers (PMFs) and single-mode fibers (SMFs), respectively, were introduced. The variable DGD generator was also used to compensate the DGD of the modulator ($\sim$17 ps) to produce null gross DGD of the setup, in addition to emulating the DGD impairment.

At the receiver side, a $LiNbO_3$-based polarization beam splitter (PBS) hybrid module (LN-PBS hybrid) was employed for homodyne detection by 90-degree polarization rotation of the pilot carrier. In the LN-PBS hybrid module, a PBS, a phase shifter (ϕ-shifter), a half-wave plate ($\lambda/2$) for pilot carrier rotation, and a 3-dB coupler for balanced detection were integrated. The performance of the QPSK homodyne detection was optimized by manually adjusting a second polarization

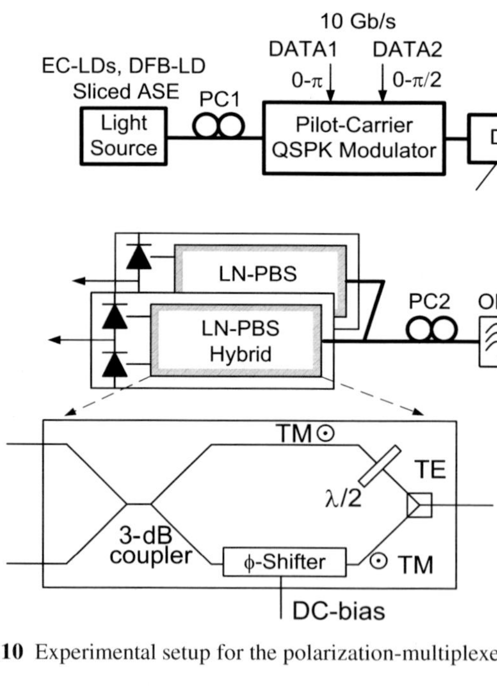

Fig. 4.10 Experimental setup for the polarization-multiplexed QPSK homodyne

controller (PC2) before the LN-PBS hybrid. Although bias-voltage control to the phase shifter (ϕ-shifter) of the LN-PBS hybrid module enables optimization of the homodyne detection, no feedback control was applied to the phase shifter.

At first, the receiver sensitivity penalty versus ratio of pilot carrier optical power (P_{pilot}) to total optical power (P_t) was evaluated to determine the optimum splitting ratio (P_{pilot}/P_t), to evaluate polarization-dependent loss (PDL) tolerance. The ratio was measured for the NRZ format with both dispersion and gross DGD set to zero, using the EC-LD with the linewidth of 200 kHz. The measurement results are shown in Fig. 4.11. We found that a 50:50 splitting ratio between the pilot carrier and the QPSK signal achieved the minimum power penalty. To attain a power penalty of less than 1 dB, the ratio should be controlled within about ±15% at around the optimum point of 50%, meaning that the PDL tolerance was less than ±0.7 dB.

Next, the phase noise tolerance was evaluated by using light sources with various spectral linewidths while keeping both DGD and dispersion at 0 ps and 0 ps/nm. Figure 4.12 shows the receiver sensitivities of the detected QPSK signal versus spectral linewidth ($\Delta\nu$) of the light sources at the transmitter side; for this measurement we used three tunable EC-LDs with linewidths of less than 500 kHz and a DFB-LD module with a linewidth of 30 MHz. The receiver sensitivities were defined at a BER of 10^{-9} for these light sources. Furthermore, we used a spectrum-sliced amplified spontaneous emission (ASE) light source with a sliced optical bandwidth of 3 nm (~380 GHz), and the receiver sensitivity was defined at a BER of 10^{-6}. Eye diagrams for the EC-LD ($\Delta\nu$ = 400 kHz), DFB-LD ($\Delta\nu$ = 30 MHz), and sliced ASE ($\Delta\nu$ = 380 GHz) are shown in the inset. We confirmed that the receiver sensitivity stayed at around −30 dBm in spite of the increasing linewidth of the light source, even for a linewidth of 30 MHz. Even in the case of the spectrum-sliced light source, although it was degraded by the excessive intensity noise, a receiver

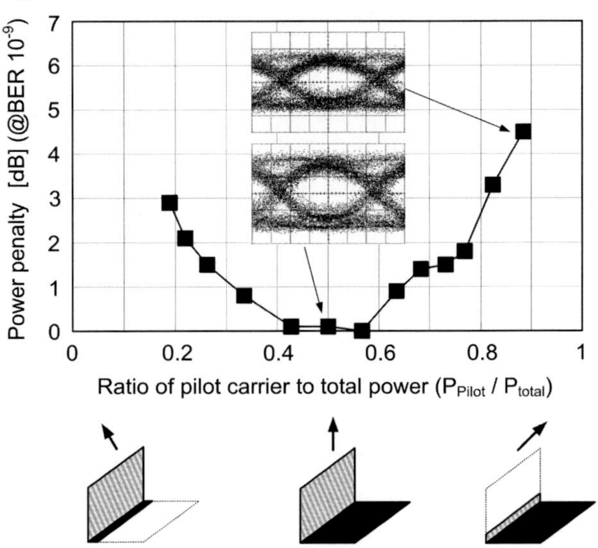

Fig. 4.11 The PDL tolerance of the polarization-multiplexed pilot carrier technique

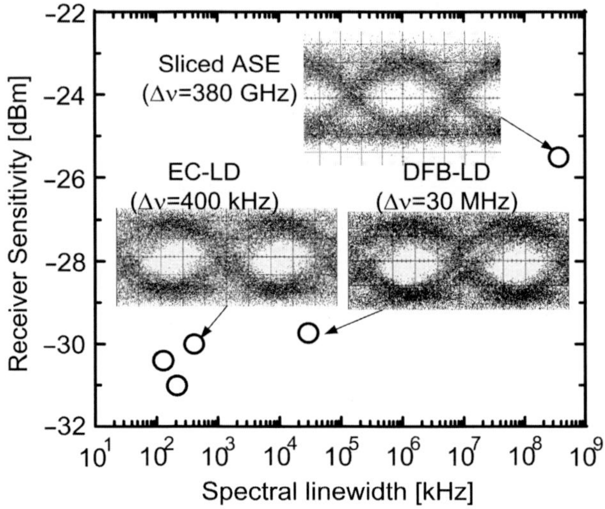

Fig. 4.12 Receiver sensitivity versus spectral linewidth

sensitivity of −25.5 dBm was obtained, due to the time-synchronized transmission of the QPSK signal and the pilot carrier.

For DGD and dispersion tolerance, use of the EC-LD light source ($\Delta\nu = 200$ kHz) and the ASE light source was investigated for comparison. To evaluate the DGD tolerance, we varied the introduced DGD and investigated the influence on the receiver sensitivity with the dispersion set to 0 ps/nm. Figure 4.13 shows the receiver sensitivity penalty of the detected QPSK signal versus the introduced DGD. It should be noted that the receiver sensitivity was almost insensitive to the DGD when using the EC-LD light source (open circles). On the other hand, for the

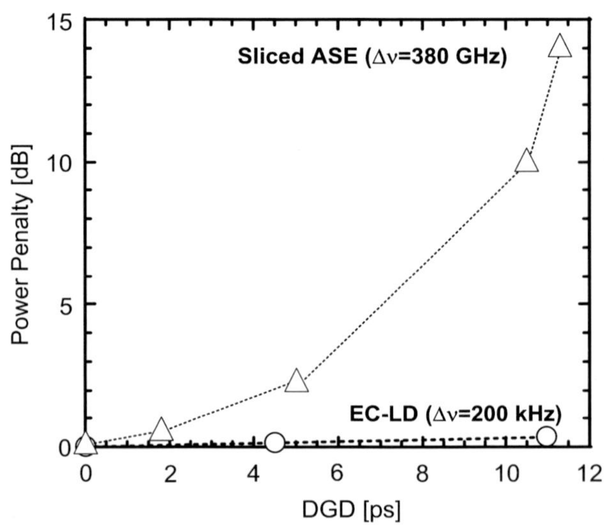

Fig. 4.13 The DGD tolerance of the polarization-multiplexed pilot carrier technique

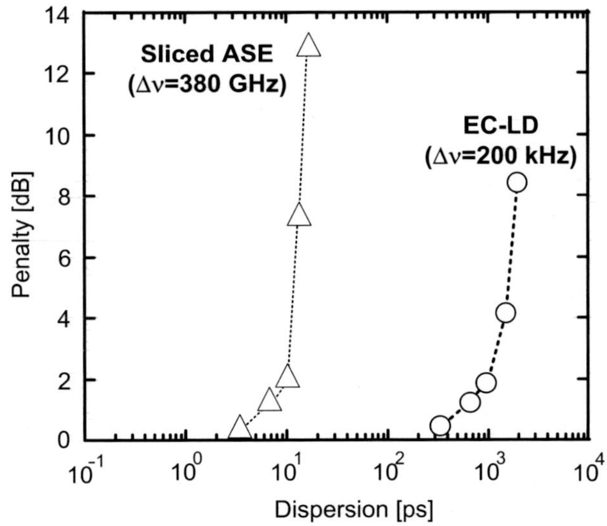

Fig. 4.14 The dispersion tolerance

ASE light source (open triangles), the DGD tolerance to attain a power penalty of less than 1 dB was only about 3 ps, which is nearly equal to the coherence time of the 3-nm sliced ASE light source. To evaluate the dispersion tolerance, we introduced dispersion while keeping the gross DGD at 0 ps. Figure 4.14 shows the receiver sensitivity penalty versus introduced dispersion. The dispersion tolerances to attain a power penalty of less than 1 dB were larger than 635 ps/nm and 7 ps/nm for the EC-LD and the ASE light sources, respectively. The dispersion tolerance of the ASE light was restrained by a spectrum-slice bandwidth as large as 3 nm.

4.3.3 8PSK Demonstration

In this section, a phase noise tolerant 8PSK demonstration using the proposed scheme is introduced [19]. The proposed scheme can effectively simplify the modulator/demodulator hardware configuration. The 8PSK signal was modulated using only one $LiNbO_3$-based integrated modulator, schematically shown in Fig. 4.15a, instead of using a cascaded configuration of multiple modulators [20]. In the modulator, the modulation of the TM component was encoded by applying DATA1 for $0-\pi$ and DATA2 for $0-\pi/2$ (in PM-TM), while the TE component was modulated by DATA3 for $0-\pi/4$ (in PM-TE), although it also acted as a pilot carrier, to generate an 8-ary polarization-shift keying (8PolSK) signal. Consequently, the modulated signal could be detected by using PBSs and balanced photo-detectors (PDs) [21] at the receiver side for detection of the I- and Q-components, as shown in Fig. 4.15b. Constellations could be observed by introducing these detected I- and Q-components to a digital oscilloscope operating in X–Y mode. Therefore, the BER was measured by an error detector (ED) while simultaneously observing the constellation using an oscilloscope.

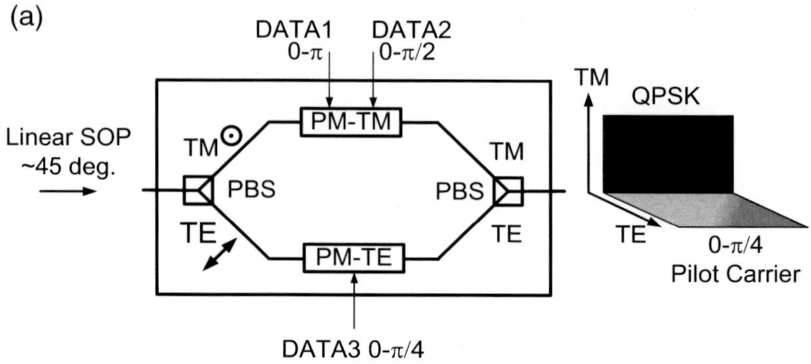

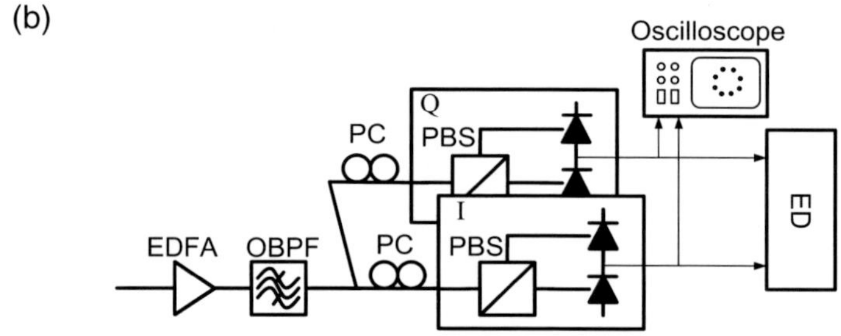

Fig. 4.15 **a** 8PSK modulator. **b** Receiver configuration for 8PSK

Figure 4.16 shows receiver sensitivities (defined at a BER of 10^{-9}) versus spectral linewidths when using an EC-LD ($\Delta\nu = 200$ kHz) and a DFB-LD ($\Delta\nu = 30$ MHz) at 1 Gsymbol/s 8PSK demodulation. The constellations are also shown in the inset. The non-uniformity of the spacing in these constellations was mainly

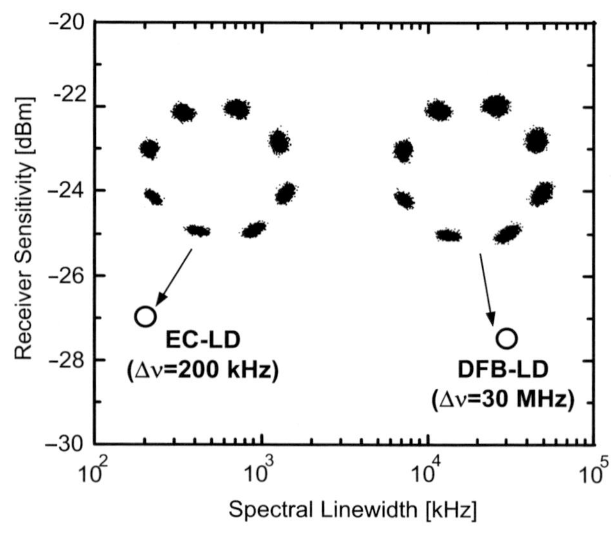

Fig. 4.16 Receiver sensitivity (at BER = 10^{-9}) versus spectral linewidth in 8PSK

caused by inter-symbol interference (ISI) due to the imperfect frequency responses of the modulator and electrical amplifiers, although it can easily be compensated for by adopting the digital signal pre-equalization technique mentioned in the next section. Here, it should be noted that no notable degradation due to the large linewidth of the DFB-LD ($\Delta\nu = 30$ MHz) was observed at this comparatively slow modulation speed of 1Gsymbol/s, where the influence of phase noise is severer for differential detection. Both 8PSK demonstrations at 1 Gsymbol/s (3 Gbit/s) and 10 Gsymbol/s (30 Gbit/s) attained error-free operation with a BER of less than 10^{-9}, even in the case using the DFB-LD. The DGD and dispersion tolerances were reported in [19]. This enormous linewidth tolerance even at the comparatively slow modulation speed is a major feature of our proposed scheme, allowing it to achieve superior phase noise cancellation performance.

4.4 ISI Digital Pre-equalization Technique for M-QAMs

4.4.1 Pre-equalization for ISI

One of the major issues in electrical multi-level modulation other than phase noise is signal impairment due to inter-symbol interference (ISI). ISI is caused by non-flat and finite frequency response characteristics in both optical and electrical components of the transmission system, such as the driver amplifiers, modulator, fiber, and receiver front-end electronics, as shown in Fig. 4.17. Especially in non-binary electrical modulation, ISI becomes a severe problem [6]. To solve this problem, a sophisticated integrated M-QAM optical encoder comprising multiple sub-Mach–Zehnder elements to allow binary electrical modulation has been proposed [6] and demonstrated in the case of 16-QAM [22, 23]. Although this technique can reduce the ISI caused in the driver electronics and optical modulator, enhancing the symbol rate, the many driver amplifiers and complex bias adjustment are technical issues especially in higher M-QAM modulation, such as 16-QAM and beyond.

Here, we focus on a digital pre-equalization technique to compensate for ISI caused in a vector modulator and electronics used in transmitter and receiver (excluding ISI caused by transmission impairment in the fiber with temporal change in Fig. 4.17), when applying electrical multi-level M-QAM encoding. The digital equalization technique employed here is achieved by a finite impulse response (FIR) filter, which is well established in wireless communications systems. First, tap coefficients of the FIR filter at the receiver side, serving as a post-equalizer in offline mode, were determined by a least mean square (LMS) algorithm, and then they were applied to the coefficients in a pre-equalizer. Because ISI caused by these elements (a vector modulator with transmitter and receiver sides electronics) is deterministic and includes negligible temporal changes, no additional offline calculation is necessary for ISI compensation once the FIR parameters are fixed. The number of taps in the FIR was tuned in the range between 16 and 128 to optimize performance. Details of the process for designing an FIR pre-equalizer are shown in [24].

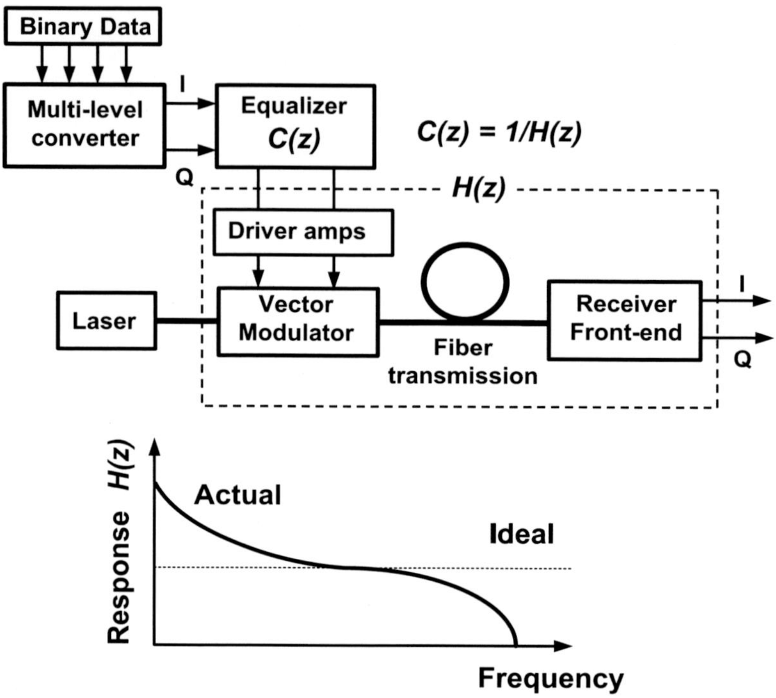

Fig. 4.17 Pre-equalization for ISI in electrical multi-level modulation

4.4.2 16-QAM and 64-QAM Demonstration

We combined the digital pre-equalization technique described in Sect. 4.4.1 with our proposed pilot carrier technique to simultaneously observe the BER and constellation performance of M-QAMs in real time. The experimental setup is shown in Fig. 4.18. Here, we employed a pilot carrier vector modulator that includes a vector modulator in the TM-PM portion in Fig. 4.15a. A detailed explanation of the setup is described in [25, 26].

Figure 4.19 shows the observed 40-Gbit/s (10 Gsymbol/s) 16-QAM (upper) and 30 Gbit/s (5 Gsymbol/s) 64-QAM (lower) constellation diagrams without (left side) and with (right side) the digital pre-equalization, respectively. Here we utilized an EC-LD with 200 kHz spectral linewidth as a light source. By applying the pre-equalization, both 16-QAM and 64-QAM constellations were drastically improved. A BER of less than 10^{-9} was attained in the 16-QAM constellation, and a BER of less than 10^{-7} was attained in the 64-QAM constellation.

In the M-QAM fiber transmission experiment, both 10-Gbit/s 16-QAM and 15-Gbit/s 64-QAM signal transmission over 120 km standard single-mode fiber (SSMF) without any optical or electrical dispersion compensation have successfully been demonstrated with the setup shown in Fig 4.18. Thanks to the combination of digital pre-equalization and phase noise canceling techniques, these transmission

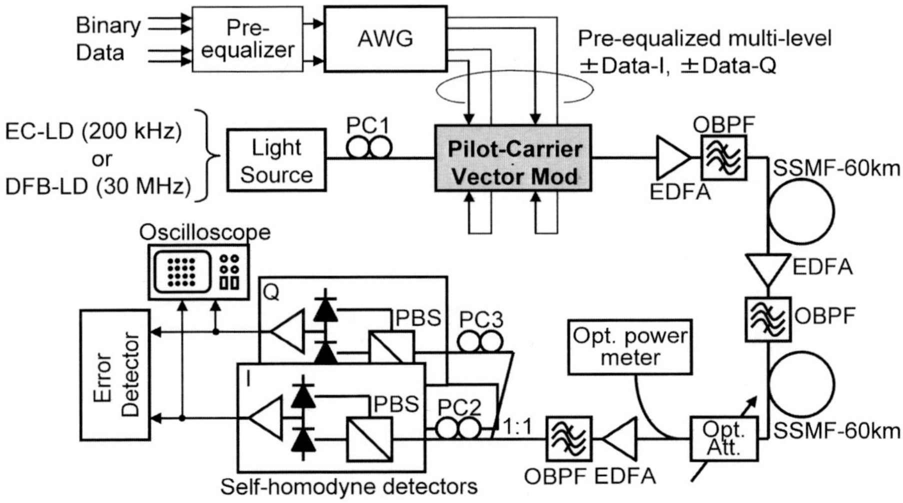

Fig. 4.18 Demonstration setup for 16-QAM and 64-QAM

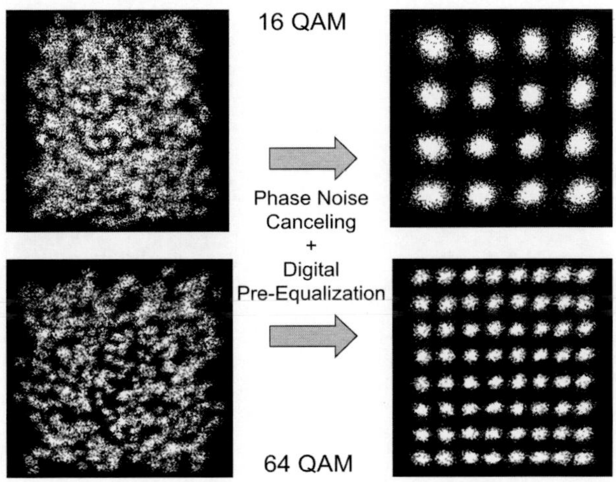

Fig. 4.19 Constellations for 40-Gbit/s 16-QAM (*upper*) and 30-Gbit/s 64-QAM (*lower*) with (*right*) and without (*left*) digital pre-equalization

demonstrations can be performed in real time to simultaneously measure symbol error rate (SER) and constellations. Figure 4.20 shows how to measure the SER in the case of 10-Gbit/s (2.5 Gsymbol/s) 16-QAM, by adjusting the threshold levels, Th_a, Th_b, and Th_c, one by one, for each I- and Q-component. The total SER of the 16-QAM transmission can be obtained by summing the measured error rates at all threshold levels of the I- and Q-components. Assuming Gray-coding, the BER is one-fourth of the total SER because 16-QAM uses four-bits-per-symbol modulation.

Figure 4.21 shows the BER characteristics of the 10-Gbit/s 16-QAM signals before (open squares) and after (closed triangles) transmission through the SSMF using a DFB-LD with 30-MHz linewidth, which were obtained from the measured SERs. Detected eye diagrams for both cases are also shown in the inset. It should

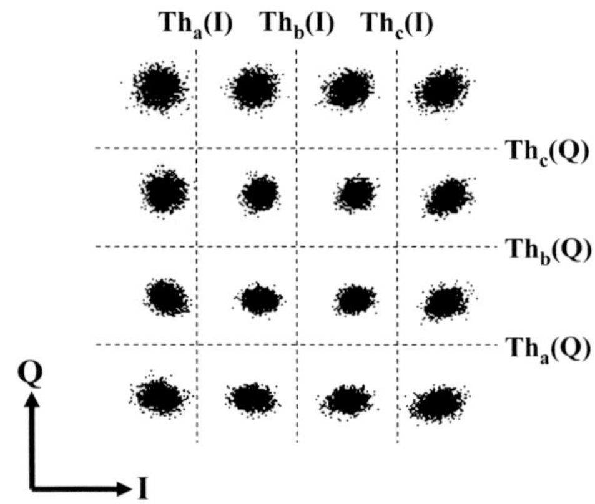

Fig. 4.20 Threshold setting for the symbol error rate (SER) measurement

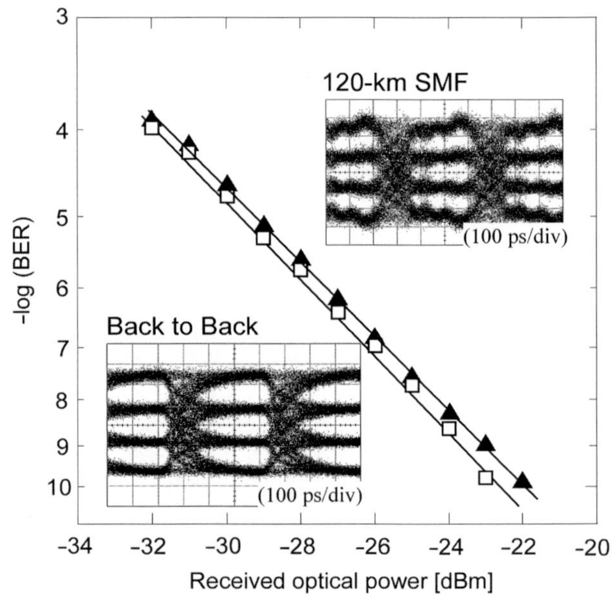

Fig. 4.21 The BER characteristics of 16-QAM. *Open circle*: back-to-back. *Closed circles*: 120-km SSMF transmission

be noted that floorless BER characteristics of less than 10^{-9} were achieved in both cases without the assumption of applying any forward error correction (FEC) technique. Furthermore, the power penalty for the 120-km SSMF transmission without any dispersion compensation was less than 1 dB at a BER of 10^{-9}. In the same way, we also experimentally demonstrated 15-Gbit/s and 30-Gbit/s 64-QAM SSMF transmission [26, 27].

Thanks to the combination of the phase noise canceling capability of the self-homodyne technique and a standard electronic pre-equalization technique, we confirmed clear M-QAM constellations and error-free characteristics with a BER of less than 10^{-9}. Furthermore, we also successfully demonstrated error-free transmission over 120-km SSMF without any dispersion compensation. Our proposed scheme will allow the use of inexpensive light sources and less-complex transmitters with fewer driver amplifiers for M-QAM multi-bit-per-symbol transmission systems.

4.5 Simulated Results in 256-QAM

In this section, we numerically show the possibility of using our proposed scheme at 80 Gbit/s for 10-Gsymbol/s 256-QAM to exhibit superior differential group delay (DGD) tolerance even with a linewidth of 1 MHz, compared with 40-Gsymbol/s DQPSK.

It is necessary to estimate the bit error rate (BER) using the received data and a limited number of symbols in a computer simulation. In the case of a binary decision, the Q-factor defined from the eye-opening based on the ratio of noise variance sigma and amplitude is normally used. However, careful consideration is required for a multi-level case.

Figure 4.22 shows the detected eye pattern for 16-QAM. In this case, we must consider three threshold levels between the four solid black lines indicating the original signal levels. If serious waveform distortion is assumed due to dispersion or nonlinearity, one of the transmitted signal levels, for example, level 1, may be received as the level shown as 1*, which is actually rather close to level 3. In the conventional eye-opening estimation, the noise variance is underestimated due to the wrong estimation of the signal distortion by the distance between level 1* and level 3. In our simulation, we considered the levels of both the original transmitted signal and the received signal to evaluate the noise variance sigma, in order to avoid underestimation in the case of large variance.

In our proposed method, first, the ratio of the received signal's energy per bit to the spectral noise density, E_b/N_0, is estimated using the variance between the transmitted modulation signal and the received signal. Then, the BER is

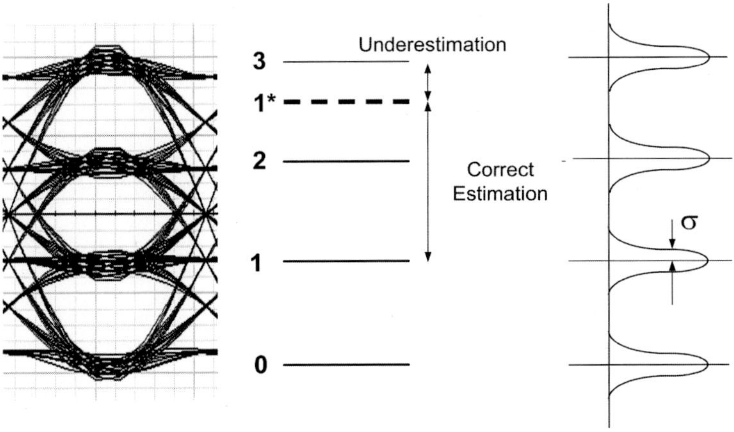

Fig. 4.22 Schematic of the BER evaluation in 16-QAM simulation

estimated from a theoretical expression for Gray-coded M-QAM using the estimated E_b/N_0 [14].

Details of the simulation procedure are described in [28]. In this simulation, tolerances against phase noise and DGD are considered by neglecting the effect of ISI mentioned in the previous section, dispersion and nonlinearity. It is assumed that the SOP at the receiver side was ideally controlled. Figure 4.23 shows the I–Q constellation of 80-Gb/s (10 Gsymbol/s) 256-QAM at a received power level of −14 dBm with DGD set to 0 ps, for spectral linewidths of 200 kHz (left) and 30 MHz (right). It should be noted that this proposed detection scheme produces no degradation due to increasing spectral linewidth. Figure 4.24 shows the BER for a 256-QAM signal vs. the rotation angle θ of the 256-QAM signal or the pilot carrier from the principal state of polarization (PSP) for the system DGD. Here, the spectral linewidth was 1 MHz, DGD was 28 ps, and the received optical power was set to obtain a BER of 10^{-9} with $\Delta\nu = 0$ Hz and DGD = 0 ps. We found that, for the 256-QAM signal, the worst SOP was $\theta = 22.5°$, as shown in Fig. 4.24. On the other hand, for DQPSK, the polarization rotation was usually set at $\theta = 45°$ for the PSP as a worst-case condition.

Figure 4.25 shows BER vs. DGD for the 256-QAM and DQPSK cases with the worst-case condition, when the received optical power was set to obtain a BER

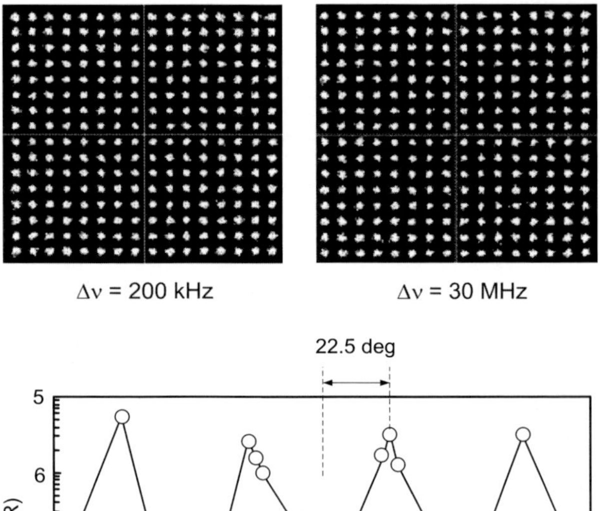

Fig. 4.23 Constellation of 80-Gb/s (10 Gsymbol/s) 256-QAM. Spectral linewidth: 200 kHz (*left*), 30 MHz (*right*)

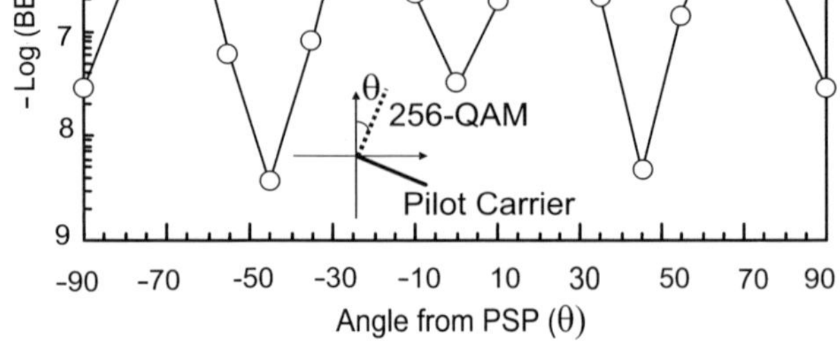

Fig. 4.24 The BER characteristics for 256-QAM signal versus the rotation angle θ

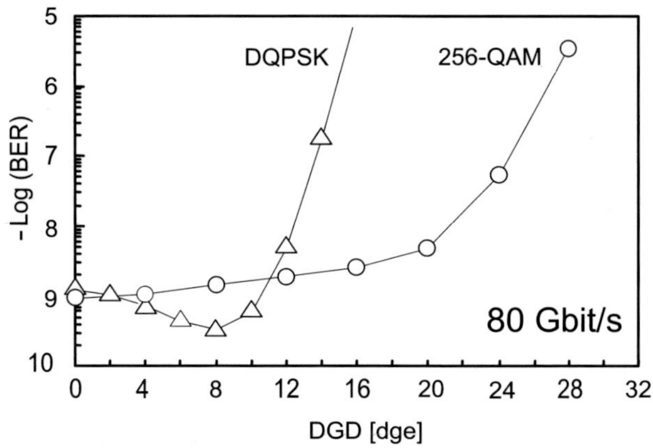

Fig. 4.25 The DGD tolerance for the 256-QAM and DQPSK at 80 Gbit/s

of 10^{-9} with spectral linewidth $\Delta\nu = 1$ MHz. Although the receiver sensitivity of 10-Gsymbol/s 256-QAM is inherently worser than that of 40 Gsymbol/s DQPSK at DGD $= 0$, the results show that the 256-QAM cases (circles) have better DGD tolerance up to 20 ps, whereas the DQPSK cases (triangles) degrade sharply above 12 ps, to attain a BER of less than 10^{-8}. It is thus feasible to use M-QAM formats in 80-Gb/s transmission with our detection scheme, without the need for an expensive narrow linewidth light source.

Acknowledgments The author wishes to acknowledge Dr. Moriya Nakamura, Dr. Guo-Wei Lu, and Dr. Yukiyoshi Kamio of NICT for their collaboration and help.

References

1. R.A. Griffin, A.C. Carter, Optical Differential Quadrature Phase-Shift Key (oDQPSK) for High Capacity Optical Transmission, OFC'02, vol. WX6, Anaheim, CA, 2002
2. Y. Konishi, K. Ishida, K. Kubo, T. Mizuochi, True PRBS Transmission of DQPSK by Differential Precoder Employing Parallel Prefix Network, Proceedings of OFC'06, OThR3, Anaheim, CA, 2006
3. L.G. Kazovsky, Balanced phase-locked loops for optical homodyne receivers: Performance analysis, design considerations, and laser linewidth requirements. IEEE J. Lightwave Technol. LT-**4**(2), 182–195 (1986)
4. S. Norimatsu, K. Iwashita, K. Noguchi, An 8 Gb/s QPSK optical homodyne detection experiment usingexternal-cavity laser diodes. IEEE Photon. Technol. Lett. **4**(7), 765–767 (1992)
5. J.R. Barry, J.M. Kahn, Carrier synchronization for homodyne and heterodyne detection of optical quadriphase-shift keying. IEEE J. Lightwave Technol. **10**(12), 1939–1951 (1992)
6. E. Ip, J.M. Kahn, Carrier synchronization for 3-and 4-bit-per-symbol optical transmission. IEEE J. Lightwave Technol. **23**(12), 4110–4124 (2005)
7. D.S. Ly-Gagnon, K. Katoh, K. Kikuchi, Unrepeated 210-km transmission with coherent detection and digital signal processing of 20-Gb/s QPSK signal. Electron. Lett. **41**(4), 206–207 (2005)
8. Y. Mori, C. Zhang, K. Igarashi, K. Katoh, K. Kikuchi, Unrepeated 200-km Transmission of 40-Gbit/s 16-QAM Signals Using Digital Coherent Optical Receiver. ECOC'08, Tu.1.E4, Sydney, 2008

9. M. Seimetz, Laser Linewidth Limitations for Optical Systems with High-Order Modulation Employing Feed Forward Digital Carrier Phase Estimation. OFC'08, OtuM2, San Diego, CA, 2008
10. L.E. Nelson, S.L. Woodward, M.D. Feuer, X. Zhou, P. D. Magill, S. Foo, D. Hanson, D. McGhan, H. Sun, M. Moyer, M. O'Sullivan, Performance of a 46-Gbps Dual-Polarization QPSK Transceiver With Real-Time Coherent Equalization Over High PMD Fiber, OFC'08, PDP9, New York, NY, 2008
11. B. Wandernoth, 1064 nm, 565 Mbit/s PSK Transmission Experiment with Homodyne Receiver Using Synchronization Bits. Electron. Lett. **27**(19), 1692–1693 (1991)
12. I. Shake, H. Takara, K. Uchiyama, T. Kitoh, T. Kitagawa, M. Okamoto, K. Magari, Y. Suzuki, T. Morioka, 160 Gbit/s full optical time-division demultiplexing using FWM of SOA-array integrated on PLC. Electron. Lett. **38**(1), 37–38 (2002)
13. K. Kikuchi, T. Okoshi, M. Nagamatsu, N. Henmi, Degradation of bit error rate in coherent optical communications due to spectral spread of the transmitter and the local oscillator. J. Lightwave Technol. **LT-2**, 1024–1033 (1984)
14. S. Sampei, *Applications of Digital Wireless Technologies to Global Wireless Communications*. (Prentice Hall, New York, NY, 1997), pp. 80–102
15. G.W. Lu, M. Nakamura, Y. Kamio, T. Miyazaki, 40-Gb/s QPSK and 20-Gb/s PSK with inserted pilot symbols using self-homodyne detection. Opt. Express **15**(12), 7660–7666 (2007)
16. T. Miyazaki, Linewidth-tolerant QPSK homodyne transmission using a polarization. multiplexed pilot carrier. IEEE Photon. Technol. Lett. **18**(2), 388–390 (2006)
17. M. Nakamura, Y. Kamio, T. Miyazaki, PMD and dispersion-tolerance of QPSK homodyne detection using a polarization-multiplexed pilot carrier, ECOC'06, Mo. 4.2.5, 2006
18. M. Nakamura, Y. Kamio, G.W. Lu, T. Miyazaki, Ultimately phase-noise tolerant QPSK homodyne using a spectrum-sliced ASE light source. IEICE Electron. Express (ELEX) **4**(13), 406–410 (2007)
19. M. Nakamura, Y. Kamio, and T. Miyazaki, Linewidth-tolerant 30-Gbit/s 8-PSK self-homodyne using a single modulator and phase-noise cancelling technique. IET Electron. Lett. **45**(7), 368–369 (2009)
20. J.B. Jensen, T. Tokle, C. Peucheret, P. Jeppesen, Transmission of Multilevel 60 Gbit/s Polarization Multiplexed RZ-D8PSK using only 10 Gbit/s equipment OFC'07, OWM4, Anaheim, CA, 2007
21. E. Hu, Y. Hsueh, K. Wong, M. Marhic, L. Kazovsky, K. Shimizu, N. Kikuchi, 4-level direct-detection of polarization shift keying (DD-PolSK) system with phase modulators, OFC'03, FD2, Anaheim, CA, 2003
22. T. Sakamoto, A. Chiba, T. Kawanishi, 50-Gb/s 16QAM by a quad-parallel Mach-Zehnder modulator, ECOC'07, PD2.8 Berlin, 2007
23. C.R. Doerr, P.J. Winzer, L. Zhang, L.L. Buhl, N.J. Sauer, Monolithic InP 16-QAM Modulator, OFC'08, PDP20, San Diego, CA, 2008
24. Y. Kamio, M. Nakamura, T. Miyazaki, ISI Pre-equalization in a Vector Modulator for 5 Gsymbol/s 64-QAM, ECOC'08, Tu.1.D.3, Brussels, 2008
25. M. Nakamura, Y. Kamio, T. Miyazaki, Linewidth-tolerant 10-Gbit/s 16-QAM transmission using a pilot-carrier based phase-noise cancelling technique. Opt. Express **16**(14), 10611–10616 (2008)
26. M. Nakamura, Y. Kamio, and T. Miyazaki, 30-Gbit/s 64-QAM transmission over 60-Km SSMF using phase-noise cancelling technique and ISI suppression based on electronic digital processing. Electron. Lett. **45**(25), 1339–1340 (2009)
27. M. Nakamura, Y. Kamio, and T. Miyazaki, Linewidth-tolerant real-time 40-Gbit/s 16-QAM self-homodyne detection using a pilot-carrier and ISI suppression based on electronic digital processing. Opt. Lett. **35**(1), 13–15 (2010)
28. Y. Kamio, M. Nakamura, T. Miyazaki, 80Gb/s-256QAM format using phase noise tolerant pilot carrier aided homodyne detection. IEEE LEOS Summer Topicals TuE3.2, pp. 250–251, (2007)

Chapter 5
High-Speed and High-Capacity Optical Transmission Systems

Peter J. Winzer and Rene-Jean Essiambre

Abstract In order to meet the exponentially growing bandwidth demands of future data applications, such as massive video services or telepresence, fiber-optic networks need to increase their spectral efficiency in order to continue to provide a cost-effective transport infrastructure. In this chapter, we review the evolution of digital optical modulation, from 10-Gb/s on/off keying systems with direct detection receivers all the way to 100-Gb/s systems using coherent detection and polarization-division multiplexing of multi-level modulation formats on one or more optical subcarriers. Finally, we review recent results on the capacity limits of single-mode optical fiber and show that current experimental records approach these theoretical limits to within a factor of 4. Over the next decade, significant efforts are expected not only in the areas of modulation and multiplexing but also in the design of low-loss and low-nonlinearity fiber to be able to further scale capacity.

5.1 The Need for Capacity and Spectral Efficiency

Over the past 50 years, digital communication technologies of various sorts have transformed our society into a true communication and information society. In many countries of the world, the vast majority of people today have access to a wealth of global and constantly updated information resources from their personal computers as well as from their handheld wireless devices. These technological advances have profoundly shaped many of our society's behavioral patterns, especially in the entertainment, commerce, and communication/information sectors. Today, data traffic is dominated by a plethora of video applications, as illustrated in Fig. 5.1(a) [1]. As a further communications paradigm, currently still in its infancy, millions of end users will be sending rich, video-centric, and real-time content between one another from wherever they are, and future corporate and personal communication products will evolve to include telepresence as their next frontier [2]. This step will significantly

P.J. Winzer (✉)
Bell Labs, Alcatel-Lucent, Paris, France; 209 Holmdel-Keyport Rd., Holmdel, NJ 07733, USA
e-mail: peter.winzer@alcatel-lucent.com

M. Nakazawa et al. (eds.), *High Spectral Density Optical Communication Technologies*, Optical and Fiber Communications Reports 6, DOI 10.1007/978-3-642-10419-0_5,

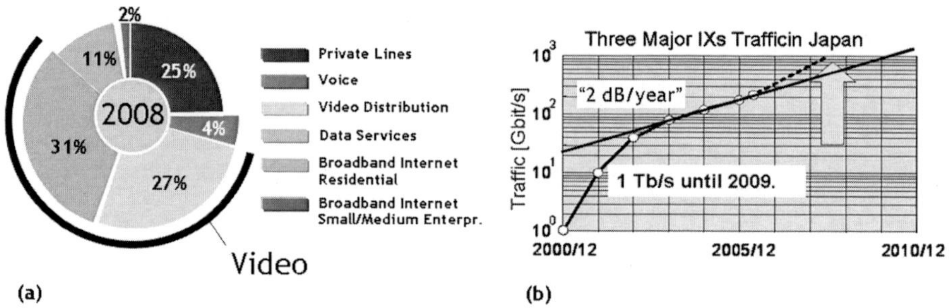

Fig. 5.1 (**a**) Data traffic is dominated by video content today (figure courtesy of J.-L. Beylat, Alcatel-Lucent [1]); (**b**) Traffic evolution and projected '2-dB/year' growth (figure courtesy of T. Miyazaki, NICT)

shape the long-distance travel patterns of our society, especially in times of limited global energy resources. Given all these developments, there is no reason to believe that the growth of data traffic should suddenly slow down from its long-term historic rate of between 1 and 2 'dB per year' (i.e., a factor of 10 every 5–10 years) [3], as visualized for the example of Japanese internet exchange (IX) points in Fig. 5.1(b).

In order to transport vast amounts of data, *optical communications* has evolved as the key enabling technology across the board, starting from interconnects in data centers and massive fiber-to-the-home (FTTH) build-outs worldwide to transparent metropolitan and regional optical networks to nation-wide long-haul optical backbones and submarine systems spanning the globe [4, 5]. Traditionally dominated by the synchronous optical network (SONET), the synchronous digital hierarchy (SDH), and the optical transport network (OTN) standards, Ethernet [6] has started to enter optical networking at the first common SONET/Ethernet bit rate of 10 Gb/s. The success of 10G Ethernet and the clear trend from circuit to data traffic has triggered the desire to scale optical networks using Ethernet. The IEEE has now standardized higher speed Ethernet at 40 Gb/s and 100 Gb/s, and the ITU-T-based OTN standard has followed the trend by departing from its classical 4× scaling in favor of taking up OTU4 at 112 Gb/s to natively support 100G Ethernet in optical transport networks [7]. Irrespective of the underlying transport formats or protocols, optical networks have to carry an exponentially increasing amount of capacity in a way that is both *cost-effective* and *energy efficient*.

Today, several technological options exist to increase the capacity of an optical transport network. In the simplest approach, one just deploys a new system on a parallel optical fiber path, which is possible as long as there are dark fiber strands available in the network. However, the capital expenditure (CAPEX) and the energy consumption per transmitted information bit remain roughly constant in this approach, and the wavelength management complexity increases, which all leads to an increased operational expenditure (OPEX). Very similar considerations apply to dual-band (C+L band) or triple-band (S+C+L band) systems [8], whose main advantage is the avoidance of the cost associated with laying new cable in fiber exhaust situations.

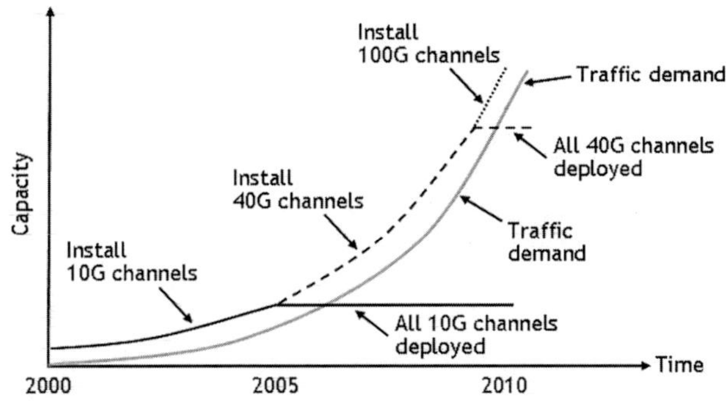

Fig. 5.2 In order to meet the exponentially growing demand for data traffic (*gray curve*), optical transport solutions must increase their spectral efficiency when proceeding to the next higher per-channel bit rate (figure courtesy of P. Magill, AT&T Research [9])

Alternatively, one strives to increase the capacity of an already installed system by reusing as much of the existing line infrastructure as possible. From a modulation and transmission standpoint, this is done by increasing the amount of information transported within the limited amplification band of optical amplifiers, quantified by the *spectral efficiency* (SE). The SE is defined as the per-channel information bit rate divided by the WDM channel spacing

$$\mathrm{SE} = \frac{R_B}{B_{\mathrm{Ch}}}. \tag{5.1}$$

Without a proportional increase in SE, single-band optical transport networks do not support the exponentially increasing traffic demand. This is visualized very clearly in Fig. 5.2 [9]. The traffic demand is given by the gray curve, while the supply is represented by the black curves. At first, installing more 10-Gb/s channels onto an existing WDM network allows the network to keep pace with the increasing traffic demand. When all WDM channels supported by the system are filled, the capacity supply stalls unless either another fiber is lit or 10-Gb/s wavelengths are exchanged by 40-Gb/s wavelengths on the same WDM channel grid (typically at 50-GHz channel spacing in today's networks). Then, system capacity continues to scale until all 40-Gb/s channels are deployed, which necessitates the need for 100-Gb/s systems, but again on a 50-GHz grid in order to further satisfy the increasing demand. This motivates the development of 100-Gb/s systems at a SE of 2 b/s/Hz. On the downside, increased SEs and higher per-channel line rates inherently reduce the transmission distance that can be obtained in an optical network, and careful trade-offs taking into account the complexity, energy consumption, and cost of opto-electronic regeneration need to be made.

5.2 Modulation at High Spectral Efficiencies

Attaining high spectral efficiencies in optical networks requires careful use of the available physical dimensions offered for signal transport within an optical fiber. Understanding how these dimensions can be used to construct spectrally efficient modulation formats and multiplexing techniques is essential to the understanding of high-capacity optical networks. The concept of orthogonal signal spaces lends itself particularly well to further this understanding. The following sections give an advanced overview on these topics. For a more introductory exposition, the interested reader is referred to [10] and references cited therein.

5.2.1 Signal Orthogonality in Optical Communications

The basic options for modulation and multiplexing in optical communication systems are given by the exhaustive list of physical properties describing a propagating optical signal shown in Fig. 5.3. Apart from the property 'code,' which is a combination of time and frequency dimensions and may be suited for optical access applications, all other properties are being used in one way or another for modulation and multiplexing in optical transport networks today. The physical properties shown here can be used to construct orthogonal signal dimensions for modulation and multiplexing [10].

Loosely speaking,[1] two signals occupy *orthogonal dimensions* if messages sent in these two dimensions can be uniquely separated from one another at the receiver without impacting each other's detection performance. The task of separating two messages can be used

- to let two users share a common transmission medium (*multiplexing*),
- to send a string of symbols from a transmitter to a receiver (*modulation*), or
- to send a packet header along with an optical payload (*label encoding*).

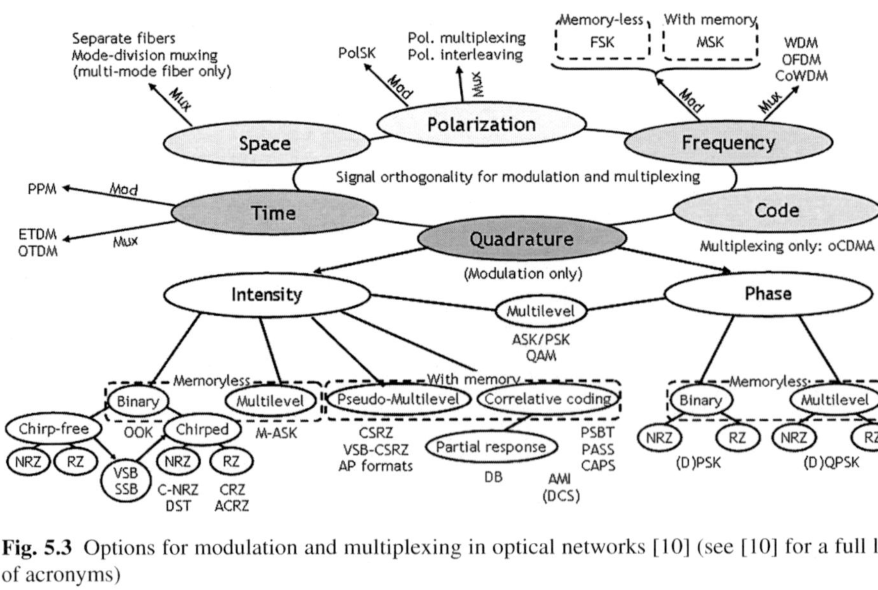

Fig. 5.3 Options for modulation and multiplexing in optical networks [10] (see [10] for a full list of acronyms)

[1] A rigorous definition of orthogonality in the context of optical communications is given in [10].

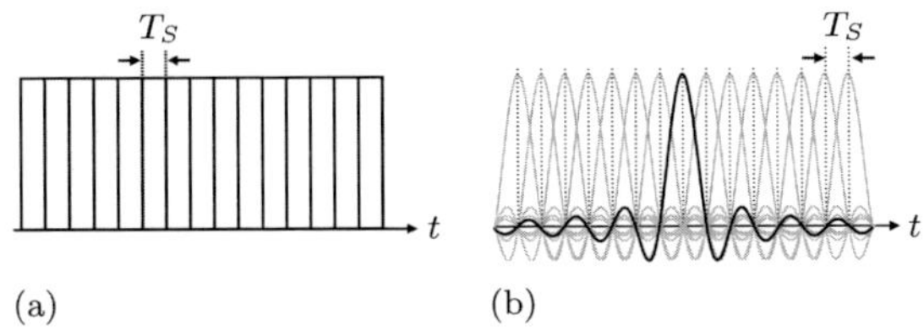

Fig. 5.4 Temporal orthogonality can be obtained by (**a**) allocating disjoint time slots or by (**b**) making sure that different pulses do not disturb each other's amplitude levels at sampling and decision

In this chapter, we will only be concerned about the first two applications of signal orthogonality.

Orthogonality in time is obtained by setting up temporally *disjoint time slots*, as shown in Fig. 5.4a, or by using *pulse shaping* such as to let different pulses not disturb each other's amplitude levels at the T_S-spaced *sampling instants* at the receiver's decision gate (Fig. 5.4b). In both cases, a receiver can clearly distinguish between the information sent every T_S seconds by proper sampling. Orthogonality in time can therefore be used to form streams of independent symbols. If two or more symbols impact each other's amplitudes at the sampling instant, orthogonality is degraded and we speak of inter-symbol interference (ISI).

Apart from generating a symbol stream belonging to a single user, temporal orthogonality can also be employed to *time division multiplex* multiple users onto a single channel. This is done, e.g., in SONET/SDH/OTN by assigning specific time slots within a well-synchronized frame to different users. Of particular practical importance are *electronically* time division multiplexed (ETDM) systems that perform all multiplexing and demultiplexing operations in the electrical domain and directly imprint the resulting high-rate signal onto an optical carrier. In contrast, *optical* time division multiplexing (OTDM) is used in research labs at data rates for which ETDM devices are not yet available.

Orthogonality in frequency is obtained by setting up disjoint frequency bins as shown in Fig. 5.5(a). This technique is used for frequency division multiplexing, which in optical systems is referred to as *wavelength division multiplexing*[2] (WDM). Any leakage from one wavelength bin into its neighbors degrades signal orthogonality, which impairs detection and is generally referred to as 'WDM crosstalk'. Usually, each WDM channel uses a single-carrier signal modulated at the full transponder bit rate. Sometimes, however, *inverse multiplexing* is performed by splitting a logical data stream at the full line rate into N lower-rate streams and frequency multiplexing them together to form an N-subcarrier signal within a single WDM frequency bin. This technique is referred to as *subcarrier multiplexing* (SCM).

[2] Note, though, that WDM for transport networks always uses a constant *frequency* spacing among channels rather than a constant *wavelength* spacing. The latter is only found in coarse WDM (CWDM) systems for access applications [11].

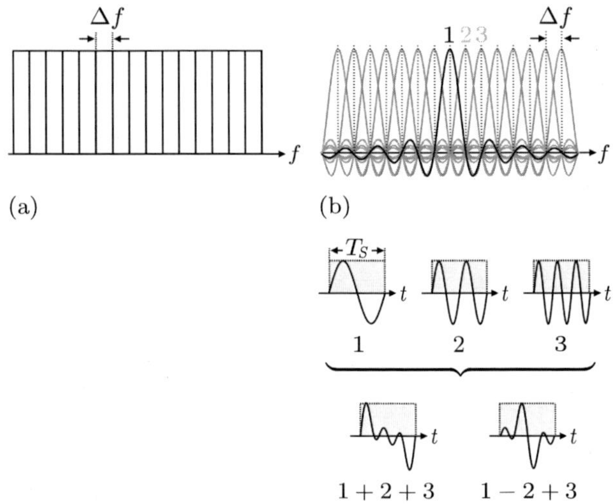

Fig. 5.5 Frequency orthogonality can be obtained by (**a**) allocating disjoint time slots or by (**b**) choosing a frequency spacing of $1/T_S$ between orthogonal (sub)carriers. Since their carrier frequencies differ by an integer number of cycles, these frequencies are orthogonal and individual symbol streams can be superimposed in time while keeping perfect separability

Alternatively, one can use rectangular[3] pulses of duration T_S and place them at carrier frequencies $f_k = k/T_S$, i.e., at a frequency separation of $\Delta f = 1/T_S$, resulting in the spectrum shown in Fig. 5.5(b). Also shown are three pulses fulfilling the above frequency separation criterion. These pulses can be individually modulated and superimposed within the same (synchronized) time slot. If the pulses are, e.g., modulated with binary phase-shift keying, a pulse is either multiplied by $+1$ or by -1 prior to superposition. The resulting waveforms for the two superpositions '1+2+3' and '1−2+3' are shown in the figure. At the receiver, each of the individual transmit pulses can be perfectly recovered from the transmitted superposition by multiplying the received signal with a sine wave at the desired subcarrier frequency and integrating the product over T_S. Due to the specific $1/T_S$ frequency spacing of the pulses, the result of the integration will be zero for all but the desired subcarrier, i.e., the receiver can perfectly filter out the information contained in each 'orthogonal' subcarrier; the signals carried by the subcarriers do not impair each other, even though they show considerable frequency overlap. This is the basic principle employed in *orthogonal frequency division multiplexing* (OFDM) and *coherent WDM* (CoWDM), as described in a dedicated section below.

Orthogonality in polarization is achieved by choosing one of two orthogonal polarization states propagating in a single-mode optical fiber. This orthogonality is exploited for *polarization division multiplexing* (PDM) in high-capacity optical communication systems. Here, one transmits different data streams in the same WDM frequency band but in orthogonal polarizations to double the spectral efficiency. Alternatively, the polarization dimension is sometimes used as an additional

[3] Other pulse shapes are possible as well, as long as the pulse spectrum after reception exhibits zeros at integer multiples of $1/T_S$ from the carrier.

degree of freedom to improve the orthogonality of signals that suffer from orthogonality degradation in another dimension. For example, in the presence of strong WDM crosstalk, orthogonality in the frequency domain is compromised and can be re-established by transmitting adjacent WDM channels on orthogonal polarizations ('polarization interleaved WDM'). In another example, ISI between adjacent pulses reduces temporal orthogonality and can be mitigated by alternating the polarization of adjacent pulses ('alternate polarization modulation').

Orthogonality of the two quadratures of a bandpass signal, i.e., the real and the imaginary parts or the amplitude and phase of the signal's complex envelope, is the basis for the most widely used modulation signal space. The resulting modulation formats are generally referred to as *quadrature amplitude modulation* formats. The wealth of possibilities encountered in optical communications is indicated in Fig. 5.3, with details and acronyms explained in Refs. [10, 12].

Orthogonality in space can be achieved by allowing no spatial overlap between co-propagating signals, e.g., by placing them on separate optical fibers. This trivial multiplexing technique is sometimes referred to as *space division multiplexing* and is frequently encountered on short optical links, e.g., for optical interconnects between racks of equipment, where fiber ribbons can be used. Much less trivially, different transverse modes propagating in a multi-mode optical fiber are all orthogonal [13] and may be assigned to different users, which leads to the notion of *mode division multiplexing* (MDM) [14]. However, multi-mode fibers are prone to random mode coupling, and the recovery of any particular transverse fiber mode from a mixture of modes is not easily achieved [15].

5.2.2 The Evolution of Optical Modulation Formats

5.2.2.1 From 10 to 40 Gb/s

Up until about a decade ago, the bandwidth provided by optical amplifiers was sufficient to carry the capacity demand in optical transport networks and spectral efficiency was of little concern. Binary intensity modulation (on/off keying OOK) had dominated fiber-optic transport since its beginning, since it is the simplest of all modulation options. OOK was implemented in a variety of flavors, including non return-to-zero (NRZ) and return-to-zero (RZ), the latter featuring improved receiver sensitivity [16] and frequently a higher tolerance to distortions from fiber nonlinearity [17]. Due to its simplicity, OOK was also the first modulation format used in early 40-Gb/s products [18] and was taken all the way to 100 Gb/s in research demonstrations using optical equalization to make up for the limited bandwidth of opto-electronic components [19, 20].

When 40-Gb/s systems started to enter optical networking at the turn of the millennium, the fourfold increase in per-channel bit rates resulted in a 6-dB higher optical signal-to-noise ratio (OSNR) requirement compared to 10-Gb/s systems. At fixed per-channel transmit powers, this reduces the reach in the linear propagation regime by a factor of four. In the nonlinear regime, the reduction depends

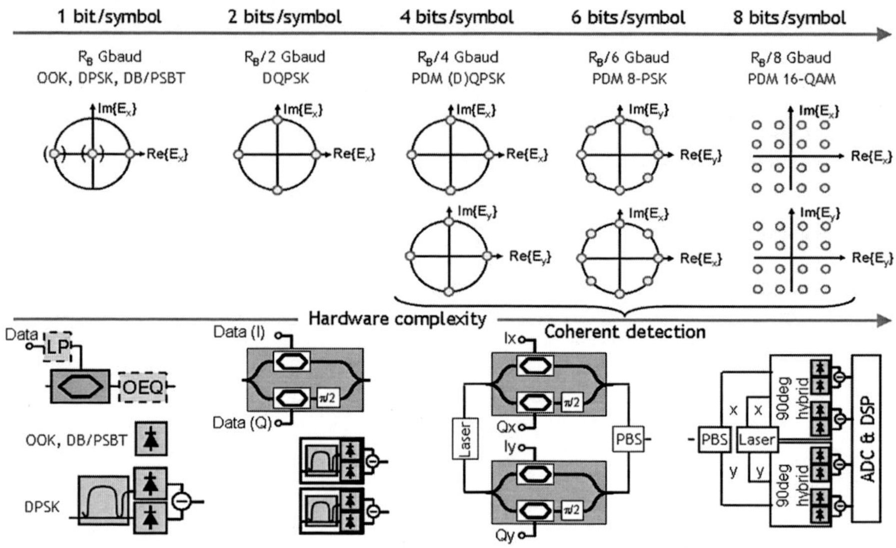

Fig. 5.6 Options for modulation and detection at high spectral efficiencies, especially those discussed for 100-Gb/s optical transport. The hardware complexity increases with the number of bits carried by each modulation symbol

significantly on the fiber type and system configuration, and a factor of 2 is more typical since one can usually launch higher per-channel powers at 40 Gb/s. To make up for the lost reach at 40 Gb/s compared to 10 Gb/s, enhanced forward error correction (FEC) [21, 22], distributed Raman amplification [23], and advanced optical modulation formats [12, 24, 25] were introduced. In particular, the development of *differential phase-shift keying* (DPSK) [26] improved the OSNR margin by about 3 dB.

As evident from Fig. 5.6, though, the benefits of DPSK come at the expense of a more complex receiver structure, involving an optical delay interferometer and a balanced pair of photodiodes as the differential demodulation element. Note that differential demodulation requires the information to be encoded in the phase *difference* between adjacent symbols, which in turn necessitates a digital precoding operation prior to modulation at the transmitter. In laboratory experiments using true pseudo-random bit sequences (PRBS), differential precoding can be omitted since a PRBS is invariant to differential precoding [27].

5.2.2.2 From a 100-GHz to a 50-GHz WDM grid

In parallel with the introduction of 40-Gb/s transmission, systems started to move from a 100-GHz to a 50-GHz WDM channel spacing and also from point-to-point operation to optically routed mesh networks [28]. This made *spectral efficiency* at 40 Gb/s a major system design consideration. Modulation formats needed not only to be resilient to optical amplifier noise, fiber nonlinearity, or chromatic and polarization-mode dispersion, but also to filter concatenation due to repeated passes through reconfigurable optical add/drop multiplexer (ROADM) nodes [12, 29, 30]. Initially, this led to a high interest in reduced bandwidth binary formats, such as NRZ-DPSK [26] or duobinary (DB) modulation[4] (also known as phase-shaped binary transmission, PSBT [31]), as well as to optimizations of the receiver optics to counteract heavy optical filtering of DPSK [32]. Subsequently, the development of quaternary modulation formats in the form of differential quadrature phase-shift keying (DQPSK) [33] allowed a bandwidth compression[5] compared to DPSK at almost the same performance in terms of its required optical signal-to-noise ratio (OSNR) at typical FEC input bit error ratios (BERs) [12]. DQPSK allows for long-haul 40-Gb/s transport on a 50-GHz grid with multiple in-line ROADMs [34].

As evident from Fig. 5.6, the hardware complexity at both transmitter and receiver is significantly higher for DQPSK than it is for DPSK, even though the bandwidth requirements of the modulation and detection electronics and opto-electronics are reduced from R_B to $R_B/2$. Furthermore, DQPSK has $\sim$6 times higher frequency stabilization requirements on the interferometer than DPSK [35]. These are some of the factors that have somewhat delayed the commercialization of 40-Gb/s DQPSK transponders. As with DPSK, DQPSK requires a digital precoding operation to be performed prior to modulation at the transmitter [36, 37]. This operation is substantially more difficult to perform than the one for DPSK, but can be parallelized and implemented in real time, even at 100 Gb/s [38].

5.2.2.3 Early 100-Gb/s DQPSK Transport Demonstrations in Lab and Field

Research on electronically multiplexed 100-Gb/s transmission systems started in 2005/06 with the first demonstrations of OOK transmitters [39, 40], receivers [41], and long-haul fiber transmission [42, 43]. Although technical difficulties regarding the available bandwidth of electronic and opto-electronic components for binary modulation were shown to be manageable, it was realized early on that the spectral efficiency of OOK, even when using vestigial-sideband modulation [44], would not be sufficient for the required transport network capacity scaling, and research on 100-Gb/s started to include DQPSK [45, 46], with a demonstration of optically routed networking over 1200 km including 6 ROADMs at a spectral efficiency of 1 b/s/Hz, i.e., on a 100-GHz WDM grid [47]. Subsequently, 100-Gb/s DQPSK was taken to the field [48] using the LambdaXtreme® platform.

LambdaXtreme® [49] is a long-haul optical transport platform manufactured by Alcatel-Lucent. It operates on a 50-GHz WDM frequency grid in the extended L-band between 1554 and 1608 nm and can carry 128 wavelengths over 4000 km at 10 Gb/s or 64 wavelengths at 10 Gb/s together with 64 wavelengths at 40 Gb/s over 2200 km. The ROADM architecture is based on *asymmetric interleavers*, as shown in Fig. 5.7. The interleavers have alternating 3-dB passband widths of 65

[4] Strictly speaking, DB/PSBT is a ternary partial-response format [12], but it can be detected using binary receiver electronics.

[5] A multi-level modulation format transmits one out of M symbols at the symbol rate $R_S = R_B/\log_2 M$. Hence, its spectrum is compressed by a factor $\log_2 M$ compared to binary modulation.

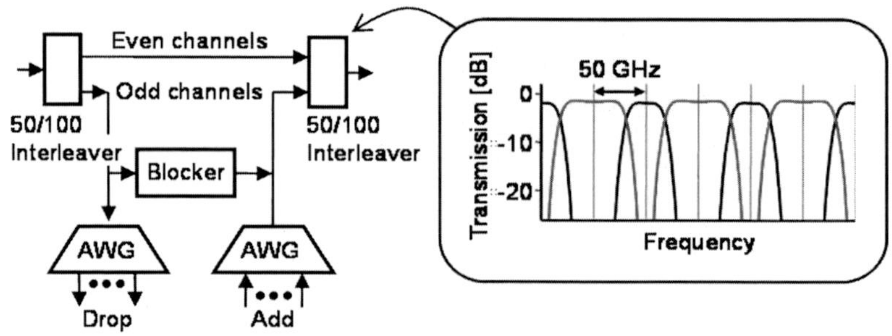

Fig. 5.7 Asymmetric interleaver structure enabling 100-Gb/s DQPSK transmission in a 50-GHz asymmetric WDM environment

and 35 GHz. The wider passband of a full ROADM (two interleavers) has a 3-dB bandwidth of 55 GHz. Optical amplification is accomplished through forward and backward Raman pumping.

Importantly, an FPGA was set up in this 100-Gb/s field trial to perform client-side aggregation from 10.7 to 107 Gb/s and real-time DQPSK precoding at the full 107-Gb/s line rate [38]. This enabled the first demonstration of live video traffic being transmitted on a single 107-Gb/s wavelength in an operational carrier network and proved the feasibility of upgrading existing optically routed networks to 100-Gb/s services.

5.2.2.4 Fiber Capacity Records Using DQPSK

In research labs, DQPSK at 43 and 56 Gbaud was used together with PDM to establish fiber transmission capacity records[6] of initially 14 Tb/s [51], then 25.6 Tb/s at a spectral efficiency of 3.2 b/s/Hz (1.6 b/s/Hz per polarization) whose optical spectra are shown in Fig. 5.8 [52]. Both groups used proven direct-detection technologies with delay demodulation to set these records. Interestingly, both experiments employed spectrally broader RZ versions of DQPSK rather than NRZ-DQPSK. The RZ-DQPSK signals were narrowly filtered prior to wavelength division multiplexing, which with today's high-speed opto-electronic technologies usually yields a better quality signal with less ISI than what would be obtained by shaping the spectrum purely electronically through NRZ-DQPSK modulation. The reason is that optical filters feature less ripple and resonances than electrical filters and amplifiers, and hence perform a better shaping of the optical transmit waveform. Ultimately, one would like to implement perfectly spectrally compact yet ISI-free pulses (i.e., pulses with a rectangular spectrum and a sinc-type temporal waveform). This is still hard to achieve at multi-Gbaud rates but may become possible in the future.

[6] Recently, the optical fiber capacity record was moved to 69.1 Tb/s using 16-QAM and, in contrast to the direct-detection real-time record of 25.6 Tb/s, used coherent detection with off-line digital signal processing [50].

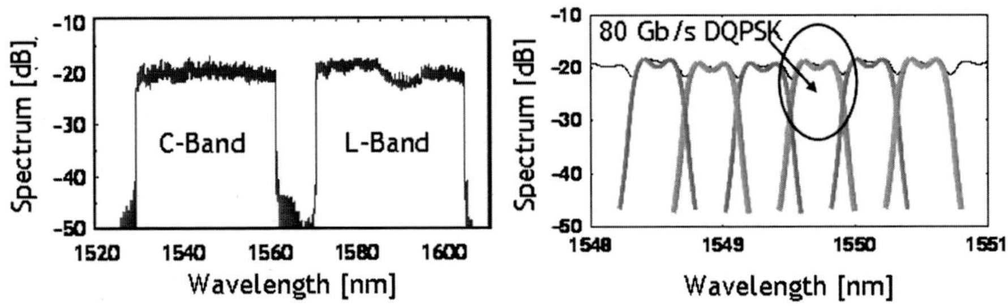

Fig. 5.8 Optical spectra of the 25.6-Tb/s capacity record experiment [52] using 86-Gb/s RZ-DQPSK in C and L bands at a spectral efficiency of 3.2 b/s/Hz (1.6 b/s/Hz per polarization). The *right figure* shows a detailed view of eight channels within the band

5.2.2.5 Coherent Detection and PDM-QPSK

One of the conclusions from the 100-Gb/s experiments described above is the fact that DQPSK would only support a 100-GHz channel spacing in an operational long-haul optically routed networking environment with multiple ROADM passes, even when using slightly asymmetric ROADM architectures. Acknowledging from Fig. 5.2 that spectral efficiency is a strong driving force for future optical networking, other approaches therefore have to be investigated. One possibility to further narrow the optical spectrum to support 112 Gb/s over a 50-GHz WDM grid would be to go to yet higher level constellations, such as 8-(D)PSK at 37 Gbaud or 16-QAM at 28 Gbaud in a single polarization. While the former would still be feasible using direct detection and delay demodulation [53], the latter requires a *coherent receiver* [54] to efficiently demodulate real and imaginary parts of the data signal.

Coherent (heterodyne or homodyne) detection schemes were intensively studied prior to the advent of the erbium-doped fiber amplifier in the early 1990s and mainly focused on increasing the shot-noise-limited sensitivity of binary modulation formats (OOK, PSK, FSK) in order to increase the spacing between opto-electronic regenerators. In contrast, modern coherent receivers are built on the *intradyne* principle [55]. As illustrated in Table 5.1, the key difference to traditional schemes is that

Table 5.1 Coherent detection options: Heterodyne detection requires a significant excess opto-electronic front-end bandwidth, and homodyne detection requires an analog PLL to lock the LO to the received signal. Intradyne detection uses a free-running LO and performs digital phase locking after A/D conversion of the beat signal

	Heterodyne	**Homodyne**	**Intradyne**
Front-end bandwidth	**~ 3 – 5 x Symbol rate**	Symbol rate	~Symbol rate
Phase/frequency locking	Frequency locking	**Analog optical PLL**	**Digital electronic PLL** (free-running LO)
Spectral sketch	0 IF f	0 f	IF f

intradyning lets the intermediate frequency fall somewhere *within the signal band*, which avoids enormous technical problems associated with opto-electronically phase locking a local oscillator (LO) laser (homodyning) but at the same time does not require the large opto-electronic front-end bandwidth of traditional heterodyning (about three to five times the symbol rate). The price to be paid for this flexibility is the necessary analog-to-digital (A/D) conversion at a sampling rate between $1.5R_S$ and $2R_S$ to convert the beat product between signal and LO into the digital domain. Once digital, negative-frequency and positive-frequency components of the intradyne signal can be brought back to overlap at baseband, which is done by digital frequency estimation together with a digital phase-locked loop (PLL). Equalization and source separation algorithms well known from wireless applications can also be implemented in a straightforward manner [58, 62–66].

Importantly, a coherent receiver relies on the polarization-aligned beating of the signal with the LO laser and hence needs *polarization diversity* in a fiber-optic environment, where the signal polarization varies in a random fashion along the propagation path. This essentially doubles the opto-electronic complexity of the receiver compared to a single-polarization structure (cf. Fig. 5.6). However, one can turn this seeming disadvantage into a benefit by using PDM in combination with less complex quadrature modulation formats that offer a higher resilience to optical amplifier noise. For example, one can use PDM QPSK instead of single-polarization 16-QAM to achieve the same bit rate at the same spectral extent of the signal. This thought process naturally results in PDM QPSK as an excellent candidate for single-carrier 100-Gb/s transport.[7] As an important added benefit, coherent detection offers the advantage of full optical field equalization in the digital domain, which dramatically increases the system's robustness to chromatic dispersion, polarization-mode dispersion, narrow-band optical filtering, and potentially also to distortions from fiber nonlinearity [58].

The benefits of intradyne detection listed above require the availability of massive digital signal processing, which necessitates the development of mixed-signal ASICs. The highest speed chip reported today operates at a symbol rate of 11.5 Gbaud to transport a line rate of 46 Gb/s using PDM QPSK [59]. Such dedicated ASIC design efforts are typically undertaken solely in the course of product development and do not provide the flexibility needed for research purposes. In research, one has to resort to FPGA implementations to develop and test signal processing algorithms in real time [60]. The flexibility provided by FPGA processing comes at the expense of a reduced bit rate, though. FPGA-based demonstrations of PDM QPSK have been demonstrated at rates of 10 Gb/s [61]. Alternatively, one can perform *off-line signal processing* to develop and test digital signal processing algorithms. Here, a continuous portion of the beat product between signal and LO is sampled using a fast digital oscilloscope (presently limited to 80 GSamples/s), and the data are transferred to a personal computer to

[7] The same conclusion is obtained by first adding the polarization dimension to a 50-Gbaud direct-detection DQPSK system with delay demodulation [56], and noting that PDM can be more practically implemented using coherent detection than using direct detection [57].

perform all the digital signal processing. This way, 111-Gb/s PDM QPSK was first demonstrated in a long-haul optical networking environment including ROADMs in [62] and has since been shown to operate over 2550 km on a 50-GHz WDM grid [63]. Note, however, that only a limited number of symbols (typically on the order of 1 million) are captured using real-time oscilloscopes. This prevents the study of certain fiber nonlinearities introducing long memory into the channel [64] and also inhibits real-time traffic demonstrations, such as the DQPSK video demonstration described above. In some respect, relying on off-line processing requires a similar leap of faith as the common procedure of measuring system performance at the FEC *input* and hoping for the FEC to correct these errors following its 'additive white Gaussian noise' characteristic, even in the presence of different channel impairments.

5.2.2.6 Going Beyond PDM QPSK

As shown in [65, 63], a PDM QPSK signal operating at a line rate of 112 Gb/s can tolerate optical filtering with a bandwidth of about 30 GHz before significant penalties start to arise. Assuming a second-order super-Gaussian ROADM filter characteristic with a 3-dB bandwidth of around 45 GHz, this allows the signal to traverse about five ROADMs on a 50-GHz WDM grid. However, some networks may require signals to pass through 15 or more ROADM nodes [28], which asks for a yet higher spectral compression using more than four symbols per polarization.

One possibility is the use of PDM 8-PSK, which has been demonstrated in a 662-km point-to-point experiment on a 25-GHz WDM grid at a spectral efficiency of 4.2 b/s/Hz [66, 67]. As seen from Fig. 5.9, 8-PSK shows a sensitivity

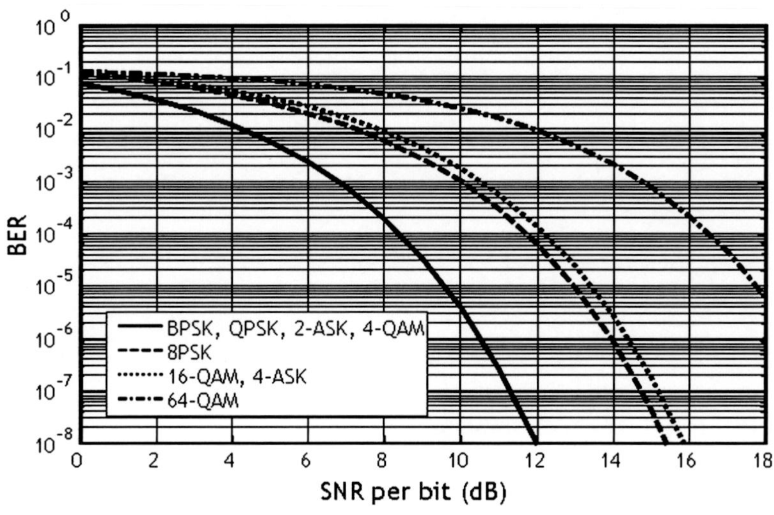

Fig. 5.9 BER performance of important higher level modulation formats

penalty of 3.25 dB compared to QPSK at a BER of 10^{-3} due to its reduced symbol spacing.[8]

Compressing the spectrum by another 25% using PDM 16-QAM, one can perform 100G optical networking with ROADMs on a 25-GHz WDM grid [69]. Even though the transmitter now operates at lower symbol rates, it must support multi-level electrical drive signals. Alternatively, one could use more advanced quad-parallel modulator structures that use exclusively binary drive signals at the expense of a more complex modulator design [70]. At present, the highest spectral efficiency reported at multi-Gbaud symbol rates and for single-carrier transmission is 8 b/s/Hz using 36-QAM, transmitted over 320 km of fiber [71].

Yet higher level (up to 256-ary [72]) QAM as well as other combinations of PSK and ASK could be an option for future high-capacity systems. At high per-channel bit rates, however, these formats place stringent requirements on the underlying modulation and detection hardware.

5.2.2.7 Multi-tone Modulation (CoWDM, OFDM)

All considerations so far were related to *single-carrier* WDM transmission, where the underlying modulation format is imprinted directly onto a single optical frequency per WDM channel. Alternatively, one can use multiple subcarriers to transport the information of a single WDM channel. In general, this *inverse multiplexing* approach leads to a reduced spectral efficiency compared to single-carrier modulation but may yield cost benefits when combined with high-density photonic integration [73]. A loss in spectral efficiency by inverse multiplexing can be avoided if the individual subcarriers are put at their theoretically minimum ('*orthogonal*,' cf. Fig. 5.5) spacing of $\Delta f = R_S = 1/T_S$, where T_S is the symbol duration used on each subcarrier; the symbol slots of all subcarriers are synchronized to maintain orthogonality. Using N information-bearing orthogonal subcarriers and M-ary modulation with symbols $a(k)$ chosen from the alphabet $\{a_1, a_2, \ldots a_M\}$ on subcarrier k, the orthogonal subcarrier-multiplexed information signal reads

$$x(t) = \sum_{k=1}^{N} a(k) e^{j2\pi k \Delta f t}, \qquad (5.2)$$

[8] The SNR per symbol (often simply called the 'SNR'), always defined as a single-polarization quantity, is related to the OSNR by

$$\text{SNR} = \text{OSNR}\frac{2B_{\text{ref}}}{(2\cdot)R_S},$$

where the factor of 2 applies for polarization-multiplexed transmission. The relation to the frequently cited 'SNR per bit' (or E_B/N_0) is given by [68]

$$\text{SNR} = \text{OSNR}\frac{2B_{\text{ref}}}{R_B},$$

where R_B is the aggregate bit rate (including possible PDM).

where we assume a perfectly rectangular pulse envelope as shown in Fig. 5.5. At times $t = lT_S$, this equation can be written as

$$x(lT_S) = \sum_{k=1}^{N} a(k)e^{j2\pi kl} = \mathcal{F}^{-1}\{a(k)\}, \tag{5.3}$$

which is identified as the (inverse) *Fourier transform* of the symbol sequence $\{a(k)\}$, lending itself to a particularly efficient digital implementation using the fast Fourier transform[9] (FFT) algorithms. If subcarrier modulation is done *electronically* [preferably by means of an FFT according to Eq. (5.3)] we speak of OFDM and of Coherent WDM 'CoWDM' or 'optical OFDM' (O-OFDM) if modulation is performed *optically* on each subcarrier according to Eq. (5.2). Table 5.2 puts the two methods in context with other optical and electronic modulation and multiplexing techniques. Figure 5.10 visualizes the transmitters for OFDM and CoWDM for the case of a four-frequency orthogonal multiplex. Note that OFDM is not a 'modulation format' in its strict sense but rather an inverse *multiplexing technique*. Within the multiplex, each subcarrier uses a certain modulation format.

Irrespective of how it is generated, the orthogonal frequency multiplex transports an aggregate bit rate of $R_B = NR_S \log_2 M$ in a channel bandwidth of $B_{\mathrm{Ch}} = NR_S = R_B/\log_2 M$ per WDM channel, exactly as a single-carrier format (with proper sinc-shaped pulses) would. In addition, however, OFDM requires the use of unmodulated pilot carriers and a substantial cyclic prefix, which reduces the transmitted information rate at fixed line rate up to 20% [74]. Techniques such as "No-Guard-Interval OFDM" [75] or "Reduced-Guard-Interval OFDM" [76] reduce this overhead and attain the same spectral efficiency as single-carrier formats.

As for the required OSNR, OFDM, and CoWDM show the same performance as single-carrier formats in the linear propagation regime provided that the same modulation format is used [74]. The tolerance to distortions from fiber nonlinearity is still under investigation [74].

Regarding the complexity of implementation, OFDM has been shown to require somewhat less digital signal processing power than single-carrier modulation [77]. On the other hand, OFDM requires a high-speed (R_S) digital-to-analog (D/A) converter at the transmitter, which together with a reasonably linear drive amplifier can pose serious practical problems. For single-carrier modulation a high-resolution

Table 5.2 The role of OFDM and CoWDM in the context of other optical and electronic modulation and multiplexing techniques

Orthogonality through	Optical multiplexing	Electronic multiplexing
Temporally disjoint time intervals	OTDM	ETDM
Disjoint frequency bins	WDM	SCM
Orthogonal subcarriers	CoWDM, O-OFDM	OFDM

[9] Note that due to the symmetry of time domain and frequency domain, the FFT is algorithmically equivalent to the inverse FFT (IFFT).

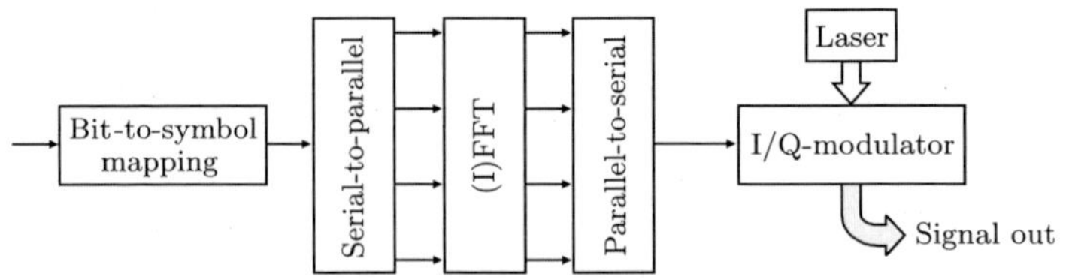

Bit stream	0 1	1 0	0 0	1 0	1 1	0 0	1 0	0 0
Symbol stream	$-j$	j	1	j	-1	1	j	1
(I)FFT	$1+j$	$-1-j$	$1-3j$	$-1-j$	$1+j$	$-1-j$	$-3+j$	$-1-j$

(a) OFDM transmitter

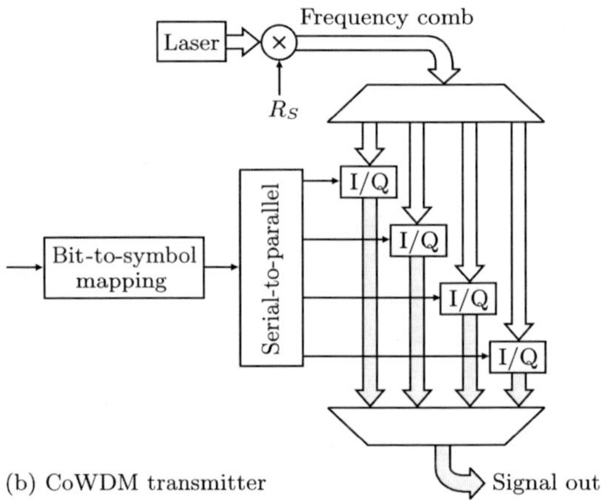

(b) CoWDM transmitter

Fig. 5.10 Transmitter implementations of OFDM and CoWDM

D/A converter is only required in the case of pre-distortion [78, 79]. Much lower D/A requirements are placed on single-carrier modulation formats without pre-distortion; for example, an M-QAM transmitter only needs a D/A resolution of $\frac{1}{2}\log_2 M$ at the transmitter.

Perhaps the only real advantage of OFDM agreed upon today is its ability to adapt both the modulation format of each subcarrier and as well as the number of subcarriers on a dynamic basis, which could allow for software-defined, modem-like optical transponders in future flexible optical networks.

Spectral efficiencies of 7.0 b/s/Hz at bit rates beyond 40 Gb/s have been demonstrated using PDM OFDM with 32-QAM per subcarrier and at a net bit rate of 56 Gb/s [80].

5.2.2.8 Coded Modulation

All considerations above assumed a strict separation between *modulation* and *coding*. This separation has the advantage of providing a clear interface between the modulation format used for data transport and the FEC used for error correction after detection of the raw (channel) data. This approach is attractive and practical at high speeds, since it allows separate design of the (individually!) highly complex coherent processing engine and the FEC chip. However, it is well known from information theory that optimum performance is only achieved when simultaneously considering modulation and coding.

In *coded modulation*, one considers symbol constellation and FEC as one design entity and introduces FEC overhead not only in the time domain but also in symbol space by allocating more symbols than necessary at a given line rate. In the context of optical communications, DQPSK at 40 Gbaud was used to transmit 40 Gb/s worth of information [81], spending the resulting 100% overhead in symbol space for trellis-coded modulation (TCM). In another example, the polarization dimension was used to implement coded modulation [82]. Coded modulation techniques are well known in radio frequency engineering and are now starting to enter optical communications, with more research to be expected over the next years. Regarding the expected performance, one can use information-theoretic techniques to evaluate the potential to be exhausted, as explained in the context of optical communications, e.g., in [83].

5.3 Theoretical Fiber Capacity Limits

Having explored the possibilities of increasing spectral efficiency in optically routed networks with the digital processing power and the high-speed opto-electronic hardware available today, we will next investigate how much spectral efficiency scaling can still be expected.

The seminal work of Shannon published in 1948 [84] gave birth to information theory. Shannon determined the capacity of memoryless channels, focusing on channels impaired by additive white Gaussian noise (AWGN) for a given SNR. The extension of Shannon's theory to the 'fiber channel' in optical core networks faces several major difficulties. Even the definition of a 'fiber channel' by itself is application specific[10] and cannot be characterized by a single SNR-like quantity, like its AWGN equivalent can. An important challenge originates from the simultaneous presence of three fundamental physical phenomena in optical fibers, as visualized in Fig. 5.11 [58]:

- *Noise* and *distortion* enter optical networking in several ways. Whether an impairment constitutes noise or distortion depends on whether one considers it random ('noise') or deterministic ('distortion'). Many impairments only *appear* random

[10] For example, a point-to-point WDM system can use joint detection to fully exploit cross-channel impairments. In contrast, in an optically-routed network one typically does not have access to the WDM channels that have shared parts of the propagation path with the channel of interest at the receiver.

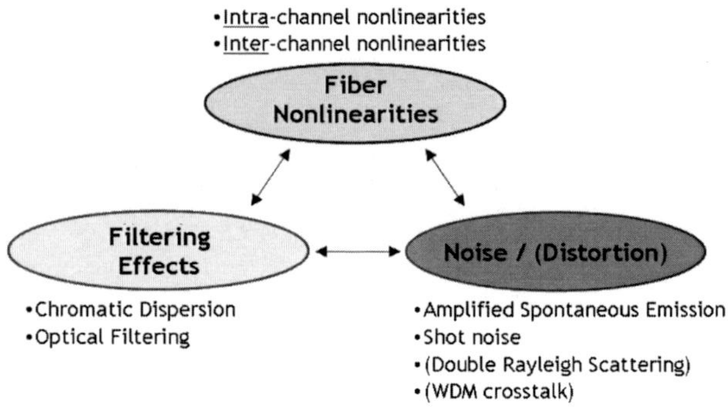

Fig. 5.11 Each fiber channel is fundamentally limited by the interplay of three classes of impairments: Noise (or uncompensated distortion), Kerr nonlinearity, and linear filtering [68]

because one is unwilling or unable (with reasonable engineering means) to measure and compensate for them.

Most fundamentally, *amplified spontaneous emission* (ASE) is noise, quantum mechanically unavoidably produced by in-line optical amplification. Optical amplification is needed to compensate for fiber loss, and hence ASE is as fundamental a system limitation as fiber loss itself. Compared to the fundamental loss limit of $\sim$ 0.16 dB/km for standard silica-core fiber, hollow-core fiber may be a way to further reduce fiber loss. Investigating the fundamental lower bounds on the loss of single-mode waveguide structures is an interesting field of future research. The optimum amplification scheme uses perfectly distributed amplification, where local gain equals loss at all points along a transmission fiber, ideally leading to a power spectral density of $N_0 = hf\alpha L$ per polarization [85, 58], where α relates to the loss coefficient expressed in decibels via $\alpha_{\mathrm{dB}} = \alpha \cdot 10\log_{10} e$. The quantity αL is the cumulative loss of the transmission fiber along its entire length L. Experimental demonstrations of nearly ideal distributed gain can be found in [86, 87].

Another fundamental source of noise is *shot noise*. However, even when using distributed amplification, shot noise is negligible compared to ASE-induced beat noise whenever the $\alpha L \gg 1$, which is the case for all practical transmission distances [71].

Double–Rayleigh backscatter (DRB) is usually considered a source of noise.[11] Since for DRB the backscattered light propagates over a significant fiber length in the backward direction before being scattered back in the forward direction, optical isolators can be inserted along the line to suppress the backward propagation of DRB [88]. Assuming the ability to put an arbitrary number of lossless isolators in-line, DRB is no longer as fundamental a source of noise as ASE or shot noise.

WDM crosstalk arises when signals spill into the neighboring WDM bands, either due to imperfectly limited spectral truncation at the transmitter or due

[11] In principle, DRB could be considered to be distortion if the locations of the frozen scatterers were measured (hence known) and the laser linewidth was small enough to avoid phase decorrelation between all possible double-reflection points.

to fiber nonlinearity-induced spectral broadening. Other nonlinear interactions between WDM channels also exist, as detailed under the topic 'fiber nonlinearity' later. Depending on the point of view, these effects may be categorized as noise or as distortion, and some parts of it may be considered fundamental while others are not.

- *Filtering* in an optically routed network is performed not only by in-line ROADMs but as importantly by fiber *chromatic dispersion*, which can be considered as an all-pass filter with a quadratic phase response. Chromatic dispersion introduces a significant amount of memory into the system, which when coupled with nonlinearity becomes an impairment that is particularly hard to deal with.
- Instantaneous *Kerr nonlinearity* is a phenomenon that results in signal distortions and imposes fundamental limits on transmission of information over fibers. One can distinguish between *intra-* and *inter-channel* nonlinearities. Intra-channel nonlinearities are nonlinear interactions confined to fields present within the frequency band of the signal that one wants to transmit from one location to another in a network. Nonlinear interactions that involve fields outside this frequency band are referred to as inter-channel nonlinearities. Intra-channel nonlinearities can be efficiently compensated by using backward propagation [68, 58] at either the transmitter or the receiver. In the results reported below, all elementary (instantaneous) Kerr nonlinear interactions in the presence of signal and noise are taken into account by direct numerical solution of the stochastic generalized nonlinear Schrödinger equation (GNSE) describing transmission in fibers.

Most importantly, all three fundamental interactions are distributed and coupled upon transmission. In the context of a fiber capacity analysis, the simultaneous presence of these effects is captured without restrictive assumptions in [58].

Using the calculation method detailed in [58], we arrive at fiber capacity predictions of the form shown in Fig. 5.12, assuming a 500-km standard single-mode transmission system in an optically routed network scenario [68]. The fiber loss

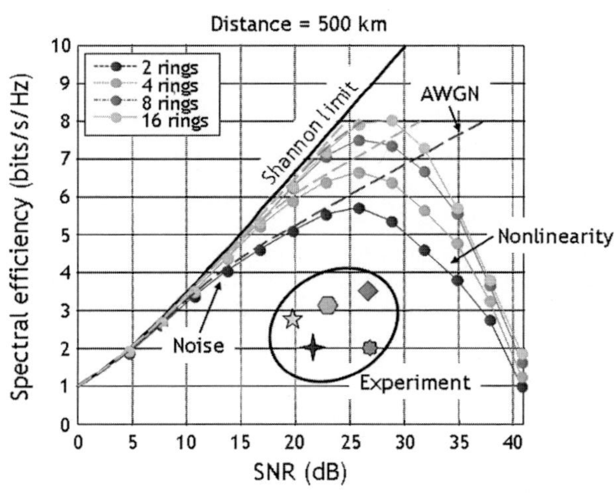

Fig. 5.12 Fundamental capacity limits predicted by the methods detailed in [58] for 500-km transmission in an optically routed network and assuming multi-ring modulation formats. The three symbols indicate experimentally achieved results over comparable distances. All results apply for a single polarization. Polarization multiplexing is expected to approximately double capacity

coefficient is 0.2 dB/km and the effective area is 80 μm^2. The underlying modulation formats are M-ASK/N-PSK, where the number M of rings is given in the legend.[12] We assume a single optical polarization. Polarization multiplexing is expected to approximately double the predicted spectral efficiency. We assume the most sophisticated electronic compensation techniques to be available at transmitter and receiver, including signaling with near Nyquist pulses for maximum spectral efficiency, reverse nonlinear propagation at both the transmitter and receiver, and coherent detection.

Since the noise in distributed amplification is given by its lower bound established above, varying the SNR implies varying the signal power, which in turn leads to different amounts of nonlinear signal distortions. At low signal powers (low SNRs), the system is linear and only limited by noise. Therefore, the numerically simulated curves closely follow those obtained for the AWGN channel (dashed) and for large constellation size approach the Shannon limit[13] [58]. For high SNR (i.e., high signal power) nonlinear interactions between different WDM channels as well as between the channel of interest and co-propagating ASE impair the signal at the receiver, and capacity drops. The optimum spectral efficiency is on the order of 8 b/s/Hz for the 16-ring constellation. Importantly, we note that the fiber channel turns the strictly monotonic Shannon capacity into a curve with a pronounced optimum SNR.

Also shown in the figure are experimentally achieved results over comparable distances [66, 69, 71, 80, 89], all referenced to a single polarization. Although these experiments do not make use of all the sophisticated methods assumed to be available in the calculation of the fundamental limits, they already fall within a factor of four from the predicted limits. This indicates that a lot of work has to be spent, e.g., on developing methods to lower fiber loss and/or nonlinearity in order to be able to further scale WDM optical transport systems.

5.4 Conclusion

In order to meet the exponentially growing bandwidth demand of future data applications such as massive video services or telepresence, fiber-optic networks need to increase spectral efficiency in order to continue to provide cost-effective transport solutions. As of today, a single optical fiber has been shown to support up to 69 Tb/s

---

[12] The number N is chosen high enough to enable the respective capacities. For example, the four-ring case predicts a maximum capacity of close to 7 b/s/Hz, which would require a total of $2^7 = 128$ constellation points in the absence of coding. Adding about 2 bits per symbol for coding, which is a typically used overhead in wireless systems, we arrive at a total requirement for $2^9 = 512$ symbols. This translates into $N = 512/4 = 128$ points per ring. Note, however, that in our numerical simulations we chose M much larger than this value in order to be limited by the true channel capacity rather than by assumptions on the exact modulation format or coding overhead.

[13] The Shannon limit is obtained through a Gaussian-shaped modulation constellation [84], which is well approximated using multi-ring formats [58].

at a spectral efficiency of 6.4 b/s/Hz. Spectral efficiencies as high as 8.0 b/s/Hz have been demonstrated at multi-GBaud rates using multi-level modulation on a single or on multiple coherent subcarriers, coherent detection, and polarization division multiplexing. Fundamental analyses of theoretical channel capacities suggest that these experimental efforts are within a factor of four from the Shannon limit of the fiber-optic channel. Over the next decade, significant efforts are expected not only in the areas of modulation and multiplexing but also in the design of low-loss and low-nonlinearity fiber to be able to further scale capacity.

Acknowledgments The work reported in this chapter is based on several original contributions accomplished together with many of our colleagues. In particular, we wish to acknowledge Greg Raybon, Alan Gnauck, Chris Doerr, Bob Tkach, Andy Chraplyvy, Tetsuya Kawanishi, Jerry Foschini, and Gerhard Kramer. We are also grateful to Jean-Luc Beylat, Pete Magill, and Tetsuya Miyazaki for permission to reuse some of their figures.

References

1. J.L. Beylat, How will optical networks continue to transform the way the world communicates? Asia-Pacific Optical Communication (APOC), Plenary Presentation (2007)
2. D.W. Schloerb, A quantitative measure of telepresence. Presence: Teleoperators Virtual Environ. **4**, 64–80 (1995)
3. R.W. Tkach, Scaling optical communications for the next decade and beyond. Bell Labs Tech. J. **14**(4), 3–10 (2010)
4. R. Mack, Global landscape in broadband: Politics and economics and applications, in *Optical Fiber Telecommunications V-B: Systems and Networks*, ed. by I.P. Kaminov, T. Li, A.E. Willner (Academic Press, New York, 2008)
5. R.E. Wagner, Fiber-based broadband access technology and deployment, in *Optical Fiber Telecommunications V-B: Systems and Networks*, ed. by I.P. Kaminov, T. Li, A.E. Willner (Academic Press, New York, 2008)
6. C.F. Lam, W.I. Way, Optical Ethernet: Protocols and management and 1–100 g G technologies, in *Optical Fiber Telecommunications V-B: Systems and Networks*, ed. by I.P. Kaminov, T. Li, and A.E. Willner (Academic Press, New York, 2008)
7. S. Trowbridge, High-speed Ethernet transport. IEEE Commun. Mag. **45**, 120–125 (2007)
8. A. Sano, Y. Miyamoto, Technologies for ultrahigh bit-rate WDM transmission. Proc. IEEE/LEOS Annual Meeting (2007)
9. P. Magill, 100 Gigabit Ethernet from a carrier's perspective. Proc. IEEE/LEOS Annual Meeting (2007)
10. P.J. Winzer, R.-J. Essiambre, Advanced optical modulation formats, in *Optical Fiber Telecommunications V-B: Systems and Networks*, ed. by I.P. Kaminov, T. Li, A.E. Willner (Academic Press, New York, 2008)
11. H.-J. Thiele, M. Nebeling (eds.), *Coarse Wavelength Division Multiplexing: Technologies and Applications* (CRC Press, Boca Rotan, 2007)
12. P.J. Winzer, R.J. Essiambre, Advanced optical modulation formats. Proc. IEEE **94**, 952–985 (2006)
13. H.G. Unger, *Planar Optical Waveguides and Fibers* (Clarendon Press and Oxford Engineering Science Series, New York, 1977)
14. H.R. Stuart, Dispersive multiplexing in multimode optical fiber. Science **289**(5477), 281–283 (2000)
15. J. Kahn, Compensating multimode fiber dispersion using adaptive optics. Proc. Opt. Fiber Commun. Conf. (OFC), paper OTuL1 (2007)

16. P.J. Winzer, A. Kalmar, Sensitivity enhancement of optical receivers by impulsive coding. J. Lightwave Technol. **17**(2), 171–177 (1999)
17. R.-J. Essiambre, G. Raybon, B. Mikkelsen, Pseudo-linear transmission of high-speed TDM signals: 40 and 160 Gb/s, in *Optical Fiber Telecommunications IV*, ed. by I. Kaminow, T. Li (Academic Press, New York, 2002), pp. 232–304
18. M. Birk, L. Raddatz, D. Fishman, S.Woodward, P. Magill, Field trial of end-to-end OC-768 transmission using 9 WDM channels over 1000 km of installed fiber. Proc. Opt. Fiber Commun. Conf. (OFC), paper TuS4 (2003)
19. P.J. Winzer, G. Raybon, C.R. Doerr, M. Duelk, C. Dorrer, 107-Gb/s optical signal generation using electronic time-division multiplexing. J. Lightwave Technol. **24,** 3107–3113 (2006)
20. J.H. Sinsky, A. Adamiecki, L. Buhl, G. Raybon, P.J. Winzer, O. Wohlgemuth, M. Duelk, C.R. Doerr, A. Umbach, H.-G. Bach, D. Schmidt, A 107-Gbit/s optoelectronic receiver utilizing hybrid integration of a photodetector and electronic demultiplexer. J. Lightwave Technol. **26**, 114–120 (2008)
21. T. Mizuochi, K. Kubo, H. Yoshida, H. Fujita, H. Tagami, M. Akita, K. Motoshima, Next generation FEC for optical transmission systems. Proc. Opt. Fiber Commun. Conf. (OFC), paper ThN1 (2003)
22. T. Mizuochi, Recent progress in forward error correction and its interplay with transmission impairments. IEEE J. Select. Topics Quantum Electron. **12**, 544–554 (2006)
23. J. Bromage, Raman amplification for fiber communications systems. J. Lightwave Technol. **22**, 79–93 (2004)
24. A.H. Gnauck, Advanced amplitude- and phase coded formats for 40-Gb/s fiber transmission. Proc. IEEE/LEOS Annual Meeting, paper WR1 (2004)
25. S. Bigo, Multiterabit DWDM terrestrial transmission with band width limiting optical filtering. IEEE J. Select. Topics Quantum Electron. **10**(2), 329–340 (2004)
26. A.H. Gnauck, P.J. Winzer, Optical phase-shift-keyed transmission. J. Lightwave Technol. **23**(1), 115–130 (2005)
27. M. Rice, S. Tretter, P. Mathys, On differentially encoded m-sequences. IEEE Trans. Commun. **49**, 421–424 (2001)
28. E.B. Basch, R. Egorov, S. Gringeri, S. Elby, Architectural tradeoffs for reconfigurable dense wavelength-division multiplexing systems. IEEE J. Select. Topics Quantum Electron. **12**, 615–626 (2006)
29. G. Raybon, S. Chandrasekhar, A.H. Gnauck, B. Zhu, L.L. Buhl, Experimental investigation of long-haul transport at 42.7 Gb/s through concatenated optical add/drop nodes. Proc. Opt. Fiber Commun. Conf. (OFC), paper ThE4 (2004)
30. G. Raybon, S. Chandrasekhar, A. Agarwal, A.H. Gnauck, L.L. Buhl, J. Sinsky, A. Adamiecki, Limitations of optical add/drop filtering on 42.7-Gb/s transmission with 50-GHz channel spacing. Proc. European Conf. on Opt. Commun. (ECOC), paper Mo4.5.1 (2004)
31. D. Penninckx, M. Chbat, L. Pierre, J.-P. Thiery, The phase-shaped binary transmission (PSBT): A new technique to transmit far beyond the chromatic dispersion limit. IEEE Photon. Technol. Lett. **9**(2), 259–261 (1997)
32. B. Mikkelsen, C. Rasmussen, P. Mamyshev, F. Liu, Partial DPSK with excellent filter tolerance and OSNR sensitivity. IEE Electron. Lett. **42**, 1363–1364 (2006)
33. R.A. Griffin, A.C. Carter, Optical differential quadrature phase shift key (oDQPSK) for high-capacity optical transmission. Proc. Opt. Fiber Commun. Conf. (OFC), paper WX6 (2002)
34. A.H. Gnauck, P.J. Winzer, S. Chandrasekhar, C. Dorrer, Spectrally efficient (0.8 b/s/Hz) 1-Tb/s (25×42.7 Gb/s) RZ-DQPSK transmission over 28 100-km SSMF spans with 7 optical add/drops. Proc. Opt. Fiber Commun. Conf. (OFC), paper Th4.4.1 (2004)
35. H. Kim, P.J. Winzer, Robustness to laser frequency offset in directdetection DPSK and DQPSK systems. J. Lightwave Technol. **21**(9), 1887–1891 (2003)
36. M. Serbay, C. Wree, W. Rosenkranz, Implementation of differential precoder for high-speed optical DQPSK transmission. IEE Electron. Lett. **40**, 1288–1289 (2004)

37. Y. Konishi, K. Ishida, K. Kubo, T. Mizuochi, True PRBS transmission of DQPSK by differential precoder employing parallel prefix network. Proc. Opt. Fiber Commun. Conf. (OFC), paper OThR3 (2006)
38. H. Song, A. Adamiecki, P.J. Winzer, C. Woodworth, S. Corteselli, G. Raybon, Multiplexing and DQPSK precoding of 10.7-Gb/s client signals to 107 Gb/s using an FPGA. Proc. Opt. Fiber Commun. Conf. (OFC), paper OTuG3 (2008)
39. P.J. Winzer, G. Raybon, M. Duelk, 107-Gb/s optical ETDM transmitter for 100G Ethernet transport. Proc. European Conf. on Opt. Commun. (ECOC), paper Th4.1.1 (2005)
40. C.R. Doerr, P.J. Winzer, G. Raybon, L.L. Buhl, M.A. Cappuzzo, A. Wong-Foy, E.Y. Chen, L.T. Gomez, M. Duelk, A single-chip optical equalizer enabling 107-Gb/s optical non-return-to-zero signal generation. Proc. European Conf. on Opt. Commun. (ECOC), paper Th4.2.1 (2005)
41. R.H. Derksen et al., Integrated 100 Gbit/s ETDM receiver in a transmission experiment over 480 km DMF. Proc. Opt. Fiber Commun. Conf. (OFC), paper PDP37 (2006)
42. G. Raybon, P.J. Winzer, C.R. Doerr. 10 × 107-Gbit/s electronically multiplexed and optically equalized NRZ transmission over 400 km. Proc. Opt. Fiber Commun. Conf. (OFC), paper PDP32 (2006)
43. P.J. Winzer, G. Raybon, C.R. Doerr, 10 × 107 Gb/s electronically multiplexed NRZ transmission at 0.7 bits/s/Hz over 1000 km non-zero dispersion fiber. Proc. European Conf. on Opt. Commun. (ECOC), paper Tu1.5.1 (2006)
44. K. Schuh et al., 8 Tbit/s (80 × 107 Gbit/s) DWDM ASK-NRZ VSB transmission over 510 km NZDSF with 1bit/s/Hz spectral efficiency. Proc. European Conf. on Opt. Commun. (ECOC), paper PD1.8 (2007)
45. M. Daikoku, I. Morita, H. Taga, H. Tanaka, T. Kawanishi, T. Sakamoto, T. Miyazaki, T. Fujita, 100 Gbit/s DQPSK transmission experiment without OTDM for 100G ethernet transport. Proc. Opt. Fiber Commun. Conf. (OFC), paper PDP36 (2006)
46. P.J. Winzer, G. Raybon, C.R. Doerr, L.L. Buhl, T. Kawanishi, T. Sakamoto, M. Izutsu, K. Higuma, 2000-km WDM transmission of 10×107-Gb/s RZ-DQPSK. Proc. European Conf. on Opt. Commun. (ECOC), paper Th4.1.3 (2006)
47. P.J. Winzer, G. Raybon, S. Chandrasekhar, C.R. Doerr, T. Kawanishi, T. Sakamoto, K. Higuma, 10 × 107-Gb/s NRZ-DQPSK transmission at 1.0 b/s/Hz over 12 × 100 km including 6 optical routing nodes. Proc. Opt. Fiber Commun. Conf. (OFC), paper PDP24 (2007)
48. P.J. Winzer, G. Raybon, H. Song, A. Adamiecki, S. Corteselli, A.H. Gnauck, D.A. Fishman, C.R. Doerr, S.Chandrasekhar, L.L. Buhl, T.J. Xia, G. Wellbrock, W. Lee, B. Basch, T. Kawanishi, K. Higuma, Y. Painchaud, 100-Gb/s DQPSK transmission: From laboratory experiments to field trials. J. Lightwave Technol. **26**, 3388–3402 (2008)
49. D.A. Fishman et al., LambdaXtreme transport system: R&D of a high capacity system for low cost, ultra long haul DWDM transport. Bell Labs Technical J. **11**, 27 (2006)
50. A. Sano et al., 69.1-Tb/s (432 x 171-Gb/s) C- and extended L-Band transmission over 240 km using PDM-16-QAM. Proc. OFC, paper PDPB7 (2010)
51. A. Sano, H. Masuda, Y. Kisaka, S. Aisawa, E. Yoshida, Y. Miyamoto, M. Koga, K. Hagimoto, T. Yamada, T. Furuta, H. Fukuyama, 14-Tb/s (140 × 111-Gb/s PDM/WDM) CSRZ-DQPSK transmission over 160 km using 7-THz bandwidth extended L-band EDFA. Proc. European Conf. on Opt. Commun. (ECOC), paper Th4.1.1 (2006)
52. A.H. Gnauck, G. Charlet, P. Tran, P.J. Winzer, C. Doerr, J. Centanni, E.C. Burrows, T. Kawanishi, T. Sakamoto, K. Higuma, 25.6-Tb/s C+L-band transmission of polarization-multiplexed RZ-DQPSK signals. Proc. Opt. Fiber Commun. Conf. (OFC), paper PDP19 (2007)
53. M. Ohm, Optical 8-DPSK and receiver with direct detection and multilevel electrical signals. IEEE/LEOS Workshop on Advanced Modulation Formats, paper 45–46 (2004)
54. K. Kikuchi, Coherent optical communication systemscoherent optical communication systems, in *Optical Fiber Telecommunications V-B: Systems and Networks*, ed. by I.P. Kaminov, T. Li, A.E. Willner (Academic Press, New York, 2008).
55. F. Derr, Optical QPSK transmission system with novel digital receiver concept. IEE Electron. Lett. **27**, 2177–2179 (1991)

56. S. Chandrasekhar, X. Liu, Experimental investigation of system impairments in polarization multiplexed 107-Gb/s RZ-DQPSK. Proc. Opt. Fiber Commun. Conf. (OFC), paper OThU7 (2008)
57. D. van den Borne, T. Duthel, C.R.S. Fludger, E.D. Schmidt, T. Wuth, C. Schulien, E. Gottwald, G.D. Khoe, H. de Waardt, Coherent equalization versus direct detection for 111-Gb/s ethernet transport. 2007 Digest of the IEEE LEOS Summer Topical Meetings, paper MA2.4 (2007)
58. R.-J. Essiambre, G.J. Foschini, G. Kramer, P.J. Winzer, Capacity limits of information transport in fiber-optic networks. Phys. Rev. Lett. **101**, 163901 (2008)
59. H. Sun, K. Wu, K. Roberts, Real-time measurements of a 40 Gb/s coherent system. Opt. Express **16**, 873–879 (2008)
60. A. Leven, N. Kaneda, Y.-K. Chen, Intradyne receivers using FPGA processing. Proc. Conf. on Lasers and Electro-optics, paper CThJJ4 (2008)
61. A. Leven, N. Kaneda, Y.-K. Chen, A real-time CMA-based 10 Gb/s polarization demultiplexing coherent receiver implemented in an FPGA. Proc. Opt. Fiber Commun. Conf. (OFC), paper OTuO2 (2008)
62. C.R.S. Fludger, T. Duthel, D. van den Borne, C. Schulien, E.-D. Schmidt, T. Wuth, E. de Man, G.D. Khoe, H. de Waardt, 10 × 111 Gbit/s and 50 GHz spaced and POLMUX-RZ-DQPSK transmission over 2375 km employing coherent equalisation. Proc. Opt. Fiber Commun. Conf. (OFC), paper PDP22 (2007)
63. G. Charlet et al., Transmission of 16.4 Tbit/s capacity over 2,550 km using PDM QPSK modulation format and coherent receiver. Proc. Opt. Fiber Commun. Conf. (OFC), paper PDP3 (2008)
64. L.K. Wickham, R.-J. Essiambre, A.H. Gnauck, P.J. Winzer, A.R. Chraplyvy, Bit pattern length dependence of intrachannel nonlinearities in pseudolinear transmission. IEEE Photon. Technol. Lett. **16**, 1591–1593 (2004)
65. C.R.S. Fludger, T. Duthel, D. van den Borne, C. Schulien, E.-D. Schmidt, T. Wuth, J. Geyer, E. De Man, Giok-Djan Khoe, H. de Waardt, Coherent equalization and POLMUX-RZ-DQPSK for robust 100-GE transmission. J. Lightwave Technol. **26**, 64–72 (2008)
66. X. Zhou, J. Yu, D. Qian, T. Wang, G. Zhang, P.D. Magill, 8 × 114 Gb/s, 25-GHz-spaced, PolMux-RZ-8PSK transmission over 640 km of SSMF employing digital coherent detection and EDFA-only amplification. Proc. Opt. Fiber Commun. Conf. (OFC), paper PDP1 (2008)
67. J. Yu, X. Zhou, M.-F. Huang, Y. Shao, D. Qian, T. Wang, M. Cvijetic, P. Magill, L.E. Nelson, M. Birk, S. Ten, H.B. Matthew, S.K. Mishra, 17 Tb/s (161 × 114 Gb/s) PolMux-RZ 8PSK transmission over 662 km of ultra-low loss fiber using C-band EDFA amplification and digital coherent detection. Proc. European Conf. on Opt. Commun. (ECOC), paper Th.3.E.2 (2008)
68. R.-J. Essiambre et al., Capacity limits of optical fiber networks. J. Lightwave Technol. **28**, 662 (2010)
69. P.J. Winzer et al., Spectrally efficient long-haul optical networking using 112-Gb/s PDM 16-QAM. J. Lightwave Technol. **28**, 547 (2009)
70. T. Sakamoto, A. Chiba, T. Kawanishi, 50-Gb/s 16 QAM by a quadparallel Mach-Zehnder modulator. Proc. European Conf. on Opt. Commun. (ECOC), paper PD2.8 (2007)
71. X. Zhou et al., 64-Tb/s (640x107-Gb/s) PDM-36QAM transmission over 320km using digital equalization. Proc. OFC, PDPB9 (2010)
72. M. Nakazawa et al., 256 QAM (64 Gbit/s) coherent optical transmission over 160 km with an optical bandwidth of 5.4 GHz. Proc. OFC, paper OMJ5 (2010)
73. D.F. Welch, F.A. Kish, S. Melle, R. Nagarajan, M. Kato, C.H. Joyner, J.L. Pleumeekers, R.P. Schneider, J. Back, A.G. Dentai, V.G. Dominic, P.W. Evans, M. Kauffman, D.J.H. Lambert, S.K. Hurtt, A. Mathur, M.L. Mitchell, M. Missey, S. Murthy, A.C. Nilsson, R.A. Salvatore, M.F. Van Leeuwen, J. Webjorn, M. Ziari, S.G. Grubb, D. Perkins, M. Reffle, D.G. Mehuys, Large-scale inp photonic integrated circuits: Enabling efficient scaling of optical transport networks. J. Lightwave Technol. **13**, 22–31 (2007)
74. S.L. Jansen, I. Morita, K. Forozesh, S. Randel, D. van den Borne, H. Tanaka, Optical OFDM, a hype or is it for real? Proc. European Conf. on Opt. Commun. (ECOC), paper Mo.3.E.3 (2008)

75. Sano et al., No-guard-interval coherent optical OFDM for 100-Gb/s long-haul WDM transmission. J. Lightwave Technol. **27**, 3705 (2009)
76. X. Liu et al., Transmission of a 448-Gb/s reduced-guard-interval CO-OFDM signal with a 60-GHz bandwidth. Proc. OFC, PDPC2 (2010)
77. B. Spinnler et al., Adaptive equalizer complexity in coherent optical receivers. Proc. European Conf. on Opt. Commun. (ECOC), paper We.2.E.4 (2008)
78. D. McGhan, M. O'Sullivan, C. Bontu, and K. Roberts, Electronic dispersion compensation. Proc. Opt. Fiber Commun. Conf. (OFC), paper OWK1 (2006)
79. M. Nakamura, Y. Kamio, T. Miyazaki, Linewidth-tolerant, ISI suppressed 15-Gbit/s 64-QAM transmission over 120-km SSMF. Proc. European Conf. on Opt. Commun. (ECOC), paper Tu1.E.5 (2008)
80. H. Takahashi, A. Al Amin, S.L. Jansen, I. Morita, H. Tanaka, DWDM transmission with 7.0-bit/s/Hz spectral efficiency using 8 $\times$ 65.1-Gbit/s coherent PDM-OFDM signals. Proc. Opt. Fiber Commun. Conf. (OFC), paper PDPB7 (2009)
81. H. Bulow, G. Thielecke, F. Buchali, Optical trellis-coded modulation (oTCM). Proc. Opt. Fiber Commun. Conf. (OFC), paper WM5 (2004)
82. H. Buelow, Polarization qam modulation (pol-qam) for coherent detection schemes. Proc. Opt. Fiber Commun. Conf. (OFC), paper OWG2 (2009)
83. G. Kramer, A. Ashikhmin, A.J. van Wijngaarden, X. Wei, Spectral efficiency of coded phase-shift keying for fiber-optic communication. J. Lightwave Technol. **21**, 2438–2445 (2003)
84. C.E. Shannon, A mathematical theory of communication. Bell System Technical J **27**, 623–656 (1948)
85. V.E. Perlin, H.G. Winful, Optimizing the noise performance of broad-band WDM systems with distributed raman amplification. IEEE Photon. Technol. Lett. **14**, 1199 (2002)
86. Bouteiller, J.-C., Brar, K., Bromage, J., Radic, S., Headley, C. Dualorder, Raman pump. IEEE Photon. Technol. Lett. **15**, 212–214 (2003)
87. T.J. Ellingham, J.D. Ania-Castanon, R. Ibbotson, X. Chen, L. Zhang, S.K. Turitsyn, IEEE Photon. Technol. Lett. **18**, 268 (2006)
88. J. Bromage, P.J. Winzer, R.-J. Essiambre, Multiple-path interference and its impact on system design, in *Raman Amplifiers and Oscillators in Telecommunications*, ed. by M.N. Islam (Springer Verlag, Berlin, 2003)
89. H. Takahashi, A. Al Amin, Sander L. Jansen, I. Morita, H. Tanaka, 8 $\times$ 66.8-Gbit/s coherent PDM-OFDM transmission over 640 km of SSMF at 5.6-bit/s/Hz spectral efficiency. Proc. European Conf. on Opt. Commun. (ECOC), paper Th3.E.4 (2008)

Part II
Advanced Modulation Formats

Chapter 6
Multilevel Signaling with Direct Detection

Nobuhiko Kikuchi

Abstract In this section, we introduce the concept of direct-detection multilevel signaling technology, in which "optical delay-differential detection" is used as a main tool of signal phase detection. Significant progress toward more than four-level signaling has been achieved by the use of an "orthogonally coupled differential receiver" with digital signal processing. We review a recent 50 Gbit/s 32-level signaling experiments and further application of advanced digital signal processing, such as the compensation of fiber chromatic dispersion and the use of QAM constellations.

6.1 Introduction

Multilevel signaling is a viable candidate of the next-generation signaling formats for high-speed and high-capacity optical fiber links, such as 40-Gbit/s, 100-Gbit/s, and beyond. One of the realizing techniques, the digital coherent signaling (for example, [1, 2]) is described in previous chapters. In this section, we review the concept of yet another multilevel signaling scheme, that is, a direct-detection (incoherent) multilevel signaling, in which "intensity detection" and "delay-differential detection" are used as the tools of detecting amplitude and (differential) phase components, respectively.

The advantages of using the direct detection lies in its potential simplicity; it does not require a receiving-side local laser nor its frequency tracking mechanism. Also the direct detection is immune to the polarization states of the received signal, therefore, it does not need to have polarization multiplexing and diversity configurations. Therefore, direct-detection multilevel receiver can be smaller, more cost-effective, and less power consuming. However, it has some restrictions to be overcome; For

N. Kikuchi (✉)
Central Research Laboratory, Hitachi Ltd., Higashi-Koigakubo 1-280, Kokubunji, Tokyo 185-8601, Japan
e-mail: nobuhiko.kikuchi.ca@hitachi.com

M. Nakazawa et al. (eds.), *High Spectral Density Optical Communication Technologies,* Optical and Fiber Communications Reports 6, DOI 10.1007/978-3-642-10419-0_6,

example, the restriction in applicable multilevel signal formats especially when number of signal levels (N) exceeds four and the restricted capability of digital chromatic dispersion compensation (CD). Each results in insufficient receiver sensitivity and less optical CD compensation-free transmission reach, respectively.

The aim of this section is to review the recent progress of direct-detection multilevel signaling to overcome these problems, especially in terms of signaling formats and detection schemes.

6.2 Combined Binary Detection

One of the significant problems of the direct-detection multilevel signaling is how to detect incoming optical field of a multilevel symbol. Straightforward solution of it is the use of combined parallel binary receivers, each detects a partial modulation component of it. The simplest form of it is a conventional differential QPSK (DQPSK) signaling as shown in Fig. 6.1. The DQPSK signal is the four-level signal with constant amplitude carrying two bits of information by modulating its phase (exactly, differential phase) into four different angles with $\pi/2$ spacings as in Fig. 6.1(a). The DQPSK receiver in Fig. 6.1(b) consists of two delay-differential receivers; each consists of a delay demodulator (the delay time T is set to be equal to the symbol time T_s), with the interference angle of $-\pi/4$ or $+\pi/4$, a balanced photo detector, and a binary-decision circuit. In the delay demodulator, a received multilevel symbol is split into two, and one of it is interfered with the other one with a one symbol delay at the specified interference angle, and then its phase modulation component is converted to binary amplitude modulation. Using binary-decision circuits, two independent binary data streams are obtained at the symbol rate from the two delay-differential receivers. Various DQPSK signaling experiments (for example, [3]) have been reported so far, because of its high OSNR sensitivity, high spectral efficiency, and simplicity.

Further increase of the number of signal levels N is a key to exploit the potential efficiency of multilevel signaling. Figure 6.2 shows the examples of the signal constellation diagrams with more than four signal levels. However, direct detection of such signals not straightforward, mainly because the use of delay-differential detection, that is, the multiplication of the optical fields of two successive received

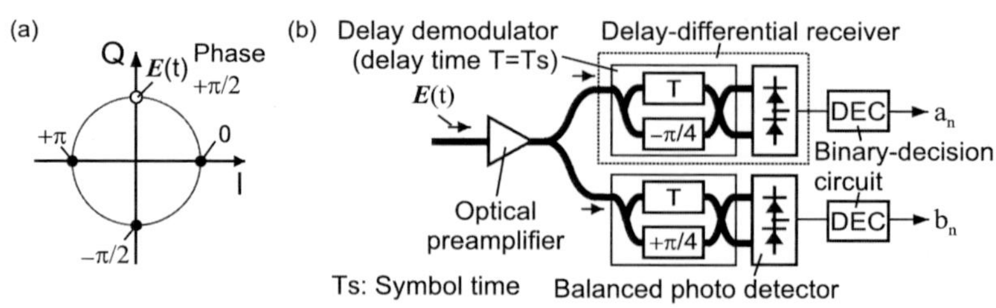

Fig. 6.1 DQPSK signaling. (**a**) Signal constellation diagram, (**b**) receiver configuration

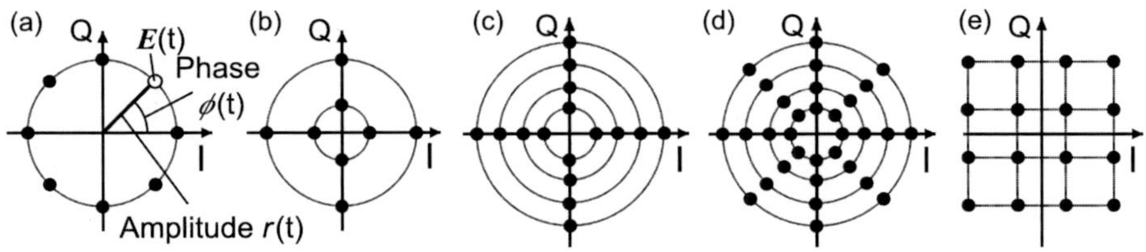

Fig. 6.2 Example of multilevel signals with more than four signal levels. (**a**) 8-DPSK, (**b**) 8-APSK, (**c**) 16-APSK, (**d**) 32-APSK, (**e**) 16-QAM

multilevel symbols causes large inter-symbol interference (ISI) and the output signal becomes complicated multilevel signal and its eye openings tend to be very small or lost.

Nonetheless, some >four-level signaling experiments are demonstrated using the combined direct-detection binary receivers, by choosing proper signal constellations and demodulation schemes; one of the example is the 30-Gbit/s eight-level DPSK signaling [4] with the signal constellation shown in Fig. 6.2(a), which uses four sets of binary delay-differential receivers as in Fig. 6.3(a) having four different interference angles and a logic operation to extract three sets of 10-Gbit/s binary data from the four receiver outputs. One of the problems of such a binary detection of N-DPSK signal is the rapid increase of the number of receivers as the increase of N.

Another approach is the use of a special set of multilevel signals having independent amplitude and phase modulations, so that the amplitude- and the differential phase-modulation components of them can be separately detected. We classify them as amplitude- and phase-shift keying (APSK) signals in this chapter. For example, eight-level APSK signaling in Fig. 6.2(b), which is the combination of two-level amplitude-shift keying (ASK) and DQPSK, is demonstrated in [5–7]. And 16-level APSK signaling in Fig. 6.2(c), in which ASK signal level is increased to four, is demonstrated in [8, 9]. Figure 6.3(b) shows the receiver configuration of the 16-APSK signaling experiment [8], in which DQPSK and quarternary ASK (QASK) components of the received signal is independently detected. Such an independent modulation and demodulation of differential PSK and ASK components relaxes the problem of increasing N, say, up to 16. However, it is almost impossible

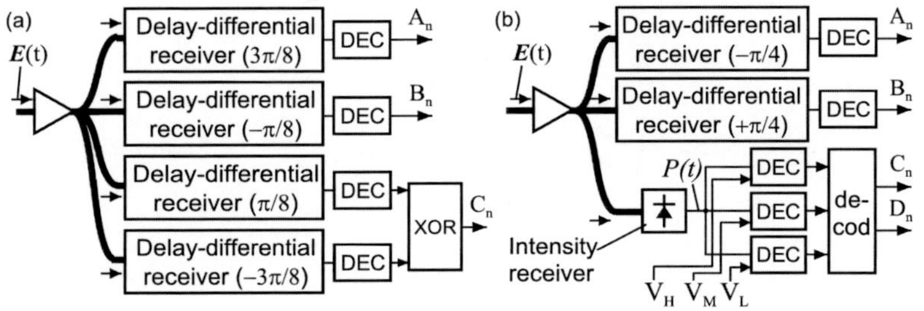

Fig. 6.3 Examples of more than four-level direct-detection multilevel receiver based on binary detections. (**a**) 8-DPSK receiver, (**b**) 16-APSK receiver

to increase it further or to use of complicated signal constellation, such as 32-APSK and 16-quadrature amplitude-shift keying (QAM) in Fig. 6.2(d) and (e).

6.3 Receiver-Side Digital Signal Processing

The limitation of the combined binary receivers can be overcome by the coupled receivers with digital signal processing. Figure 6.4 shows its basic configuration. A received signal is divided into three paths: two of them are received by orthogonally coupled differential receivers (output signal: $dI(t)$ and $dQ(t)$), that is, two pairs of an optical delay demodulator (delay time: T) with orthogonal interferometric phase 0 and $\pi/2$ and a balanced receiver. The third path is received by an intensity receiver (output: $P(t)$).

$$\boldsymbol{E}(t) = I(t) + Q(t) = r(t)\exp(j\phi(t)). \tag{6.1}$$

$$\Delta\phi(t) = \phi(t) - \phi(t-T). \tag{6.2}$$

$$dI(t) \propto r(t)r(t-T)\cos(\Delta\phi(t)). \tag{6.3}$$

$$dQ(t) \propto r(t)r(t-T)\sin(\Delta\phi(t)). \tag{6.4}$$

$$P(t) = r(t)^2. \tag{6.5}$$

$$dI_n = dI(t)_{t=nT},\ dQ_n = dQ(t)_{t=nT},\ P_n = P(t)_{t=nT}. \tag{6.6}$$

$$\Delta\phi_n = \phi_n - \phi_{n-1} = \tan^{-1}(dQ_n, dI_n). \tag{6.7}$$

$$r_n = \sqrt{P_n}. \tag{6.8}$$

The received field $\boldsymbol{E}(t)$ can be expressed as (6.1) both in the Cartesian ($I(t)$ and $Q(t)$) or in the polar coordinate (amplitude $r(t)$, phase $\phi(t)$). Meanwhile, the differential phase $\Delta\phi(t)$ is defined as in (6.2), where T is the delay time of the delay demodulators, typically set to be equal to the symbol time T_s. The output signals $dI(t)$ and $dQ(t)$ is approximated as the cosine and sine functions of $\Delta\phi(t)$ as in

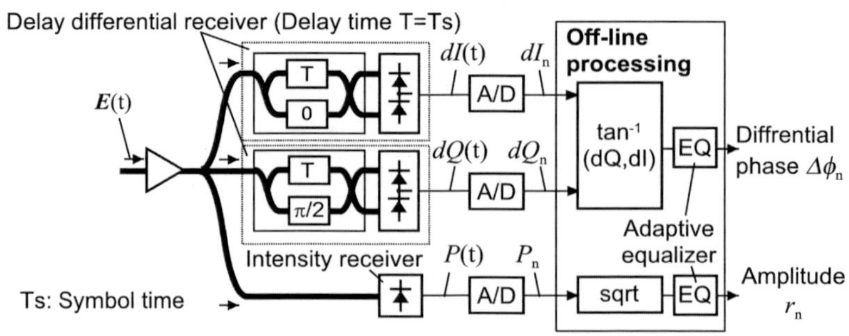

Fig. 6.4 Direct-detection multilevel receiver with digital signal processing

(6.3) and (6.4). When all the signal path lengths are the same, $\Delta\phi(t)$ is obtained with high resolution by arctangent of $(dI(t), dQ(t))$. Such a non-linear operation can be effectively implemented by using digital signal processing. Using high-speed A/D converters and representing nth sampled A/D values at the center of a symbol time as in (6.6), the differential phase $\Delta\phi_n$ can be obtained as in (6.7). Note that it should be a two-argument arctangent function, in order to avoid quadrant uncertainty. Also the sampled amplitude r_n is obtained by the square root of P_n as in (6.8). The advantage of the digital receiver in Fig. 6.4 is the high-resolution detection of amplitude r_n and differential phase $\Delta\phi_n$ (not absolute phase ϕ_n) of a received multilevel symbol. Therefore, it is very suitable for the precise detection of N-level APSK signals with large N.

So far, the highest number of signal levels N with direct detection is attained by the 32-level signaling [10] with the constellation diagram in Fig. 6.2(d). The setup of a 32-level transmitter is shown in Fig. 6.5. A CW light from a laser diode (LD) is first modulated by a zero-chirp ISI-suppressed QASK modulator [11] in which two 10-Gbit/s binary-PSK (BPSK) signals and an CW light are interfered in phase to generate a clear QASK signal. The ratio of four optical intensity levels of the QASK signal is set to be 7:5:3.5:2. Next, DQPSK modulation is applied by a fiber IQ modulator comprising of two dual-arm lithium-niobate (LN) Mach–Zehnder (MZ) modulator in push–pull driving. Next an LN phase modulator (voltage swing $V_{\pi/4}$) is used to add $\pi/4$ phase modulation, and a 50-Gbit/s NRZ-32-APSK signal is obtained. Finally, a pulse carver is used to convert it to RZ modulation to reduce transient frequency chirp. The receiver configuration is the same as in Fig. 6.4. The orthogonally coupled differential receivers are used to detect the 8-DPSK component from $\Delta\phi_n$, and the intensity receiver is used to detect the QASK component from r_n. In the experiment, A/D converters are emulated by a high-speed real-time digital oscilloscope with the sampling rate of 10 GSa/s. No clock extraction is implemented and only the sample sequences (with the length of 100 Kbits) taken at the center of symbols are used in off-line signal processing. Adaptive equalizers (EQ) with complex taps (10 FFE and 10 DFE) and real taps are used to reduce signal ISI for both 8-DPSK and ASK components, respectively. The first 500 bits of a received pattern are used for the tap adaptation with the LMS algorithm, and the rest is evaluated with fixed taps. Three- and two-bits gray decoders

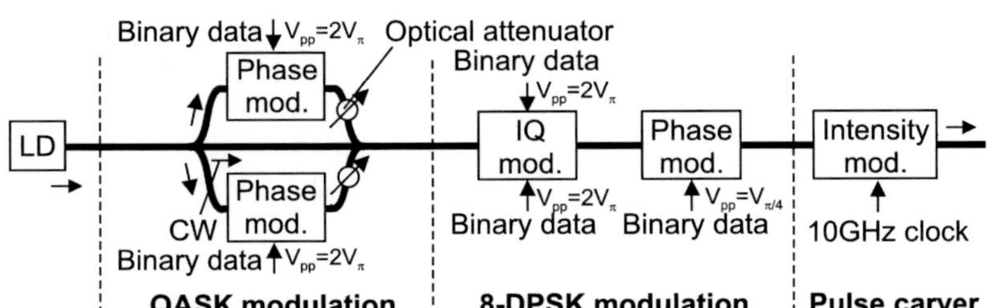

Fig. 6.5 Experimental 50-Gbit/s 32-level APSK transmitter. All the binary data streams are 10-Gbit/s $2^{15}-1$ PRBSs

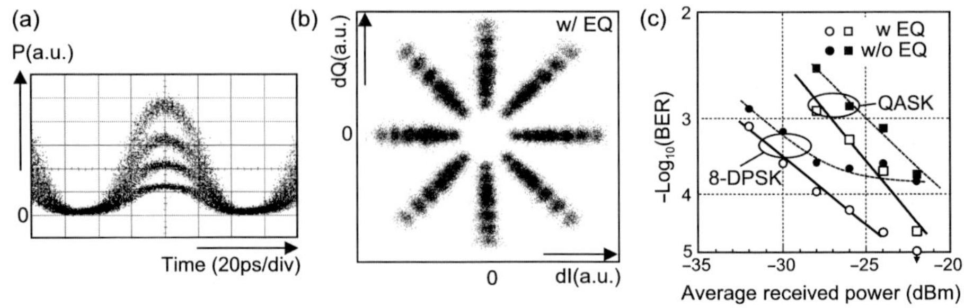

Fig. 6.6 Results of 50-Gbit/s 32-level APSK signaling experiment. (**a**) Intensity eye-pattern, (**b**) differential constellation diagram, (**c**) back-to-back BER performances

are implemented to evaluate multi-bit error occurrences in 8-DPSK and QASK components.

Figure 6.6(a) shows an intensity eye pattern of the RZ-32-APSK signals at back to back, which shows four signal levels and three clear eye openings of its QASK component. The two-dimensional outputs (dI_n, dQ_n) of the orthogonally coupled differential receiver are plotted as a "differential" constellation diagram as in Fig. 6.6(b). They are clearly divided into the radial patterns with eight distinct angles, which is the proof of a highly accurate 8-DPSK modulation/demodulation process. Fig. 6.6(c) shows the back-to-back bit error ratio (BER) performances of the 20-Gbit/s QASK and the 30-Gbit/s 8-DPSK components of 50-Gbit/s RZ-32-APSK signals. Without adaptive equalizers, a BER floor around 10^{-4} appears on the 8-DPSK component (filled circles). By using it, the BER floor is vanished and the error-free operations are confirmed. The receiver sensitivities at BER $= 10^{-4}$ of the QASK and 8-DPSK components are −23 and −27 dBm, respectively, by the help of 1- to 2-dB sensitivity improvement by the adaptive equalizers. The corresponding OSNR sensitivities at BER $= 10^{-3}$ are 24.6 and 19.7 dB, respectively, with the ASE resolution of 0.1 nm. Using the 8-DPSK parts of the transmitter and receiver, a 16-channel WDM transmission of 30-Gbit/s 8-DPSK signals over 1040-km fiber has also been reported [12]. These results show that the application of digital signal processing is very effective to increase signal levels and to improve the performance of multilevel signaling.

6.4 Transmitter-Side Digital Signal Processing

The direct-detection multilevel signaling described so far still have severe restrictions: (1) applicable signal constellation is limited to PSK and APSK signals and (2) it lacks digital CD compensation capability. As to (2), a digital field reconstruction schemes have been reported [13, 14] for the receiver-side digital CD compensation, but its compensation performance is limited by the relatively large field reconstruction error by the optical front-end noise. Recently, the use of transmitter-side digital signal processing is proposed to overcome these restrictions: (1) phase pre-integration [15] for the use of arbitrary signal constellation design by canceling the phase-differential operation of optical delay demodulation process and (2) the signal pre-distortion for chromatic dispersion compensation [16, 17].

$$\boldsymbol{E}_n = I_n + Q_n = r_n \exp(j\theta_n). \tag{6.9}$$

$$\boldsymbol{E'}_n = r_n \exp(j\phi_n) = r_n \exp(j \sum_{i=-\infty}^{n} \theta_i). \tag{6.10}$$

$$\Delta\phi_n = \phi_n - \phi_{n-1} = \sum_{i=-\infty}^{n} \theta_i - \sum_{i=-\infty}^{n-1} \theta_i = \theta_n. \tag{6.11}$$

Figure 6.7 shows the experimental setup with the proposed transmitter-side digital signal processing. At first, random complex multilevel symbol sequences $\boldsymbol{E}_n$ in (6.9) where $n = 1, 2, 3\ldots$ at the symbol rate of 10 GSymbol/s and the length of 32,768 symbols are generated. Next, the phase pre-integration is applied to them, in which the phase component θ_n of each signal point is numerically integrated on symbol-by-symbol basis as in (6.10). After twice up-sampling, CD pre-distortion is performed with a 53-stage digital FIR filter emulating the transfer function of fiber CD, and in-phase (I'_m) and quadrature-phase (Q'_m) components are passed to two high-speed DA converters (sampling speed: 20 GSa/s). A CW light is converted to a multilevel optical signal by using an LN-IQ MZ modulator driven by the $I'(t)$ and $Q'(t)$ signals.

Since the CD pre-distortion cancels out the pulse distortion caused by fiber CD, the receiver input signal can be assumed to be identical to the analog version of $\boldsymbol{E'}_n$ in (6.10). When it is received by the direct-detection multilevel receiver as in Fig. 6.4, the original absolute phase θ_n can be obtained from the sampled differential phase $\Delta\phi_n$ as in (6.11), since the phase pre-integration acts as differential pre-coding and cancels out the effect of optical delay-differential detection. By combining θ_n

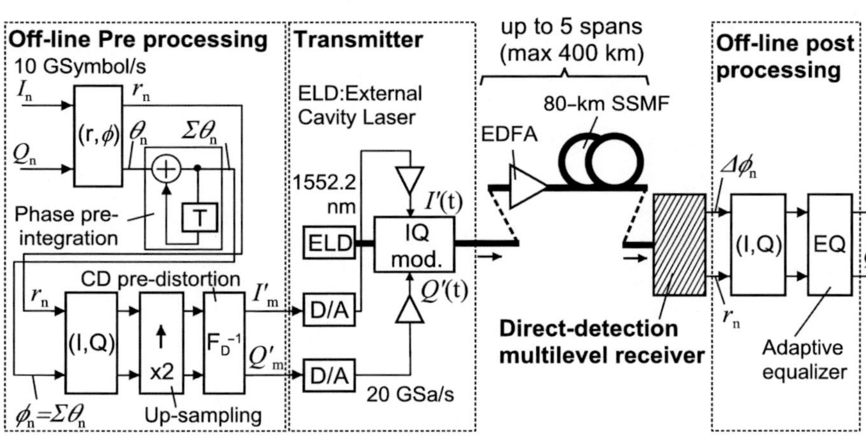

Fig. 6.7 Experimental setup of direct-detection multilevel signaling with the transmitter-side digital phase pre-integration and CD pre-distortion

and separately detected amplitude components r_n, original multilevel symbols $\boldsymbol{E}_n$ can be reconstructed. Therefore, almost arbitrary two-dimensional multilevel signal can be used even in direct-detection signaling by the phase pre-integration technique. However, it should be noted that the signal point at origin ($r_n = 0$) should be avoided since it causes the loss of differential phase of succeeding symbols.

Signaling experiments are performed using the fiber transmission line consisting of EDFA repeaters and three or five 80-km standard single-mode fiber (SSMF) spans (average loss and CD of 17.2 dB/span and 16.7 ps/nm/km) with the fiber input power of −3 dBm [17]. The CD pre-distortion of −4080 and −6700 ps/nm are applied to 240- and 400-km transmissions, respectively. At first, a 30-Gbit/s 8-QAM signal is transmitted through a 400-km SSMF line. Without CD pre-distortion, the received differential signal constellation diagram (dI_n, dQ_n) is totally distorted as in Fig. 6.8(a). With the CD pre-distortion, the original 8-QAM constellation is reconstructed as in Fig. 6.8(b). The back-to-back OSNR sensitivity is 20.3 dB(@BER = 10^{-3}) using the adaptive equalizer. After the 400-km transmission, the BER of 2.7×10^{-4} is obtained. Next, a 35.8-Gbit/s 12-QAM signal is applied to a 240-km SSMF transmission, and it is reconstructed as in Fig. 6.8(c). The back-to-back OSNR sensitivity is 28.1 dB and the BER is 8.2×10^{-4} after 240-km transmission, which is also well below the threshold of a super forward error correction (FEC) circuit even considering the overhead penalty. Finally, a 40-Gbit/s 16-QAM

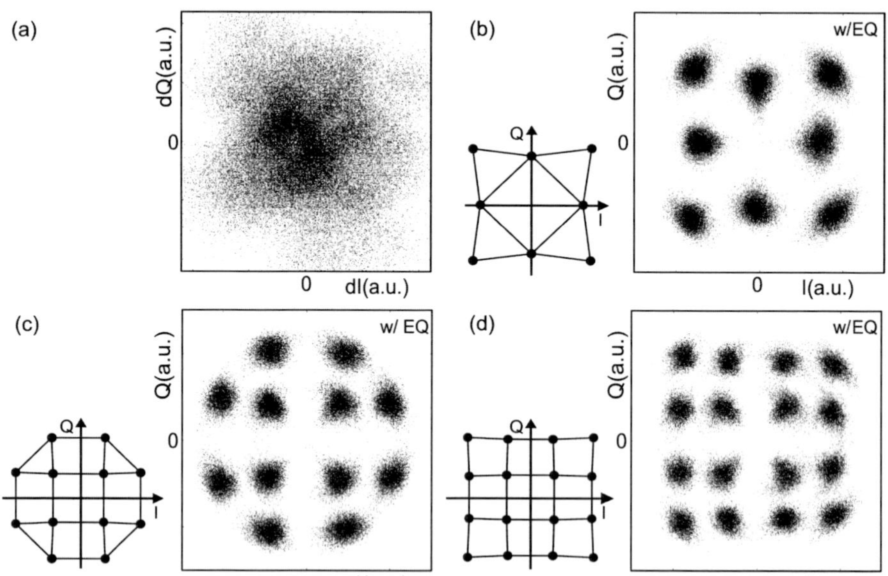

Fig. 6.8 Original and experimentally reconstructed constellation diagrams of direct-detected QAM signals with phase pre-integration and CD pre-distortion. (**a**) 8-QAM differential constellation diagram after 400-km SSMF transmission (without CD pre-distortion and phase pre-integration), (**b**) 8-QAM signal after 400-km SSMF transmission, (**c**) 12-QAM signal after 240-km SSMF transmission, (**d**) 16-QAM signal at back-to-back

signaling is performed and Fig. 6.8(d) shows its reconstructed constellation diagram, in which its four-by-four constellation is clearly observed. The bit error ratio (BER) is measured to be 1.7×10^{-3}. It should be noted that the experimental performances such as OSNR sensitivities shown above are not quite good, mainly because of the experimental error of transmitter-side digital signal synthesis. Recently, with the improved signal synthesis and advanced receiver-side signal processing, the experimental OSNR sensitivities of 8-QAM, 12-QAM, and 16-QAM signals are improved to 13.5, 16.0, and 18.1 dB, respectively at BER= 10^{-3}, which is within 0.5–2.5 dB from those of coherent detection [18].

6.5 Conclusions

In this section, we have shown the recent advance and experimental results of direct-detection multilevel signaling, especially in terms of signal formats and detection schemes. With the use of APSK signals and combined binary receivers, up to 16-level APSK signaling have been reported. It is improved to 32 level with the adoption of receiver-side digital signal processing thanks to its accurate amplitude and differential phase detection capability. Further adoption of the transmitter-side digital signal processing opens up a way to the OSNR sensitivity improvement with the use of efficient signal constellations like 16-QAM and the optical CD compensation free long-distance SSMF transmission. These advancements show the potential possibility and feasibility of future direct-detection multilevel signaling.

References

1. M.G. Taylor, Coherent detection method using DSP for demodulation of signal and subsequent equalization of propagation impairments. IEEE Photonics Technol. Lett. **16**(2), 674–676 (2004)
2. S. Tsukamoto, K. Katoh, K. Kikuchi, Unrepeated 20-Gbit/s QPSK transmission over 200-km standard single-mode fiber using homodyne detection and digital signal processing for dispersion compensation, in *Proceedings of the Optical Fiber Communication Conference (OFC/NFOEC)*, Anaheim, CA, Mar. 2006, paper OWB4
3. M. Daikoku, I. Morita, H. Taga, H. Tanaka, T. Kawanishi, T. Sakamoto, T. Miyazaki, T. Fujita, 100 Gbit/s DQPSK transmission experiment without OTDM for 100G Ethernet transport, in *Proceedings of the Optical Fiber Communication Conference (OFC/NFOEC)*, Anaheim, CA, Mar. 2006, post-deadline paper PDP36
4. M. Serbay, C. Wree, W. Rosenkranz, Experimental Investigation of RZ-8DPSK at 3×10.7Gb/s, in *Proceedings of the 18th Annual Meeting of the IEEE Lasers & Electro-Optics Society*, Sydney, Australia, Oct. 2005, paper WE3
5. S. Hayase, N. Kikuchi, K. Sekine, S. Sasaki, Proposal of 8-state per symbol (binary ASK and QPSK) 30-Gbit/s optical modulation / demodulation scheme, in *Proceedings of the 29th European Conference on Optical Communication (ECOC/IOOC 2003)*, Rimini, Italy, Sep. 2003, paper Th2.6.4
6. T. Miyazaki, F. Kubota, Superposition of DQPSK over inverse-RZ for 3-bit/Symbol modulation-demodulation, IEEE Photonics Technol. Lett. **16**(12), 2643–2645 (2004)

7. T. Tokle, M. Serbay, Y. Geng, J.B. Jensen, W. Rosenkranz, P. Jeppesen, Penalty-free transmission of multilevel 240 Gbit/s RZ-DQPSK-ASK using only 40 Gbit/s equipment, in *Proceedings of the 31st European Conference on Optical Communication (ECOC 2005)*, Glasgow, Scotland, Sep. 2005, paper We3.2.5
8. K. Sekine, N. Kikuchi, S. Sasaki, S. Hayase, C. Hasegawa, Proposal and demonstration of 10-Gsymbol/sec 16-ary (40 Gbit/s) optical modulation/demodulation scheme, in *Proceedings of the 30th European Conference on Optical Communication (ECOC 2004)*, Stockholm, Sweden, Sep. 2004, paper We3.4.5
9. M. Serbay, T. Tokle, P. Jeppesen, W. Rosenkranz, 42.8 Gbit/s, 4 bits per symbol 16-ary inverse-RZ-QASK-DQPSK transmission experiment without Polmux, in *Proceedings of the Optical Fiber Communication Conf. (OFC/NFOFEC)*, Anaheim, CA, Mar. 2007, paper OThL2
10. N. Kikuchi, K. Mandai, K. Sekine, S. Sasaki, First experimental demonstration of single-polarization 50-Gbit/s 32-level (QASK and 8-DPSK) incoherent optical multilevel transmission, in *Proceedings of the Optical Fiber Communication Conference (OFC/NFOEC)*, Anaheim, CA, Mar. 2007, paper PDP21
11. N. Kikuchi, Inter-symbol interference (ISI) suppression technique for optical binary and multilevel signal generation, IEEE J. Lightwave Technol. **25**(8), 2060–2068, Aug. (2007)
12. K. Mandai, N. Kikuchi, S. Sasaki, First long distance (1040-km) WDM transmission of 16-channels 30-Gbit/s (480-Gbit/s) 8-ary differential phase-shift-keying (8-DPSK) signals, in *Proceedings of the 12th Opto Electronics and Communications Conference 16th International Conference on Integrated Optics and Optical Fiber Communication (OECC/IOCC 2007)*, Yokohama, Japan, July, 2007, post-deadline paper P1.2
13. N. Kikuchi, K. Mandai, S. Sasaki, K. Sekine, Proposal and first experimental demonstration of digital incoherent optical field detector for chromatic dispersion compensation, in *Proceedings of the 32nd European Conference on Optical Communication (ECOC 2006)*, Cannes, France, Sep. 2006, post-deadline paper Th.4.4.4
14. X. Liu, X. Wei, Electronic dispersion compensation based on optical field reconstruction with orthogonal differential direct-detection and digital signal processing, in *Proceedings of the Optical Fiber Communication Conference (OFC/NFOEC)*, Anaheim, CA, Mar. 2007, OTuA6
15. N. Kikuchi, S. Sasaki, Incoherent 40-Gbit/s 16QAM and 30-Gbit/s Staggered 8APSK (amplitude- and phase-shift keying) signaling with digital phase pre-integration technique, IEEE LEOS Summer Topical Meeting, Acapulco, Mexico, July. 2008, paper WD3.2
16. D. McGhan, C. Laperle, A. Savchenko, C. Li, G. Mak, M. O'Sullivan, 5120 km RZ-DPSK transmission over G652 Fiber at 10 Gb/s with no optical dispersion compensation, in *Proceedings of the Optical Fiber Communication Conference (OFC/NFOEC)*, Anaheim, CA, Mar. 2005, post-deadline paper PDP27
17. N. Kikuchi, S. Sasaki, Optical dispersion-compensation free incoherent multilevel signal transmission over standard single-mode fiber with digital pre-distortion and phase pre-integration techniques, in *Proceedings of the 34th European Conference on Optical Communication (ECOC 2008)*, Brussels, Belgium, Sep. 2008, paper Tu.1.E.2
18. N. Kikuchi, S. Sasaki, Sensitivity improvement of incoherent multi-level (30-Gbit/s 8QAM and 40-Gbit/s 16QAM) signaling with non-Euclidean metric and MSPE (multi symbol phase estimation), in *Proceedings of the Optical Fiber Communication Conference (OFC/NFOFEC)*, San Diego, CA, March. 2009, paper OWG1

Chapter 7
High Spectral Efficiency Coherent Optical OFDM

William Shieh and Xingwen Yi

Abstract OFDM has quickly gained its attraction in optical communications that are evolving toward software-enhanced optical transmissions. Coherent optical OFDM (CO-OFDM) takes advantage of software capabilities of electronic digital signal processing (DSP) to perform sophisticated operations and has demonstrated its easiness of realizing high spectral efficiency and combating various distortions at the same time. This chapter presents the signal processing of coherent optical MIMO-OFDM and elucidates its capability. In the first part, centered on high spectral efficiency, CO-OFDM is investigated to implement high-order QAM modulation. In the second part, orthogonal-band-multiplexed OFDM (OBM-OFDM) is discussed to alleviate the analog bandwidth requirement of DACs/ADCs and to eliminate the frequency guard band between the optical channels. With the theoretical analysis and experimental demonstrations in both parts, CO-OFDM presents itself as an attractive candidate for high spectral efficiency optical transmissions.

7.1 Overview

7.1.1 Background

Optical and wireless communications were once world apart. Wireless communications in need of adapting to usually fast deep fading environments have employed sophisticated modulation, reception, and coding concepts that form the foundation of the modern communication theory, whereas optical communications have been incorporating rather straightforward modulation and reception methods for a much more static fiber optical channel. However, the recent drive toward dynamically reconfigurable optical networks with the transmission speed beyond 100 Gb/s has brought the two forms of communications ever close to each other. In particular,

---

W. Shieh (✉)
The University of Melbourne, Victoria 3010 Australia
e-mail: w.shieh@ee.unimelb.edu.au

M. Nakazawa et al. (eds.), *High Spectral Density Optical Communication Technologies*, Optical and Fiber Communications Reports 6, DOI 10.1007/978-3-642-10419-0_7,

digital signal processing (DSP) for optical and wireless communications is fundamentally similar as both can be modeled as band-pass systems. DSP has been playing a vital role in wireless communications and is on its way to enable the ultimately flexible software-defined radio (SDR), once only theoretically possible [1, 2]. On the other hand, the application of DSP to optical transmissions was hampered due to lack of enormous computation power required for optical communications. Recent advances in microelectronics, such as analog-to-digital converter (ADC), digital-to-analog converter (DAC), and digital signal processor (DSP), have brought about many novel optical subsystems and systems via electronic DSP up to 40 Gb/s [3]. It is anticipated that more and more DSP functionality will be applied to optical communications, fueled by the ever-advancing silicon technology underlined by the Moore's law.

In response to the emergence of plethora of analog and digital wireless standards in 1980s, the concept of software-defined radio (SDR) has been proposed as a practical solution [1, 2], namely, a software implementation of the radio transceiver that can be dynamically adapted to the user environment, instead of relying on the dedicated hardware. SDR itself does not mean a specific technique and instead it promotes the migration trend from analog to digital in wireless communications. On the other hand, a similar challenge arises in modern optical communications, where multiple advanced modulation formats [4–6] have been proposed for next generation 100-Gb/s Ethernet transport. This may signal the trend toward software-enhanced optical transmission (SEOT) in which the transponder can be reconfigured to multiple standards, multiple modulation formats, or dynamic link environment. In particular, among other proposed modulation formats, this chapter focuses solely on coherent optical OFDM (CO-OFDM) that has recently gained significant interests from the optical communication community [7, 8].

OFDM is a multi-carrier transmission technique by which a data stream is carried by many lower-rate orthogonal subcarriers [9]. Although OFDM has been proposed for about 40 years, it only gained popularity in early 1990s after being implemented in electronic DSP. Since then, OFDM has been incorporated into many wireless network standards (802.11 a/g WiFi, HiperLAN2, and 802.16 WiMAX) and digital audio and video broadcasting (DAB and DVB-T) in Europe, Asia, Australia, and other parts of the world. Therefore, software implementation brought about the success of OFDM. Multiple input and multiple output (MIMO) is another recent example greatly benefited from software implementation. MIMO in wireless communications uses multiple transmit and multiple receive antennas to improve communication performance. Similar to OFDM, MIMO also requires sophisticated signal processing at transmitter and receiver for timing synchronization, channel estimation, and compensation [10]. In addition, MIMO is generally combined with OFDM as MIMO-OFDM to further boost the performance of OFDM systems [10, 11]. Several emerging standards, including 4G wireless communications will be based on MIMO-OFDM. In short, the great impact of MIMO and OFDM on wireless communications is largely realized by using DSP.

The conceptual diagram of SEOT is proposed in Fig. 7.1. A salient difference from conventional optical transmissions is the presence of DAC/ADC and DSP. For

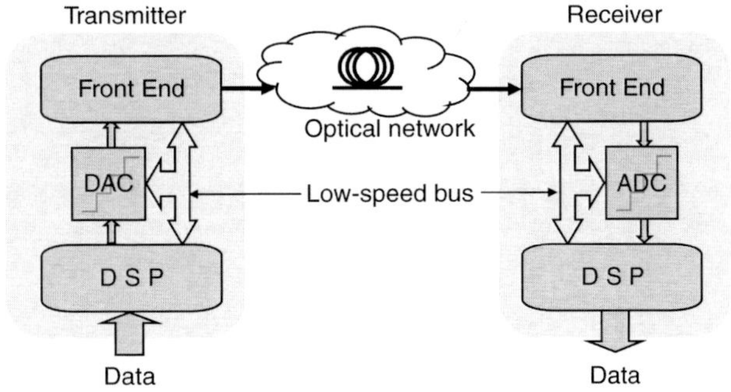

Fig. 7.1 Conceptual diagram of software-enhanced optical transmission (SEOT)

optimization and application purposes, there are low-speed interactions among DSP, DAC, or ADC, and the front end. Again, SEOT promotes the migration from analog to digital to improve optical transmissions of any front end via dynamic adaptation to the transmission channel and reconfiguration to an appropriate modulation format.

The analog-to-digital migration can be seen from the development of electronic dispersion compensation (EDC). The very early approaches of EDC are based on hardware, such as feed-forward equalizer (FFE) and decision feedback equalizer (DFE), with limited performance improvement [12]. However, the ensuing EDC by DSP has shown much more significant performance improvement [13]. SEOT in Fig. 7.1 provides a generic architecture for various EDC by DSP. For instance, for conventional front ends of intensity modulation/direct detection (IM/DD) systems, maximum likelihood sequence estimation (MLSE) can be used [14]; for an optical in-phase/quadrature (IQ) modulator and direction detection, pre-compensation can be used [15]; for a coherent detection front end, digital phase estimation can be used to replace conventional optical phase-locked loops (OPLL) [16]; for an optical IQ modulator and coherent detection front end, CO-OFDM can be realized [7, 8, 17]. In these examples, their front ends are quite different but they all take advantage of DSP to achieve significant performance enhancements of chromatic dispersion tolerance.

OFDM was merely of peripheral interests to optical communities a few years ago when most of the research studies on optical communications were dominated by IM/DD systems that are difficult to carry bi-polar signals such as OFDM. It is the recent revival of coherent optical detection by DSP that paves the way for OFDM to be applied to optical communications. In fact, since the proposal of CO-OFDM [7], the number of publications on both the theories and the experiments of CO-OFDM has been fast increasing. MIMO-OFDM has also naturally been studied to further exploit the advantage from MIMO [18].

In a generic coherent optical MIMO-OFDM system, the transmitter front end includes an optical IQ modulator and polarization multiplexer, and the receiver front end includes a polarization-diversified coherent receiver. Consequently, the transmitter fully controls the optical carrier and the receiver accesses all the optical information, which considerably enhances the capabilities of DSP. Coherent optical

MIMO-OFDM systems based on the SEOT platform present numerous flexibilities and advantages:

- Without any hardware change, each subcarrier can have different modulation formats, including quadrature amplitude modulation (QAM). In essence, the channel data rate can be adjusted according to the optical channel condition [19].
- Some hardware imperfections of front end can be addressed by software, such as the laser phase drift [20], the non-linear transfer function of optical modulator [21], and the imperfect frequency response of the electronic components.
- The transmission channel characteristics can be monitored by means of receiver signal processing [22, 23]. Subsequently, all the OFDM parameters, such as subcarrier number and individual subcarrier modulation format, can be optimized [9].
- The fiber linear dispersion and non-linear phase noise can be mitigated by DSP [24, 25].

In short, the development of coherent optical MIMO-OFDM shows the essence of SEOT, which is to shift more and more analog implementation of the physical layer to the digital domain and take advantage of DSP to perform sophisticated operations.

It is noted that there are some varieties of optical OFDM using direct detection [26–28]. They could be cost-effective but they do not have all the advantages of coherent optical MIMO-OFDM intrinsic to coherent detection.

7.1.2 Organization of the Chapter

The attraction of coherent optical MIMO-OFDM arises from its impressive capabilities to respond to several existing challenges in optical communications, such as distortion compensation from chromatic dispersion and polarization-mode dispersion (PMD), and high spectral efficiency. This chapter focuses on high spectral efficiency and its interplay with the distortion compensation.

To achieve high spectral efficiency, one of the most popular approaches is to use multi-level formats. The other way is to reduce the frequency spacing between optical carriers. This chapter is organized as follows. Section 7.2 will present the general architecture and signal processing in coherent optical MIMO-OFDM. Sects. 7.5 and 7.6 will discuss how to achieve high spectral efficiency by advanced modulation formats and orthogonal-band multiplexing, via theoretical analysis and experimental demonstrations.

7.2 Signal Processing in Coherent Optical MIMO-OFDM

OFDM has been proposed and extensively studied for several decades and been well documented in wireless communications. Compressive knowledge of OFDM may be found in [9, 29]. In this section, signal processing of OFDM will be introduced with a focus on its key features. CO-OFDM and coherent optical MIMO-OFDM will be subsequently introduced.

7.2.1 Representation of OFDM

OFDM is a special form of a broader class of multi-carrier modulation (MCM), a generic implementation of which is depicted in Fig. 7.2. The structure of a complex mixer (IQ modulator/demodulator), which is commonly used in MCM systems, is also shown in the figure. The MCM-transmitted signal $s(t)$ is represented as

$$s_{\mathrm{B}}(t) = \sum_{i=-\infty}^{\infty} \sum_{k=1}^{N_{\mathrm{sc}}} C_{ki} s_k(t - iT_{\mathrm{s}}), \tag{7.1}$$

$$s_k(t) = \Pi(t) \exp(j2\pi f_k t), \tag{7.2}$$

$$\prod(t) = \begin{cases} 1, & -T_{\mathrm{g}} < t < t_{\mathrm{s}} \\ 0, & \mathrm{t} \leq -T_{\mathrm{g}}, t > t_{\mathrm{s}} \end{cases}, \tag{7.3}$$

where C_{ki} is the ith transmitted information symbol at the kth subcarrier, s_k is the waveform for the kth subcarrier, N_{sc} is the number of subcarriers, f_k is the

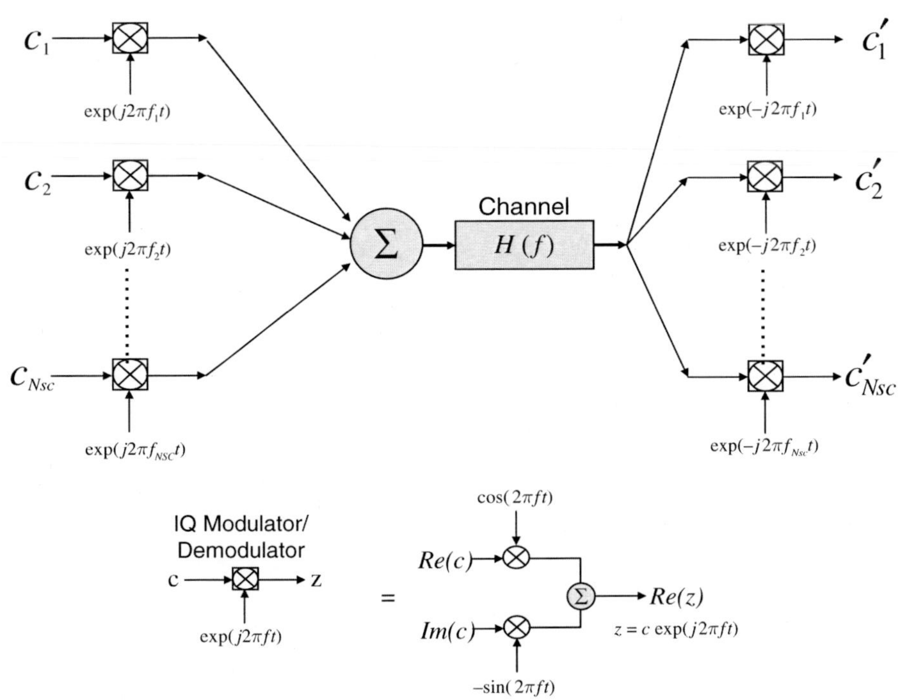

Fig. 7.2 Conceptual diagram for a generic multi-carrier modulation (MCM) system

frequency of the subcarrier, and T_s, T_g, and t_s are the OFDM symbol period, guard interval length, and observation period (or FFT window length), respectively, with a relationship of $T_s = T_g + t_s$, and $\Pi(t)$ is the rectangular pulse-shaped waveform of the OFDM symbol.

It has been shown that if the following condition:

$$f_k - f_l = m\frac{1}{T_s} \tag{7.4}$$

is satisfied, where m is a non-zero integer, then the two subcarriers are orthogonal to each other. This signifies that these orthogonal subcarrier sets, with their frequencies spaced at multiple of inverse of the symbol rate, can be recovered without inter-carrier interference (ICI) in spite of strong signal spectral overlapping.

Weinsten and Ebert first revealed that OFDM modulation/demodulation can be implemented by using inverse discrete Fourier transform (IDFT)/discrete Fourier transform (DFT) [30]. The digital implementation effectively eliminates the need for a bank of oscillators and modulators. It can be shown that the modulation can be performed by IDFT of the input information symbol C_k, and the demodulation by DFT of the sampled received signal $r(t)$. As a result, the orthogonality condition (7.4) can be strictly satisfied through IDFT/DFT. Because OFDM modulation and demodulation can be done in software instead of hardware, OFDM has gained its popularity once it was implemented in electronic DSP in early 1990s. The corresponding architecture based on SEOT platform is shown in Fig. 7.3. At the transmit end, the input serial data bits are converted to parallel by serial-to-parallel converter (S/P) and mapped onto corresponding information symbols of the subcarriers within one OFDM symbol, and the digital time-domain signal is

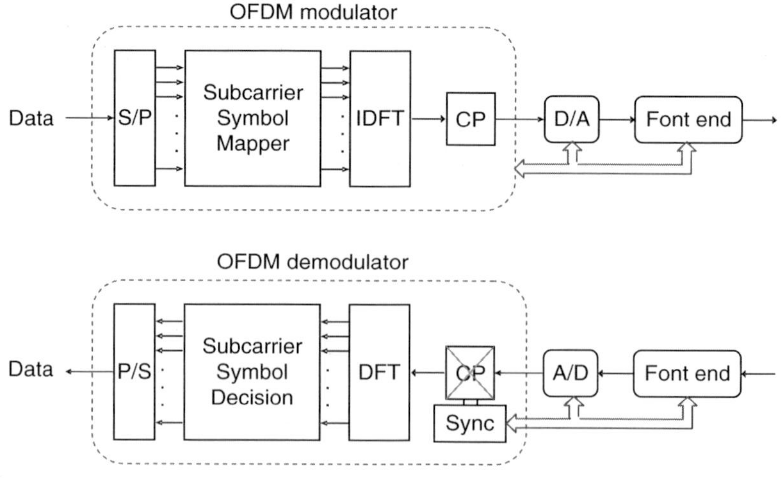

Fig. 7.3 Conceptual diagram for OFDM transmitter (*top*) and receiver (*bottom*)

obtained by using IDFT, which is subsequently inserted with cyclic prefix (CP) and converted into real-time waveform through DAC. The CP is inserted to prevent inter-symbol interference (ISI). The baseband signal can be up-converted to an appropriate RF or optical carrier with an IQ modulator. At the receive end, the OFDM signal is first down-converted to baseband with an IQ demodulator, and sampled with an ADC, synchronized to remove the CP, and demodulated by performing DFT and baseband signal processing to recover the data. The synchronization is crucial to guarantee the orthogonality condition (7.4), including time and frequency synchronization. In addition, as the block diagrams shown in Fig. 7.3, DSP in transmitter is generally called OFDM modulation and in receiver, OFDM demodulation.

A generic OFDM transmission model in the continuous time domain can be expressed as

$$r_{\mathrm{B}}(t) = s_{\mathrm{B}}(t) \otimes h(t) + n, \tag{7.5}$$

where $r_{\mathrm{B}}(t)$ is the received baseband signal, $h(t)$ is the lumped channel response in the time domain, $n(t)$ is the noise source, and $\otimes$ stands for convolution. The $h(t)$ accounts for the frequency responses from both the OFDM transceiver front end and the optical channel.

Assuming a perfect DFT window and frequency synchronization, the received signal is sampled and DFT is performed to recover the received OFDM information symbol, which is given by

$$C'_k = H_k \cdot C_k + n_k \tag{7.6}$$

where C'_k and C_k are respectively the received and transmitted information symbol for the kth subcarrier OFDM symbol, H_k is the lumped channel response in the frequency domain, and n_k is the optical noise. Equation (7.6) serves as the OFDM transmission model in the frequency domain. In this chapter, OFDM transmission models in the frequency domain will be used for discussion.

One of the enabling techniques for OFDM is the insertion of CP/guard interval to combat inter-symbol interference (ISI) [9]. ISI-free OFDM transmission condition is given by

$$T_{\mathrm{ISI}} \leq T_{\mathrm{g}} \tag{7.7}$$

which means the time duration of ISI, T_{ISI}, is no larger than CP. Therefore, the minimum length of CP can be determined by computing the T_{ISI} for the fiber channel including chromatic dispersion and polarization-mode dispersion (PMD) [8].

For convenience, the mathematical symbols used in this chapter are summarized in Table 7.1.

Table 7.1 List of mathematical symbols

i, k	Indices of OFDM symbol, subcarrier
$C_{ki}, C'_{ki},$	Transmitted/received data at ith OFDM symbol and kth subcarrier
N_{sc}, N_f	Number of OFDM subcarriers/OFDM symbols in one OFDM block
H, h	Channel response in the frequency domain/time domain
$s(t), r(t)$	Continuous time-domain OFDM signal
B	Combined laser linewidth of transmitter and receiver lasers
Γ	Ratio of combined laser linewidth to symbol rate
M	Constellation points of square QAM
T	Sampling time interval of ADC
T_g	Time interval of cyclic prefix/guard interval
T_s	Time duration of OFDM symbol
r_{cp}	Ratio of T_g to T_s
Φ_k	Phase shift of chromatic dispersion
N	Noise
$\varphi_{LD1}, \varphi_{LD2}, f_{LD1}, f_{LD2}$	Phase angle and frequency of transmit/receive laser
D_t	Total chromatic dispersion

7.3 Implementation of CO-OFDM

At both transmitter and receiver ends, the frequency conversion between the optical domain and the RF domain can be either intermediate frequency (IF) conversion or direct conversion. The detailed description of the various CO-OFDM implementations can be found in [31]. Figure 7.4 shows a typical CO-OFDM architecture. The main optical devices of transmitter front end are a laser source and an optical IQ modulator. An ideal electrical-to-optical (E/O) IQ up-conversion is realized with the so-called pre-distortion in DSP. The main devices of receiver front end are a local laser source, an optical 90° hybrid, and photodiodes, hence a direct down-conversion. The block arrows in Fig. 7.5 represent the communication interface between various components within the OFDM transponder to facilitate the SEOT. The DSP in the transmitter includes OFDM modulation. Note that only main functions are illustrated and explained inside DSP, where other functions such as

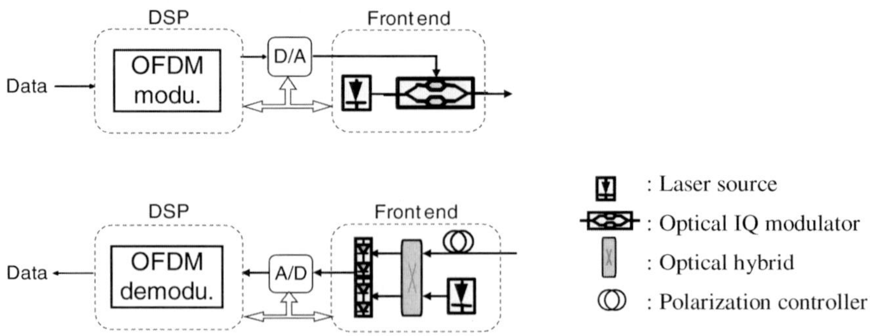

Fig. 7.4 A typical CO-OFDM architecture: transmitter (*top*) and receiver (*bottom*)

pre-distortion, not always needed, can be easily added due to its software nature. The channel response is modeled as the linear cascades of three elements including transmitter front end, optical fiber link, and receiver front end.

Following the same procedure in [7, 32], the received OFDM baseband signal in the time domain $r_{\mathrm{B}}(t)$ is given by

$$r_{\mathrm{B}}(t) = \exp(\Delta\varphi) \cdot s_{\mathrm{B}}(t) \otimes h(t) + n(t)$$

$$\Delta\varphi = \varphi_{\mathrm{LD2}} - \varphi_{\mathrm{LD1}} \tag{7.8}$$

$$h(t) = h_t(x) \otimes h_l(x) \otimes h_r(x), \tag{7.9}$$

which is the CO-OFDM transmission model in the time domain, with the frequency responses of transmitter and receiver front ends and the linear optical channel, $h_l(x)$.

For better illustration, an OFDM block structure is shown in Fig. 7.5. It shows the two-dimensional time/frequency structure for one OFDM block, which includes N_{sc} subcarriers in frequency and N_f OFDM symbols in time. The preamble is added at the beginning to realize DFT window synchronization and channel estimation.

Assuming a perfect DFT window and frequency synchronization, the received signal is sampled and DFT is performed to recover the received OFDM information symbol, which is given by [33, 34]

$$\begin{aligned} C'_{ki} &= I_{i0} \cdot H_{ki} \cdot C_{ki} + \varepsilon_{ki} + n_{ki} \\ \varepsilon_{ki} &= \sum_{m=-N_{\mathrm{sc}}/2, m\neq k}^{N_{\mathrm{sc}}/2-1} C_{mi} H_{mi} I_{i(m-k)} \\ I_{im} &= \frac{1}{N_{\mathrm{sc}}} \sum_{n=-N_{\mathrm{sc}}/2}^{N_{\mathrm{sc}}/2-1} \exp(j2\pi nm/N_{\mathrm{sc}}) \exp(j\Delta\varphi_{in}), \end{aligned} \tag{7.10}$$

where C'_{ki} and C_{ki} are respectively the received and transmitted information symbol for the kth subcarrier in the ith OFDM symbol, ε_{ki} is the inter-carrier interference (ICI) noise induced by the phase noise, and I_{im} is the ICI coupling coefficient between two subcarriers with distance of m.

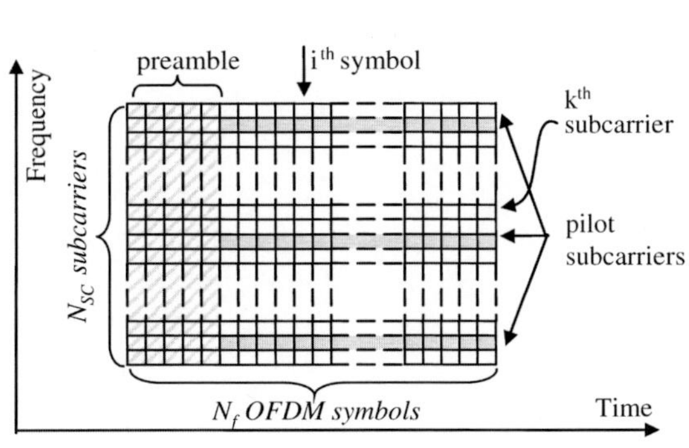

Fig. 7.5 Time and frequency representation of a block of CO-OFDM signal

By assuming slow phase change of $\Delta\varphi_{in}$ and $\phi_i = \frac{1}{N_{sc}} \sum_{n=-N_{sc}/2}^{N_{sc}/2-1} \Delta\varphi_{in}$, then $I_{i0} \cong \exp(j\phi_i)$ and (7.10) can be approximated as

$$C'_{ki} \cong \exp(j\phi_i) \cdot H_{ki} \cdot C_{ki} + n_{ki}, \tag{7.11}$$

where ϕ_i is the common phase error (CPE), and ε_{ki} can be approximated as Gaussian noise when $N_{sc} >> 1$ and is lumped into n_{ki}.

From the transmission model of (7.11), it is apparent that the channel response and phase noise need to be estimated and compensated to recover the transmitted data information. There are various methods of channel estimation such as time-domain pilot-assisted approach and frequency-domain-assisted approach [9, 35]. This chapter will focus on the techniques based upon frequency-domain pilot carriers or pilot symbols [20]. The channel response can be estimated as

$$\hat{H}_k = \sum_{i=1}^{p} \frac{\exp\left[-j\arg(C'_{k_l i})\right] \cdot C'_{ki}}{C_{ki}}, \tag{7.12}$$

where C_{ki}, C'_{ki} are respectively transmitted and received pilot subcarriers, p is the number of OFDM symbols, or preamble, used for channel estimation, and $\arg(C'_{k_l i})$ is the angle for the k_lth carrier (an arbitrary reference carrier) in the ith OFDM symbol. The purpose of the average is to remove the noise of the received symbols.

The phase estimation can be expressed as

$$\hat{\phi}_i = \arg\left(\sum_{k=1}^{Np} C'_{ki} \cdot \hat{H}^*_k \cdot C^*_{ki}/\delta^2_k\right), \tag{7.13}$$

where the asterisk stands for complex conjugate and δ_k is the standard deviation of the constellation spread for the kth subcarrier. When δ_k is assumed to be constant across all the subcarriers, (7.13) will reduce to the least square (LS) method [32].

The laser phase noise is generally a zero-mean sequence and related to laser linewidth β by

$$\sigma_c^2 = \frac{1}{N_f} \sum_{i=1}^{N_f} \left(\hat{\phi}_i - \hat{\phi}_{i-1}\right)^2 = 2\pi\beta T_s. \tag{7.14}$$

After phase estimation and channel estimation, the estimated transmitted data from received data can be finally written as

$$\hat{C}'_{ki} = \frac{C'_{ki} \cdot \hat{H}^*_k \cdot \exp(-j\hat{\phi}_i)}{\left|\hat{H}^*_k\right|^2}, \tag{7.15}$$

which can be de-mapped into the transmitted data. Subsequently, a BER calculation can be conducted to evaluate the transmission performance.

7.4 Representation of Coherent Optical MIMO-OFDM

Figure 7.6 shows a typical coherent optical MIMO-OFDM architecture. In the transmitter, two optical IQ modulators and a pair of polarization beam splitter/combiner are used for polarization multiplexing. In the receiver, two polarization beam splitters, two 90° optical hybrids, and four balanced receivers are used for polarization-diversified detection. The polarization de-multiplexing is conducted by MIMO-OFDM signal processing in DSP in the receiver, without the conventional optical polarization stabilizer.

It is well known that the single-mode fiber supports two modes in the polarization domain. Each polarization mode travels with a slight different speed due to polarization-mode dispersion (PMD) and experiences different loss due to polarization-dependent loss (PDL). Therefore, a complete CO-OFDM model requires the mathematical description of the two-polarization coupling including these polarization effects as well as the fiber chromatic dispersion. As shown in [24], such a model in essence is a 2 × 2 MIMO-OFDM model. The time-domain signal at the transmit is in a similar form of (7.1) but with the substitution of Jones matrix by

$$\vec{s}(t) = \begin{pmatrix} s_x \\ s_y \end{pmatrix}, \vec{C}_{ki}(t) = \begin{pmatrix} C_{ki}^x \\ C_{ki}^y \end{pmatrix}. \tag{7.16}$$

Similar to (7.11), the coherent optical MIMO-OFDM model can be conveniently expressed in the frequency domain as

$$\vec{C}'_{ki} = \exp(j\phi_i) \cdot H_{ki} \cdot \vec{C}_{ki} + \vec{n}_{ki} \tag{7.17}$$

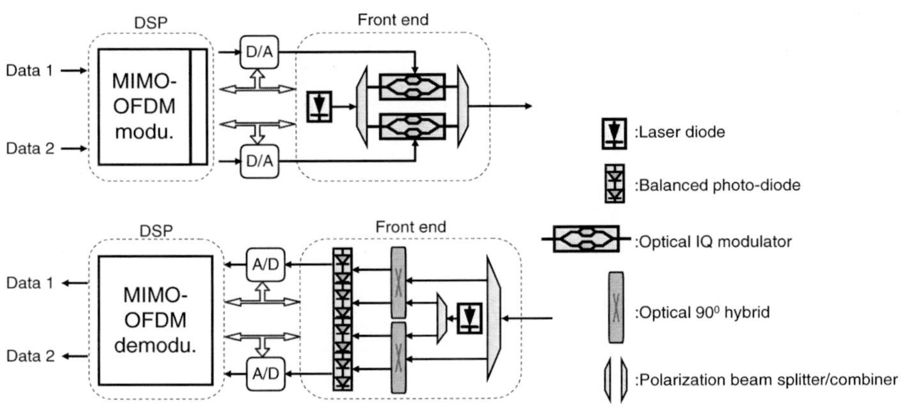

Fig. 7.6 A typical coherent optical MIMO-OFDM architecture: transmitter (*top*) and receiver (*bottom*)

Again, channel response H_{ki} needs to be re-written as

$$H_{ki} = H_{ki}^{t} \cdot H_{ki}^{l} \cdot H_{ki}^{r} \tag{7.18}$$

where H_{ki}^{t} and H_{ki}^{r} are respectively the frequency responses of the optical OFDM transmitter and receiver including the responses of DAC/ADC, RF post/preamplifier, optical IQ modulator, and balanced photo-detectors. H_{ki}^{l} is the frequency response of the fiber including chromatic dispersion, PMD, and PDL effects given by

$$H_{ki}^{l} = \exp(j\Phi_k) \prod_{p=1}^{M_0} \exp\left[\left(-\frac{1}{2}j \cdot \vec{\beta}_{pi} \cdot f_k + \frac{1}{2}\vec{\alpha}_{pi}\right) \cdot \vec{\sigma}\right] \tag{7.19}$$

$$\Phi_k = \pi \cdot c \cdot D_t \cdot f_k^2 / f_{\mathrm{LD1}}^2 \tag{7.20}$$

where Φ_k is the phase shift due to the chromatic dispersion, f_k is the frequency for the kth subcarrier, M_0 is the number of the PMD/PDL cascading elements in the entire fiber link with each (e.g., pth) section represented by its birefringence vector $\vec{\beta}_p$ and PDL vector $\vec{\alpha}_p$ [36], $\vec{\sigma}$ is the Pauli matrix vector, and D_{t} is the total chromatic dispersion assuming quadratic dependence on frequency but in general can be an arbitrary function of the frequency.

The channel response H_{ki} is a 2×2 matrix and its estimation may use a special preamble as follows. It consists of $\left[C_{ki}^{x}\ 0\right]^{T}$ and $\left[0\ C_{ki}^{y}\right]^{T}$. The basic idea is to fill zeros to one of the two orthogonal polarizations alternately so that the four elements of H_{ki} can be separately estimated.

$$H_{ki} = \begin{bmatrix} H_{11} & H_{12} \\ H_{21} & H_{22} \end{bmatrix} \tag{7.21}$$

$$\begin{bmatrix} C_{ki}^{xx} \\ C_{ki}^{xy} \end{bmatrix} = H_{ki} \cdot \begin{bmatrix} C_{ki}^{x} \\ 0 \end{bmatrix}, \begin{bmatrix} C_{ki}^{yx} \\ C_{ki}^{yy} \end{bmatrix} = H_{ki} \cdot \begin{bmatrix} 0 \\ C_{ki}^{y} \end{bmatrix}, \tag{7.22}$$

where superscript x and y stand for two-polarization states and (7.22) shows the received data when the special two alternate polarization preambles are sent.

The estimated results of the four elements of H_{ki} are

$$\begin{bmatrix} \hat{H}_{11} \\ \hat{H}_{21} \end{bmatrix} = \begin{bmatrix} \sum_{i=1}^{p} \frac{C_{ki}^{xx}}{C_{ki}^{x}} \\ \sum_{i=1}^{p} \frac{C_{ki}^{xy}}{C_{ki}^{x}} \end{bmatrix}, \begin{bmatrix} \hat{H}_{12} \\ \hat{H}_{22} \end{bmatrix} = \begin{bmatrix} \sum_{i=1}^{p} \frac{C_{ki}^{yx}}{C_{ki}^{y}} \\ \sum_{i=1}^{p} \frac{C_{ki}^{yy}}{C_{ki}^{y}} \end{bmatrix}. \tag{7.23}$$

The common phase noise is the same for the two-polarization components and therefore one polarization or both can be used for the phase estimation, similar to (7.13). The equalized signal after channel compensation and phase compensation is in a similar expression to (7.15) but the channel response here takes the form of Jones matrix.

7.5 High-order Modulation in CO-OFDM

The modulation format for optical fiber transmission has evolved to meet the ever-increasing requirement of high spectral efficiency. One way is to exploit all possible dimensions of optical modulation, i.e., polarization, amplitude, frequency, and phase. The other way is to use multi-level modulation on each dimension. OFDM on a SEOT platform greatly simplifies the implementation of different modulation formats through its software feature. Quadrature amplitude modulation (QAM) is a popular multi-level format and square QAM, or M-QAM, is the most practical member of QAM family, especially for higher order constellations [37, 38]. Hence the high-order M-QAM implementation in CO-OFDM is discussed in detail in this section.

7.5.1 BER Performance of Advanced Modulation Formats in AWGN

The BER performance of M-QAM-OFDM has been documented in [9, 29, 37] in terms of SNR (per bit unless otherwise stated). However, optical SNR (OSNR) is widely used in the optical community. If the total transmission bit-rate is fixed, it is of interest to know the BER performance of different M-QAM-OFDM in terms of OSNR. Note the BER performance versus SNR is bit-rate independent, whereas the BER performance versus OSNR depends on bit-rate and spectral efficiency because OSNR is commonly defined in 0.1-nm bandwidth. For CO-OFDM, SNR and OSNR are linearly related due to coherent detection.

Figure 7.7 shows the required OSNR at 10^{-3} and 10^{-5} BER for CO-QAM-OFDM systems at a fixed 10-Gb/s NET bit-rate with ideal conditions, i.e., there is no cyclic prefix, or $T_g = 0$, and the detection corresponds to direct down-conversion or homodyne detection. The BER performance of M-QAM-OFDM is calculated with (7.29) in Appendix. The BER performance of BPSK is also included for a comparison. As shown in Fig. 7.7, the lowest OSNR is achieved by both 4-QAM-OFDM and BPSK-OFDM. Compared with conventional IM/DD systems [14], 16-QAM-OFDM still has a better OSNR performance. Therefore, both 4-QAM-OFDM and 16-QAM-OFDM are favorable for optical transmissions. Note that the results in Fig. 7.7 are linearly scalable to higher bit-rate systems. For instance, the OSNR curves can be shifted 6 dB upward for 40-Gb/s systems.

In practical CO-QAM-OFDM implementations, cyclic prefix and pilot subcarriers are necessary overheads and require more bandwidth. Consequently, there is an OSNR penalty in addition to the results in Fig. 7.7. The penalty in the linear scale can be written as [25]

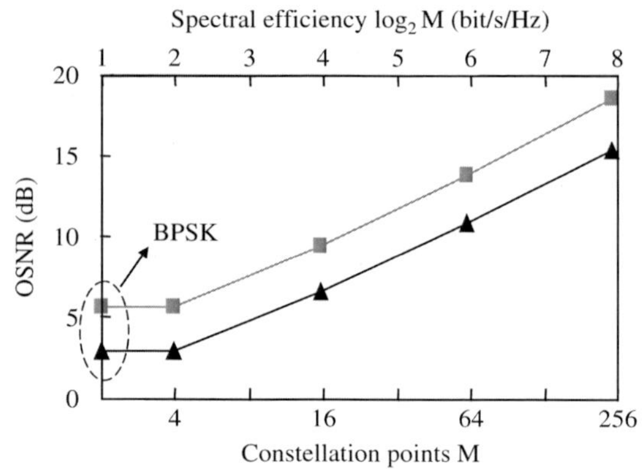

Fig. 7.7 Required OSNR at 10^{-3} and 10^{-5} BER for CO-QAM-OFDM at a fixed total bit-rate of 10 Gb/s. The curve with triangles is for 10^{-3} BER and the curve with squares is for 10^{-5} BER. OSNR is defined in 0.1-nm bandwidth

$$\text{Pen} = \frac{N_{\text{scu}}}{(N_{\text{scu}} - N_{\text{p}})(1 - r_{\text{cp}})} \tag{7.24}$$

where N_{scu} is the number of the filled subcarriers for both data transmission and pilot, and N_{p} is the number of pilot subcarriers.

7.5.2 Simulation Results on Laser Phase Noise

The laser phase noise is characterized by its linewidth. Therefore, the system design of CO-OFDM systems needs to take account of laser linewidth. Specifically, the constraints in terms of the number of subcarriers N_{sc} and the constellation points M need to be investigated for a given laser linewidth. On the other hand, it may be desirable to maximize these two parameters for ISI tolerance and high spectral efficiency. The trade-off of these two considerations has to be made to choose appropriate N_{sc} and M for a specific application, which is the aim of this subsection.

The guard interval T_{g} is used to combat ISI but it consumes bandwidth. Therefore, r_{cp} should be kept small to limit the overhead. In this section, r_{cp} is fixed as 1/17 and therefore T_{g} is proportional to N_{sc}. N_{sc} is related to tolerable chromatic dispersion D_{t} and ratio of combined laser linewidth to subcarrier symbol rate Γ as [7]

$$\frac{c}{f^2}\frac{|D_{\text{t}}|}{T} \leq T_{\text{g}} = \frac{r_{\text{cp}} N_{\text{sc}} T}{1 - r_{\text{cp}}} \tag{7.25}$$

$$\Gamma = \beta T_{\text{s}} = \frac{\beta N_{\text{sc}} T}{1 - r_{\text{cp}}} \tag{7.26}$$

where c is the light speed and f is the optical carrier frequency. Note $1/T$ is the total bandwidth of N_{sc} subcarriers and $1/T_{\text{s}}$ is the subcarrier symbol rate. The ISI source is limited to chromatic dispersion in (7.25) but the PMD-induced ISI is simply a substitution of the left side by maximum DGD in (7.25). The ratio of combined

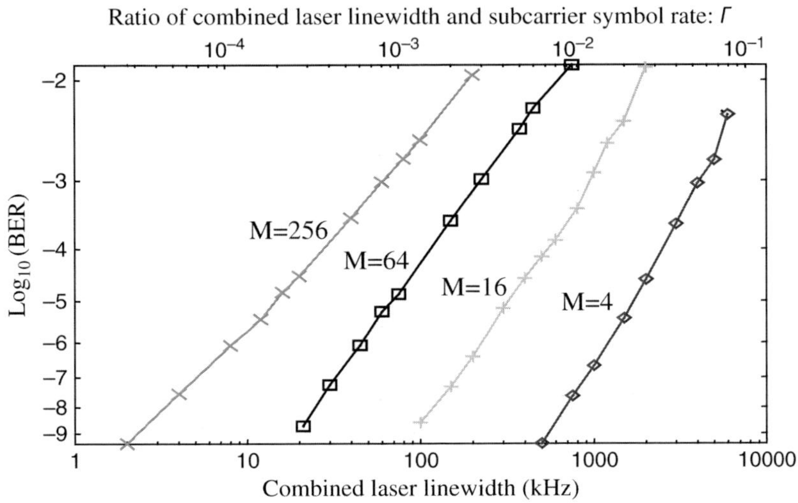

Fig. 7.8 BER floors of M-QAM-OFDM from finite laser linewidth. Symbol rate for each subcarrier is 73.5 MSymbol/s

laser linewidth to subcarrier symbol rate defined by (7.26) determines the BER performance of M-QAM-OFDM.

Figure 7.8 shows the BER floors of varying laser linewidth and QAM constellation using extensive Monte Carlo simulations. The OFDM frame and digital signal processing of the simulation are almost identical to those used in the following experiment. In simulation, the OFDM trace from the transmitter program is directly sent to the receiver program. In experiment, the OFDM trace from the transmitter program is sent to the waveform generator and the sampling trace from the scope is fed to the receiver program. The laser phase noise is simulated by white Gaussian frequency noise. Each BER point is recorded when there are more than 100 errors. The main parameters are $N_{sc} = 128$, $r_{cp} = 1/17$. The symbol rate for each subcarrier is fixed to 73.5 MSymbol/s.

From Fig. 7.8, 16-QAM-OFDM requires a combined laser linewidth of 322 kHz at 10^{-5} BER. Since the commercially available semiconductor lasers can have linewidth about 100 kHz, coherent optical 16-QAM-OFDM seems feasible.

7.5.3 Experimental Investigations of Phase Noise Effects

The phase noise in CO-QAM-OFDM transmissions is mainly contributed from laser sources, optical amplifiers, and non-linear effects in optical fibers. The ASE noise itself induced phase noise is well-behaved as AWGN and is not discussed here. Therefore, the experimental investigations are separated into two parts including laser phase noise and non-linear phase noise, respectively.

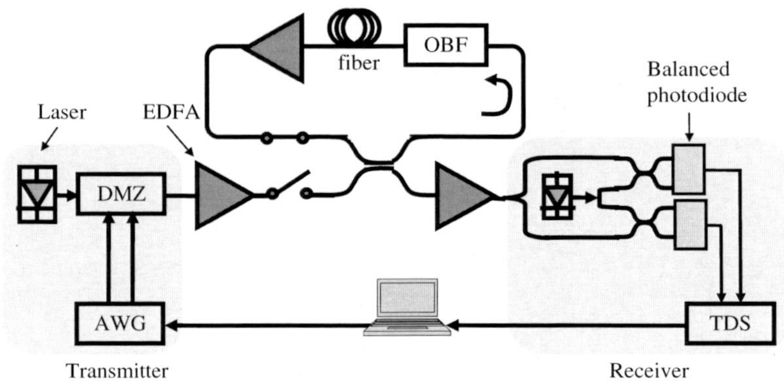

Fig. 7.9 Experimental setup for higher order QAM-OFDM transmission. *DMZ* dual Mach–Zehnder modulator, *AWG* arbitrary waveform generator, *TDS* time-domain scope, *OBF* optical band-pass filter

Figure 7.9 is a simplified illustration of experimental setup with an emphasis on hardware structure. Due to the software re-configurability of SEOT, the optical hardware structures of transmitter and receiver can scale to any M-QAM-OFDM. This is advantageous over the conventional single-carrier coherent systems where the signal processing elements such as DSP and DAC are not employed. Recall the conceptual diagram in Fig. 7.1, the transmitter front end in this experiment includes an external cavity laser, an optical IQ modulator, and its RF amplifiers. The receiver front end includes another external cavity laser, two polarization beam splitters (not shown in Fig. 7.9), two optical couplers, and two balanced receivers with built-in RF amplifiers. The configuration of the front ends shows that it is a single-input two-output (SITO) MIMO system [18]. The DAC is emulated by an arbitrary waveform generator (AWG) operated at 10 GSa/s per channel while the ADC is emulated by a time-domain sampling scope (TDS) at 20 GSa/s due to the IF down-conversion. The DAC in AWG and ADC in TDS both have 8-bit resolution. The DSP is carried out with a computer.

All the major changes to the previous work [24] are in software to accommodate the M-QAM-OFDM format. M-QAM modulation uses Gray coding and demodulation uses rectangular decision areas. The pilot-aided phase estimation is based on the description in [20] with a modification of QAM modulated pilot subcarriers.

In this work, 128 OFDM subcarriers are used, or FFT length, but only 36 subcarriers are used for data transmission. Eight more subcarriers are reserved as pilot subcarriers for laser phase drift estimation, and the remaining subcarriers are padded with zeros to avoid aliasing noise. The time duration of one OFDM symbol is $T_s = 13.6$ ns, and the guard interval is $T_g = 0.8$ ns. Note that the 44 subcarriers only occupy 3.5-GHz optical bandwidth, and the 36 data subcarriers correspond to 2.65 GSymbol/s. Therefore, the effective bit-rates of 16-QAM-OFDM and 64-QAM-OFDM are 10.59 and 15.89 Gb/s, respectively. Without polarization multiplexing, their spectral efficiencies are 3.0 and 4.5 bit/s/Hz, correspondingly. Consequently, CO-QAM-OFDM can achieve high spectral efficiency.

7.5.4 Laser Linewidth Effects

The re-circulating loop in Fig. 7.9 is by-passed to investigate the laser linewidth effects. Figure 7.10a shows the constellations from 14,400 QAM symbols. The SNR per symbol is about 25 dB for both constellations. From several measurements, the combined laser linewidth β is estimated using (7.14) as about 101 kHz, which is almost half of the specification of the lasers. The estimated linewidth is subsequently used in the simulation and the resultant BER curve is compared with the experiment results in Fig. 7.10b. The BER curves of ideal lasers with zero linewidth are also included. Compared to the results in Fig. 7.8, the ideal laser bounds in Fig. 7.10b increase about 4.1 dB to account for the heterodyne detection and the OSNR penalty described by (7.24). The heterodyne detection in this work is without image filters, and it is expected that the system performance can be improved by 3 dB if narrow optical filters are applied to reject the image noise. The received OSNR is 11.1 dB at 10^{-3} BER for 16-QAM-OFDM. This result shows that 16-QAM-OFDM can achieve similar OSNR performance to conventional IM/DD systems [14]. 64-QAM-OFDM is strongly limited by the laser phase noise and has an apparent BER floor, confirmed by both the simulation and the experiment. In Fig. 7.10b, the experimental curves approach closer to the simulation curves for the higher OSNR. The main reason is that the estimated 101-kHz linewidth includes small frequency jitters, so-called $1/f$ frequency noise. This kind of noise leads to overestimation of the laser linewidth but does not degrade BER in high bit-rate systems [39]. Therefore, the actual combined linewidth of the two lasers in this work should be smaller than 101 kHz. In other words, the simulation BER curves should be lower.

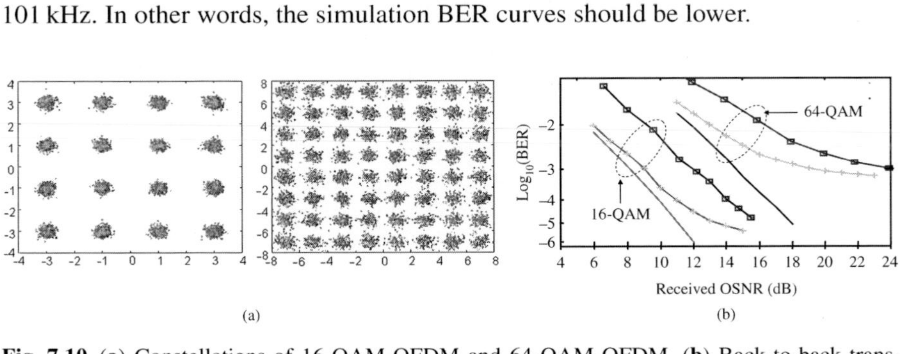

(a) (b)

Fig. 7.10 (**a**) Constellations of 16-QAM-OFDM and 64-QAM-OFDM. (**b**) Back-to-back transmission performance. The *solid lines* without symbols are for ideal lasers with zero linewidth. The *curves* with crosses are simulated results with 101-kHz combined linewidth. The *curves* with *squares* are experimental results

7.5.5 Non-linear Phase Noise from Optical Fiber Transmissions

From Fig. 7.10, 64-QAM-OFDM is strongly limited by the laser phase noise. Therefore, only 16-QAM-OFDM is used to investigate the non-linear phase noise from optical fiber transmissions in this work. The non-linear optical fiber transmissions include two scenarios: (1) single-span transmission with high launch optical power and (2) multi-span transmission with long distance. The two scenarios are investigated in the following two experiments.

In the single-span transmission, 50-km standard single-mode fiber (SSMF) is used to investigate the non-linear phase noise arising from high launch powers. The experimental setup is based on Fig. 7.9 but the re-circulating loop is replaced with the 50-km SSMF fiber. Figure 7.11 shows the Q-factor and phase drift variance versus launch optical power. The phase drift variance is defined in (7.14). Figure 7.11 also shows the non-linear phase noise, self-phase modulation (SPM) in particular, and the mitigation results given by

$$\vec{s}_{\mathrm{nr}}(t) = \vec{s}_{\mathrm{r}}(t)\exp\left\{-j\beta_{\mathrm{n}}|\vec{s}_{\mathrm{r}}(t)|^2\right\},$$
$$|\vec{s}_{\mathrm{r}}(t)|^2 = \left[s_{\mathrm{r}}^x(t) \quad s_{\mathrm{r}}^y(t)\right]\cdot\left(\left[s_{\mathrm{r}}^x(t) \quad s_{\mathrm{r}}^y(t)\right]^T\right)^*, \tag{7.27}$$

where $|\vec{s}_{\mathrm{r}}(t)|^2$ is the total power of two-polarization states, β_{n} is an empirical non-linearity parameter related to the strength of SPM, and $\vec{s}_{\mathrm{nr}}(t)$ is the non-linearity mitigated signal.

For the high-order M-QAM, the Q-factor is converted from BER by (7.30) in Appendix. There are two ways to calculate the BER, consequently Q-factor. The first way is based on AWGN assumption and calculated from SNR by (7.29) in Appendix, which is essentially the constellation spreading [22, 38]. The second way is the result of data decision and error computation. The resultant two Q-factors are both plotted in Fig. 7.11. For clarity, the Q-factor from BER decision is named calculated-Q, and the Q-factor from noise spreading, monitored-Q.

In Fig. 7.11, the Q-factors barely change from −6 to 0 dBm despite the varying received OSNR. This is because the system performance is capped by the imperfection in the electrical components. When the launch power is higher than 0 dBm, the phase drift variance begins to increase, whereas the Q-factor decreases. The difference between two Q-factors is obvious but they both show the same trend. It is apparent that the system is corrupted by the phase noise and cannot be approximated

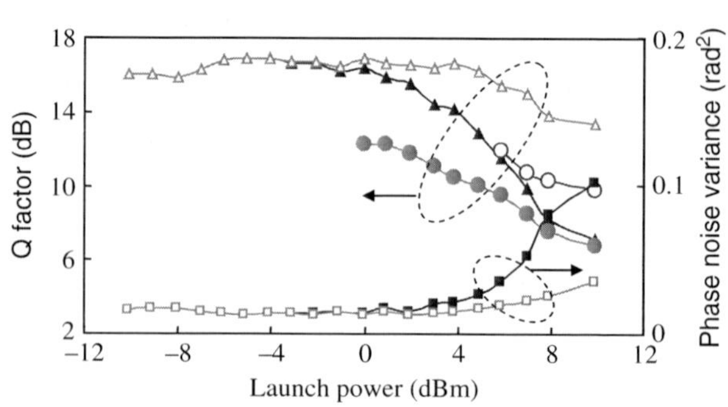

Fig. 7.11 Q-factor and phase drift variance of 16-QAM-OFDM versus launch power. The *curves* with *open symbols* are after fiber non-linearity mitigation. The *curves* with *triangles* and *circles* are monitored-Q and calculated-Q, respectively

by AWGN. The Q-factor based on AWGN overestimates the system performance. The accurate system performance analysis needs further study. The calculated-Q is not shown for the low launch power because it is difficult to obtain the meaningful BER at lower values. For the same reason, Q-factor is still used in the following experiment, even though it may overestimate the system performance. The fiber non-linearity mitigation improves the system performance significantly, manifested by the difference between the curves with solid and open symbols. The phase drift variance is reduced and Q-factor is increased about 3 dB. The tolerable launch power into fiber is also shifted to about 5 dBm.

In Fig. 7.11, the optimum launch power is about 3 dBm per channel with negligible fiber non-linearity. Therefore, the optimum launch power for 10 loops can be roughly estimated as −7 dBm. In the multi-span long-haul transmission, the launch power into SSMF fiber is −6.6 dBm. Figure 7.12a shows the Q-factor evolution with the distance up to 1,000 km. The Q-factor at the same OSNR in the back-to-back transmission is also recorded as the reference to compute system penalty. The Q-factor is mainly decided by the OSNR when the transmission distance is short. However, the non-linear phase noises occur after 500 km and the Q-factor penalty at 1,000 km become significant. As a result, there is an error floor at 10^{-4} BER. The improvement of the fiber non-linearity mitigation is not apparent because the non-linearity in the system is week due to the low launch power. Furthermore, as explained in [24], the mitigation method in this work is accurate for the single-span transmission but only an approximation for the multi-span transmission. More advanced signal processing techniques are needed to mitigate the fiber non-linearity for the multi-span transmission [33].

Figure 7.12b shows the BER performance after 500-km and 1,000-km transmissions. The inset is the spectrum of the OFDM signal after 1,000-km transmission. The spectrum is still tightly bounded within 3.5 GHz. The OSNR penalties at 10^{-3} BER for 500-km and 1,000-km transmissions are 0.1 dB and 2 dB, respectively. The fiber non-linearity mitigation results are not shown in Fig. 7.12b because the improvements are insignificant, as explained in Fig. 7.12a.

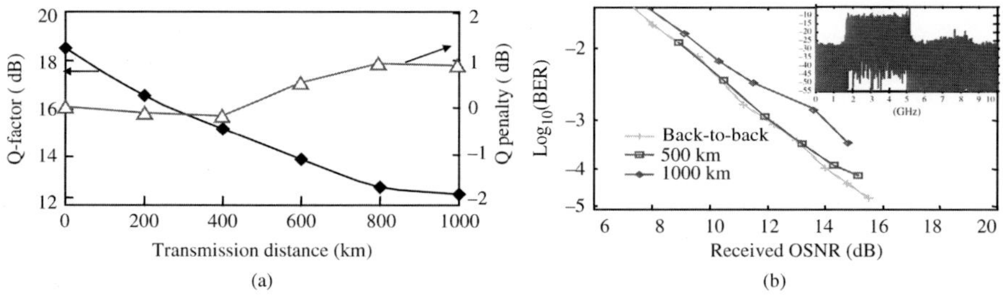

Fig. 7.12 (**a**) Q-factor and Q-penalty versus transmission distance. The Q-penalty *curve* with *squares* is after fiber non-linearity mitigation. (**b**) System performance at 500-km and 1000-km transmission

7.6 Orthogonal-Band Multiplexing Using CO-OFDM

Wavelength division multiplexing (WDM) is widely used to improve the overall system capacity. However, the conventional WDM systems have a relatively large frequency guard band between wavelength channels, which leads to poor spectral efficiency. This section will focus on reducing the frequency guard band between two adjunct wavelength channels by using orthogonal-band multiplexing [40, 41].

7.6.1 Principle of Orthogonal-Band-Multiplexed OFDM (OBM-OFDM)

The principle of OBM-OFDM is to divide the entire OFDM spectrum into multiple orthogonal OFDM (sub) bands. As shown in Fig. 7.13, the entire OFDM spectrum comprises N OFDM bands, each with the subcarrier spacing of Δf, and band frequency guard spacing of Δf_{G}. The subcarrier spacing Δf is identical for each band due to the use of the same sampling clock within one circuit. The orthogonality condition between the different bands is given by

$$\Delta f_{\mathrm{G}} = m\Delta f, \qquad (7.28)$$

which means that the guard band is multiple times ($m = 1, 2, 3, \ldots$) of subcarrier spacing. This is to guarantee that each OFDM band is an orthogonal extension of another. As such, the orthogonality condition is satisfied not only for the subcarriers inside each band, but it is also satisfied for any two subcarriers from different bands, for instance, f_i from band 1 and f_j from band 2 are orthogonal to each other (Fig. 7.13), despite the fact that they originate from different bands. The interesting scenario is that m equals 1 in (7.28) such that the OFDM bands can be multiplexed/de-multiplexed even without guard band. Note OBM-OFDM here is identical to cross-channel OFDM (XC-OFDM) in [33].

Upon reception, each OFDM (sub) band can be de-multiplexed using an anti-alias filter slightly wider than the bands to be detected. Figure 7.13 shows two approaches for OBM-OFDM detections. The first approach is to tune the receiver laser to the center of each band and to use an anti-alias low-pass filter I that filters out only one-band RF signal, such that each band is detected separately. The second

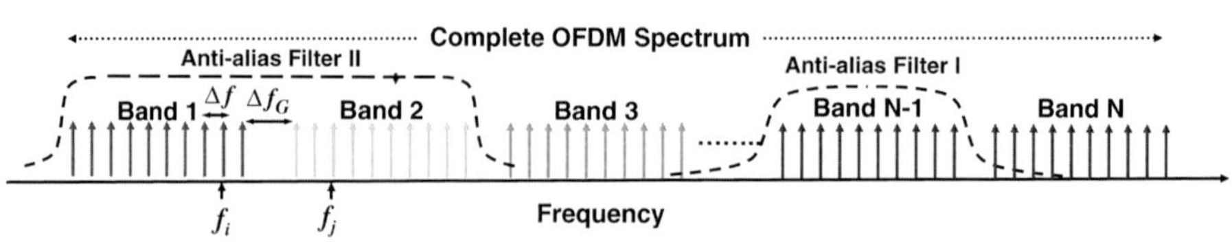

Fig. 7.13 Conceptual diagram of orthogonal-band-multiplexed OFDM (OBM-OFDM). Anti-alias filters I and II correspond to two detection approaches illustrated in the text

approach is to tune the receive laser to the center of the guard band and to use an anti-alias low-pass filter *II* that filters out two-band RF signal, such that two bands are detected simultaneously. In either case, the inter-band interference is avoided because of the orthogonality between the neighboring bands, despite the *leakage* of the subcarriers from neighboring bands.

7.6.2 Experimental Setup and Description

The OBM-OFDM could be realized using either subcarrier multiplexing [8] or wavelength multiplexing to patch multiple orthogonal bands into a complete OFDM spectrum as shown in Fig. 7.13. The transmission performances such as OSNR sensitivity, non-linearity, and phase noise impact are independent of the means of OBM-OFDM implementation. Figure 7.14 shows the experimental setup for the 107-Gb/s CO-OFDM transmission by using optical multiplexing. The five tones multi-frequency optical source spaced at 7.5 GHz is generated using cascaded intensity modulator and phase modulator architecture [42]. The tone spacing and AWG sampling clock are locked by a frequency standard of 10 MHz from the frequency synthesizer. The OFDM signal in each individual band is generated by using an AWG. The time-domain OFDM waveform is first generated with a Matlab program with the parameters as follows: total number of subcarriers is 128 with QPSK encoding, cyclic prefix is 1/8 of the observation period, middle 87 subcarriers out of 128 are filled, from which 10 pilot subcarriers are used for phase estimation. The real and imaginary parts of the OFDM waveforms are uploaded into the AWG operated at 10 GSa/s to generate two analog signals, which are then fed into I and Q ports of an optical I/Q modulator, to impress the baseband OFDM signal onto five optical tones. The optical output of the I/Q modulator consists of five-band OBM-OFDM signals, each band carrying 10.7 Gb/s. Although the five bands are filled with the same data, this will not affect the performance of the system studied and the subsequent conclusion. Unlike conventional link design, no dispersion compensation module is used in this transmission experiment, leading to fast phase walk-off and de-correlation between neighboring bands. In fact, this experimental

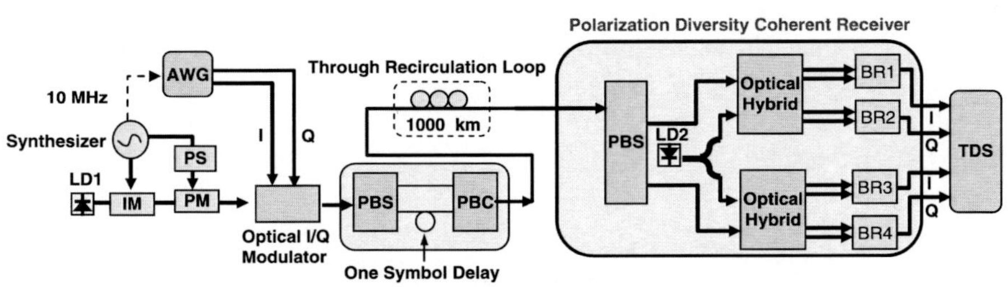

Fig. 7.14 Experimental setup for the 107-Gb/s CO-OFDM system

setup represents a worse scenario because the correlation between bands increases peak-to-average power ratio (PAPR) and consequently fiber non-linearity [41].

The optical OFDM signal from I/Q modulator is then split into two branches that are delay-mismatched by one OFDM symbol period (14.4 ns), and then polarization combined. This is to emulate the polarization diversity transmitter with a data rate of 21.4 Gb/s per band. The resultant two-polarization components are completely independent. The signal is further input into a re-circulating loop comprising 100-km fiber and an EDFA to compensate the loss. The signal is coupled out from the loop and received with a polarization diversity coherent receiver comprising a receive laser, a polarization beam splitter, two optical 90° hybrids, and four balanced receivers. The receive laser is tuned to the center of each band, and the RF signals from the four balanced detectors first pass through a 3.8-GHz low-pass filter, as the 'alias filters *I*' in Fig. 7.13, such that each band is measured independently. The RF signals are then input into a TDS, acquired at 20 GSa/s, and processed with a Matlab program using 2×2 MIMO-OFDM models. The 2×2 MIMO-OFDM signal processing involves (1) FFT window synchronization using Schmidl format to identify the start of the OFDM symbol [9], (2) software estimation and compensation of the frequency offset, (3) channel estimation in terms of Jones Matrix H, (4) phase estimation for each OFDM symbol, and (5) constellation construction for each carrier and BER computation. The channel matrix H is estimated by sending 30 OFDM symbols using alternate polarization launch, as described by (7.23). The total number of OFDM symbols evaluated is 1,000. The measurements of low BER in the order of 10^{-5} are run multiple times.

7.6.3 Experimental Results and Discussion

Figure 7.15a shows the RF spectrum after 1,000-km transmission measured with the polarization diversity coherent receiver. Note that frequency guard band ($m = 1$) is used in this transmission measurement. Therefore, five OFDM bands with 6.4-GHz bandwidth each are closely patched together and the entire OFDM spectrum occupies about 32-GHz bandwidth. Hence a spectral efficiency of 3.3 bit/s/Hz is achieved. The out of band components are due to the multi-frequency source generation not tightly bounded at five tones. This artifact will not exist in the real application using either subcarrier multiplexing or optical multiplexing OBM-OFDM. Figure 7.15b shows the detected electrical spectrum after using a 3.8-GHz electrical anti-alias filter for one-band detection. The anti-alias filter is critical for OBM-OFDM implementation. As shown in Fig. 7.15a, without electrical anti-alias filter, the single-side electrical spectrum will be as broad as 16 GHz, indicating that at least 32-GSa/s ADC has to be used. However, the filtered spectrum in Fig. 7.15b is readily sampled with 20 GSa/s, or even at a lower speed of 10 GSa/s. Additionally, despite the fact that there are some spurious components from neighboring band that is leaked at the edge of the 3.8-GHz filter, they do not contribute the interference degradation since they are orthogonal subcarriers to the interested OFDM subcarriers.

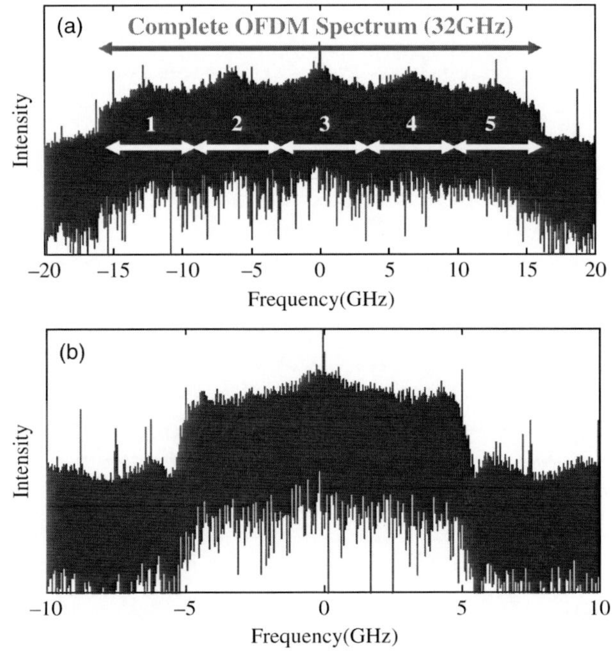

Fig. 7.15 (**a**) RF spectrum for the 107-Gb/s signal using a polarization diversity coherent receiver, the band numbers are depicted next to the corresponding bands. (**b**) The electrical spectrum at the receiver after the 3.8-GHz anti-alias filter

Figure 7.16a shows the BER sensitivity performance for the entire 107-Gb/s CO-OFDM signal at the back-to-back and 1,000-km transmission with the launch power of -1 dBm. Each BER point is the average of all five bands and two polarizations. The OSNR required for a BER of 10^{-3} is respectively 15.8 and 16.8 dB for back-to-back and 1,000 km transmission. Figure 7.16b shows the system Q performance of the 107-Gb/s CO-OFDM signal as a function of reach up to 1,000 km. It can be seen that the Q-factor decreases from 17.2 to 12.5 dB when reach increases from back-to-back to 1,000-km transmission. By improving the balance between the polarization diversity coherent receivers, the disparity between the two polarizations has been improved from 1.5 dB in [40] to 0.49 dB in this experiment.

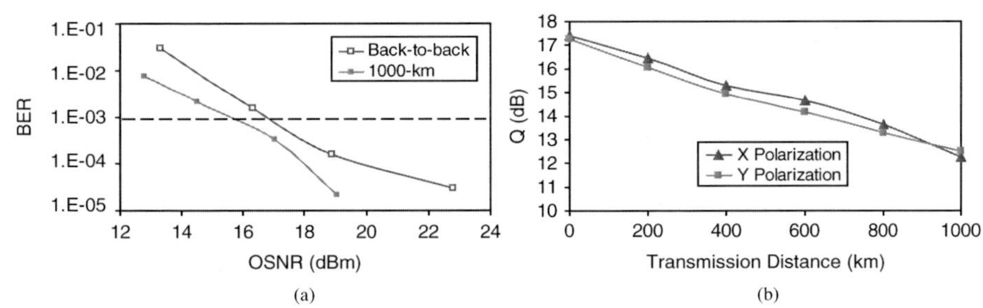

Fig. 7.16 (**a**) BER sensitivity of 107-Gb/s CO-OFDM signal at the back-to-back and at 1,000 km. (**b**) Q-factor of 107-Gb/s CO-OFDM signal as a function of reach

7.7 Conclusion

Optical transmissions have started the trend of migration from analog to digital and evolved toward software-enhanced optical transmission (SEOT). On such a platform, coherent optical OFDM has made full advantage of DSP to conduct sophisticated operations and demonstrated numerous merits over conventional optical transmissions. This chapter has particularly shown that coherent optical OFDM is a flexible and feasible solution to overcome various distortions and achieve high spectral efficiency. Without much hardware change, coherent optical OFDM can implement different advanced modulation formats. With single polarization, 16-QAM-OFDM and 64-QAM-OFDM have been experimentally realized with 3.0 and 4.5 bit/s/Hz spectral efficiency, respectively. With an extension of the OFDM concept, orthogonal-band multiplexing OFDM has been demonstrated to minimize or eliminate the conventional frequency guard band. The experimental system employs 2×2 MIMO-OFDM signal processing and has achieved 3.3 bit/s/Hz spectral efficiency at a data rate of 107 Gb/s.

Appendix

In the AWGN channel, single carrier and OFDM have about the same performance in terms of SNR and the BER of M-QAM-OFDM is given by [37, 38]

$$P_e = \frac{1 - \left[1 - 2\left(1 - \frac{1}{\sqrt{M}}\right) Q\left(\sqrt{\frac{3R_s}{M-1}}\right)\right]^2}{\log_2 M} \qquad (7.29)$$

$$Q(x) = \frac{1}{\sqrt{2\pi}} \int_x^{\infty} e^{-t^2/2} dt$$

where $M = 2^k$ and k is a positive even number, and R_s is the SNR per symbol. Note that (7.29) is a tight upper bound of a more accurate expression in [37].

In some experiments, R_s can be estimated from the symbol constellation and then BER by (7.29). The Q-factor can be inversely calculated from BER P_e by

$$Q_f = \sqrt{2} erf^{-1}(1 - 2P_e) \qquad (7.30)$$

$$y = \mathrm{erf}(x) = \frac{2}{\sqrt{\pi}} \int_0^x e^{-t^2} dt. \qquad (7.31)$$

$$x = \mathrm{erf}^{-1}(y)$$

References

1. J. Mitola, The software radio architecture. IEEE Commun. Mag. **33**(5), 26–38 (1995).
2. A.A. Abidi, The path to the software-defined radio receiver. IEEE J. Solid-State Circuits **42**(5), 954–966 (2007).
3. H. Sun, K.T. Wu, K. Roberts, Real-time measurements of a 40 Gb/s coherent system. Opt. Express **16**(2), 873–879 (2008).
4. C.R. Fludger, T. Duthel, D. van den Borne, C. Schulien, E.-D. Schmidt, T. Wuth, E. de Man, G.D. Khoe, H. de Waardt, 10 × 111 Gbit/s, 50 GHz Spaced, POLMUX-RZ-DQPSK Transmission over 2375 km Employing Coherent Equalisation. in OFC. 2007, paper PDP22.
5. P.J. Winzer, G. Raybon, M. Duelk. 107-Gb/s optical ETDM transmitter for 100G Ethernet transport. in ECOC. 2005, paper Th4.1.1.
6. I.B. Djordjevic, B. Vasic, 100-Gb/s transmission using orthogonal frequency-division multiplexing. IEEE Photon. Technol. Lett. **18**(13/16), 1576 (2006).
7. W. Shieh, C. Athaudage, Coherent optical orthogonal frequency division multiplexing. Electron. Lett. **42**(10), 587–589 (2006).
8. S.L. Jansen, I. Morita, N. Takeda, H. Tanaka, 20-Gb/s OFDM transmission over 4,160-km SSMF Enabled by RF-Pilot Tone Phase Noise Compensation. in OFC. 2007, paper PDP15.
9. S. Hara, R. Prasad, *Multicarrier techniques for 4G mobile communications.* (Artech House, Inc. Norwood, 2003).
10. A. Gupta, A. Forenza, R.W. Health, Rapid MIMO-OFDM software defined radio system prototyping. In *Proceedings of the IEEE Workshop on Signal Processing Systems*, pp. 182–187, Austin, 13–15 October 2004.
11. G.L. Stuber, J.R. Barry, S.W. Mclaughlin, Y.G. Li, M.A. Ingram, G. Pratt, Broadband MIMO-OFDM wireless communications. Proc. IEEE **92**(2), 271–294 (2004).
12. F. Buchali, H. Bulow, Adaptive PMD compensation by electrical and optical techniques. J. Lightwave Technol. **22**(4), 1116 (2004).
13. H. Bulow, Electronic dispersion compensation. in OFC. 2007, Tutorial OMG5.
14. A. Faerbert, Application of digital equalization in optical transmission systems. in OFC. 2006, paper OTuE5.
15. D. McGhan, C. Laperle, A. Savchenko, C. Li, G. Mak, O'Sullivan Maurice, 5120-km RZ-DPSK transmission over G. 652 fiber at 10 Gb/s without optical dispersion compensation. IEEE Photon. Technol. Lett. **18**(2), 400 (2006).
16. D. Ly-Gagnon, S. Tsukamoto, K. Katoh, K. Kikuchi, Coherent detection of optical quadrature phase-shift keying signals with carrier phase estimation. J. Lightwave Technol. **24**(1), 12 (2006).
17. W. Shieh, X. Yi, Y. Tang, Transmission experiment of multi-gigabit coherent optical OFDM systems over 1000 km SSMF fiber. Electron. Lett. **43**(3), 183–184 (2007).
18. W. Shieh, X. Yi, Y. Ma, Q. Yang, Coherent optical OFDM: has its time come? J. Opt. Network. **7**(3), 234–255 (2008).
19. Q. Yang, W. Shieh, Y. Ma, Bit and power loading for coherent optical OFDM. IEEE Photon. Technol. Lett. **20**(15), 1305–1307 (2008).
20. X. Yi, W. Shieh, Y. Tang, Phase estimation for coherent optical OFDM. IEEE Photon. Technol. Lett. **19**(9/12), 919 (2007).
21. Y. Tang, K.P. Ho, W. Shieh, Coherent optical OFDM transmitter design employing predistortion. IEEE Photon. Technol. Lett. **20**(11), 954–956 (2008).
22. W. Shieh, R.S. Tucker, W. Chen, X. Yi, G. Pendock, Optical performance monitoring in coherent optical OFDM systems. Opt. Express **15**(2), 350–356 (2007).
23. X. Yi, W. Shieh, Y. Ma, Y. Tang, G.J. Pendock, Experimental demonstration of optical performance monitoring in coherent optical OFDM Systems. in OFC. 2008, paper OThW3.
24. W. Shieh, X. Yi, Y. Ma, Y. Tang, Theoretical and experimental study on PMD-supported transmission using polarization diversity in coherent optical OFDM systems. Opt Express **15**(16), 9936–9947 (2007).

25. X. Yi, W. Shieh, Y. Ma, Phase noise effects on high spectral efficiency coherent optical OFDM Transmission. J. Lightwave Technol. **26**(10), 1309–1316 (2008).
26. I.B. Djordjevic, B. Vasic, Orthogonal frequency division multiplexing for high-speed optical transmission. Opt. Express, **14**(9), 3767–3775 (2006).
27. A.J. Lowery, L. Du, J. Armstrong. Orthogonal frequency division multiplexing for adaptive dispersion compensation in long haul WDM systems. in OFC. 2006, paper PDP39.
28. K. Yonenaga, A. Sano, E. Yamazaki, F. Inuzuka, Y. Miyamoto, A. Takada, T. Yamada, 100 Gbit/s All-Optical OFDM Transmission Using 4×25 Gbit/s optical duobinary signals with phase-controlled optical sub-carriers. in OFC. 2008, paper JThA48.
29. L. Hanzo, T. Keller, *OFDM and MC-CDMA: A Primer*. (Wiley, New York, 2006).
30. S. Weinstein, P. Ebert, Data transmission by frequency-division multiplexing using the discrete Fourier transform. IEEE Trans. Commun. **19**(5 Part 1), 628–634 (1971).
31. Y. Tang, W. Shieh, X. Yi, R. Evans, Optimum design for RF-to-optical up-converter in coherent optical OFDM systems. IEEE Photon. Technol. Lett. **19**(5/8), 483 (2007).
32. S. Wu, Y. Bar-Ness, A phase noise suppression algorithm for OFDM-based WLANs. IEEE Commun. Lett. **6**(12), 535–537 (2002).
33. W. Shieh, H. Bao, Y. Tang, Coherent optical OFDM: theory and design. Opt. Express **16**(2), 841–859 (2008).
34. S. Wu, Y. Bar-Ness, OFDM systems in the presence of phase noise: consequences and solutions. IEEE Trans. Commun. **52**(11), 1988–1996 (2004).
35. S. Hara et al., Transmission performance analysis of multi-carrier modulation in frequency selective fast Rayleigh fading channel. Wirel. Pers. Commun. **2**(4), 335–356 (1995).
36. N. Gisin, B. Huttner, Combined effects of polarization mode dispersion and polarization dependent losses in optical fibers. Opt. Commun. **142**(1–3), 119–125 (1997).
37. F. Xiong, *Digital Modulation Techniques*, 2nd edn. (Artech House, Boston, 2006).
38. J.G. Proakis, *Digital Communications*, 4th edn. (McGraw-Hill Higher Education, New York, 2001).
39. K. Kikuchi, Effect of 1/f-type FM noise on semiconductor-laser linewidth residual in high-power limit. IEEE J. Quantum Electron. **25**(4), 684–688 (1989).
40. W. Shieh, Q. Yang, Y. Ma, 107 Gb/s coherent optical OFDM transmission over 1000-km SSMF fiber using orthogonal band multiplexing. Opt. Express **16**(9), 6378–6386 (2008).
41. Q. Yang, Y. Tang, Y. Ma, W. Shieh, Experimental demonstration and numerical simulation of 107-Gb/s high spectral efficiency coherent optical OFDM. J. Lightwave Technol. **27**, 168–176 (2009).
42. Y. Wang, Z. Pan, C. Yu, T. Luo, A. Sahin, and A.E. Willner, A multi-wavelength optical source based on supercontinuum generation using phase and intensity modulation at the line-spacing rate. in ECOC. 2003, paper Th3.2.4.

Chapter 8
Polarization Division-Multiplexed Coherent Optical OFDM Transmission Enabled by MIMO Processing

Sander L. Jansen and Itsuro Morita

Abstract In this chapter, the realization and performance of polarization division-multiplexed orthogonal frequency division multiplexing (PDM-OFDM) for long-haul transmission systems is discussed. Polarization de-multiplexing of the PDM signal at the receiver is realized with a multiple input multiple output (MIMO) processing. With the help of PDM, continuously detectable transmission is reported of 10×121.9 Gb/s (112.6 Gb/s without OFDM overhead) at 50-GHz channel spacing over 1,000 km of standard single-mode fiber (SSMF) without any in-line dispersion compensation.

8.1 Introduction

Polarization division multiplexing (PDM) is a very effective method to double the spectral efficiency of a transmission system [1, 2]. However, it has been shown that direct-detected PDM has a reduced tolerance to polarization mode dispersion (PMD), because the polarization de-multiplexing introduces cross talk between the polarization tributaries [3]. An elegant way to overcome this PMD sensitivity is by using polarization diverse coherent detection with digital equalization instead of direct detection. In [4] it has been shown that such a system can effectively be described as a polarization multiple input multiple output (MIMO) system in which any space-time coding algorithm can be applied. This has enabled the demonstration of various single-carrier MIMO experiments with bit rates up to 111 Gb/s and a superior tolerance toward chromatic dispersion and PMD [5].

Orthogonal frequency division multiplexing (OFDM) presents an alternative to digital coherent systems that inherently offers a virtually unlimited chromatic dispersion and PMD [6] tolerance and is furthermore easier scalable to higher level modulation formats [7–9]. Unlike most single-carrier MIMO algorithms,

I. Morita (✉)
KDDI R&D Laboratories, Japan
e-mail: morita@kddilabs.jp

M. Nakazawa et al. (eds.), *High Spectral Density Optical Communication Technologies*, Optical and Fiber Communications Reports 6, DOI 10.1007/978-3-642-10419-0_8,

PDM-OFDM uses training symbols for MIMO processing. In this chapter, we give a detailed explanation of MIMO processing for PDM-OFDM transmission systems.

8.2 PDM Receiver with MIMO Processing

8.2.1 PDM-OFDM

The concept of a PDM transmission system with coherent optical detection and MIMO processing at the receiver is shown in Fig. 8.1.

At the transmitter two baseband signals, generated by transmitter 1 (Tx 1) and transmitter 2 (Tx 2), are modulated onto an optical carrier and multiplexed onto orthogonal polarizations by a polarization beam splitter. At the receiver first the received signal is split into two random polarizations. The two detected signals consist of an arbitrary mix of each transmitted signal. In order to preserve the phase information of the optical signal, two coherent optical detectors (COD) are used for conversion from the optical to the electrical domain. MIMO processing is then applied to de-rotate the polarization and separate the two received signals. After MIMO processing, the two de-rotated signals are fed to the baseband receivers.

Figure 8.2 illustrates the block diagram of a PDM-OFDM transmitter and receiver. For simplicity, a perfectly synchronized system is assumed in this illustration, meaning that the carrier frequency offset (CFO) [10] and local oscillator phase noise [11] is compensated for.

The PDM-OFDM transmitter consists effectively of two conventional OFDM transmitters. The notation vectors shown in Fig. 8.2 consist of two elements, one for each transmitter, defined for data stream **b** as

$$\mathbf{b} = \begin{pmatrix} b_1 \\ b_2 \end{pmatrix}, \tag{8.1}$$

where the index denotes the number of the transmitter. Vectors and matrices are denoted by bold lowercase and uppercase symbols, respectively. The binary data stream **b** is converted from serial to parallel and symbol mapping is applied. Periodically, training symbols are then inserted resulting in the complex signal vector **s**. Subsequently, the signal is modulated onto orthogonal carriers by taking the inverse fast Fourier transform (IFFT). After the IFFT, cyclic prefix is added to mitigate impairments caused by, for instance, chromatic dispersion or polarization mode

Fig. 8.1 Concept of PDM enabled by MIMO processing at the receiver, with Tx baseband transmitter, Opt. Mod. optical modulator, COD coherent optical detector, and Rx baseband receiver

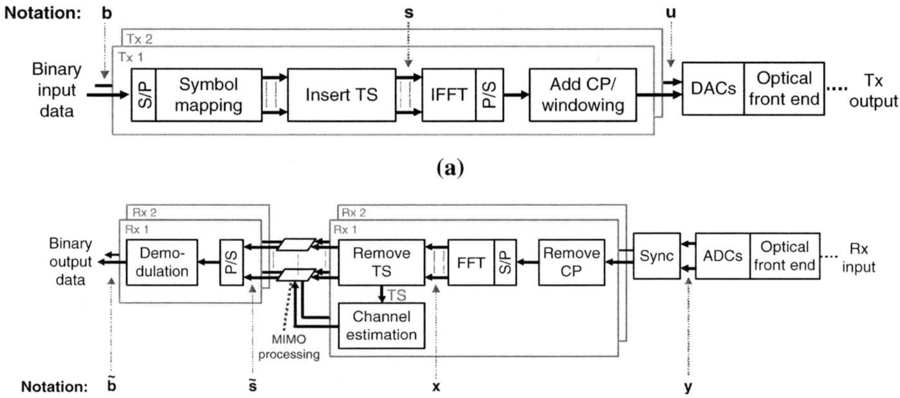

Fig. 8.2 Block diagram of a PDM-OFDM transmitter and receiver, with S/P serial to parallel converter, P/S parallel to serial converter, CP cyclic prefix, TS training symbol, DAC digital-to-analogue converter, ADC analogue-to-digital converter, CFO corr. carrier frequency offset correction, and Sync OFDM symbol synchronization. (**a**) Transmitter, (**b**) receiver

dispersion (PMD), yielding the OFDM baseband vector $\mathbf{u}$. The baseband signals of both transmitters are then converted to analogue signals using digital-to-analogue converters (DAC) and fed to the optical front end (shown in Fig. 8.1). At the receiver, the outputs of the optical front end are first digitalized using analogue-to-digital converters (ADC), yielding vector $\mathbf{y}$. After the ADCs, symbol synchronization is applied as described in [10, 12]. Both transmitters operate at the same symbol rate, so the symbol timing needs only to be recovered once. The OFDM signal is converted back to the frequency domain by taking the fast Fourier transform (FFT), yielding signal vector $\mathbf{x}$. After the FFT, the training symbols are removed from the payload and used for channel estimation. Subsequently, MIMO processing per subcarrier is applied to de-rotate the polarization of the received symbols, yielding estimate vector $\tilde{\mathbf{s}}$. Finally, these symbols are demodulated (or demapped) resulting in the estimate vector of the transmitted data $\tilde{\mathbf{b}}$. Numerous system parameters are critical for the design of an OFDM system and in [6, 11] some of the most important parameters and trade-offs for OFDM-based fiber-optic transmission systems are discussed. In this section the enabling techniques for MIMO processing will be reviewed: MIMO processing and channel estimation.

8.2.2 MIMO Processing

In a single-carrier PDM system, a butterfly structure of complex-valued, multi-tap adaptive filters has proven to be an effective equalizer for MIMO processing [5]. This equalizer can theoretically be applied to multi-carrier systems such as OFDM as well, but would result in an immense computational complexity as a separate adaptive filter has to be applied for each individual OFDM subcarrier. It has been shown that for a multi-carrier system, the computational complexity of a MIMO detector can greatly be reduced by using knowledge about the channel at the receiver [13]. Such a MIMO detector will be discussed in this section.

Assuming a perfectly synchronized system, the received signal before MIMO processing $\mathbf{x}(k)$ of the kth subcarrier can be expressed as:

$$\mathbf{x}(k) = \mathbf{H}(k)\,\mathbf{s}(k) + \mathbf{n}(k), \tag{8.2}$$

where

$$\mathbf{H}(k) = \begin{pmatrix} h_{11} & h_{12} \\ h_{21} & h_{22} \end{pmatrix} \tag{8.3}$$

represents the 2×2 channel matrix (Jones matrix) and vector $\mathbf{n}(k)$ represents the frequency-domain noise within subcarrier k for the two received polarizations. The most straightforward MIMO detector employs a zero-forcing (ZF) algorithm. Basically, ZF-MIMO detection is realized by multiplying the received signals $\mathbf{x}$ with the pseudo-inverse of the estimated channel matrix to find an estimate of the transmitted vector [10]. For subcarrier k, let $\tilde{\mathbf{H}}(k)$ be the estimation of channel matrix $\mathbf{H}(k)$, the estimate of the transmitted stream after ZF processing $\tilde{\mathbf{s}}(k)$ can then be described by

$$\tilde{\mathbf{s}}(k) = \tilde{\mathbf{H}}^{+}(k)\mathbf{x}(k), \tag{8.4}$$

where superscript + denotes the pseudo-inverse operation, defined for a matrix $\mathbf{A}$ as

$$\mathbf{A}^{+} = \left(\mathbf{A}^{H}\,\mathbf{A}\right)^{-1}\mathbf{A}^{H}. \tag{8.5}$$

Here superscripts H and -1 denote the conjugate transpose and matrix inverse, respectively. Assuming perfect channel estimation, i.e., $\tilde{\mathbf{H}}^{+}(k) = \mathbf{H}^{+}(k)$, equation (8.4) can be written as:

$$\tilde{\mathbf{s}}(k) = \mathbf{s}(k) + \mathbf{H}^{+}(k)\mathbf{n}(k). \tag{8.6}$$

From equation (8.6), we see that even with perfect channel estimation an error term will occur due to the noise added by optical amplifiers and other noise sources. More advanced MIMO detectors can be employed to improve the performance in MIMO detection, such as parallel interference detection [14] and maximum-likelihood detection (MLD) [15, 16]. However, this performance improvement comes at the cost of a significant increase in computational complexity.

8.2.3 MIMO OFDM Channel Estimation

In order for MIMO processing to work, an estimation of the MIMO channel is required in the receiver. A commonly used method to enable channel estimation is by periodically inserting a piece of known data, i.e., a training symbol, in the symbol stream at the transmitter. At the receiver, channel estimation can then be realized by comparing the received symbol per OFDM subcarrier with the transmitted training symbol. As shown in Fig. 8.2, channel estimation is performed in the frequency domain, e.g., after the FFT operation at the receiver. For a MIMO system, it is essential that the training symbols of the transmitters are orthogonal so that they can be uniquely identified at the receiver [17]. Additionally the frequency selective behavior, e.g., caused by chromatic dispersion, must be characterized. For this purpose the training symbols should also be shift orthogonal. To achieve this, several training symbol structures have been proposed in literature. In this chapter, only the time-multiplexed training symbols, illustrated in Fig. 8.3, will be discussed.

Each training period consists of two OFDM training symbols, $s_{1,\mathrm{t}1}$ and $s_{2,\mathrm{t}2}$, that are transmitted by Tx 1 and Tx 2, respectively. Indices t1 and t2 refer to the timeslot number within the training period. In the time-multiplexed symbol configuration, orthogonality between the two transmitters is realized by sending the OFDM training symbols one after the other.

The received signals on the kth subcarrier during the training period can now, using equation (8.2), be written as the 2×2 matrix.

$$\begin{aligned}\mathbf{X}_\mathrm{t}(k) = [\mathbf{x}_{\mathrm{t}1}(k)\ \mathbf{x}_{\mathrm{t}2}(k)] &= \mathbf{H}(k)\mathbf{S}_\mathrm{t}(k) + \mathbf{N}_\mathrm{t}(k) \\ &= \mathbf{H}(k)\,[\mathbf{s}_{\mathrm{t}1}(k)\ \mathbf{s}_{\mathrm{t}2}(k)] + [\mathbf{n}_{\mathrm{t}1}(k)\ \mathbf{n}_{\mathrm{t}2}(k)] \\ &= \mathbf{H}(k)\begin{pmatrix} s_{1,\mathrm{t}1}(k) & 0 \\ 0 & s_{2,\mathrm{t}2}(k)\end{pmatrix} + [\mathbf{n}_{\mathrm{t}1}(k)\ \mathbf{n}_{\mathrm{t}2}(k)],\end{aligned} \tag{8.7}$$

where $\mathbf{x}_{\mathrm{t}1}(k)$ denotes the 2×1 received signal vector for the kth subcarrier during time slot t1. The notation for the noise and transmitted signal vectors is the same. Consequently, an estimate can be derived from the received signal matrix using different estimation algorithms, such as least squares estimation (LSE) and minimum

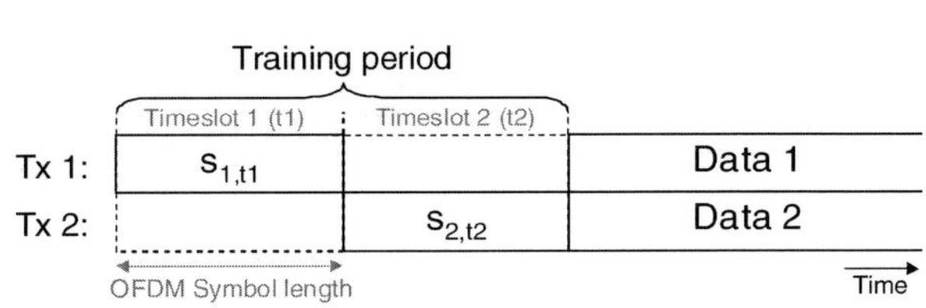

Fig. 8.3 Time-multiplexed training symbols for MIMO channel estimation

mean squared error estimation (MMSE). Here we will review the former, for which an estimate of the MIMO channel matrix is found as

$$\tilde{\mathbf{H}}(k) = \mathbf{X}_{\mathrm{t}}(k)\mathbf{S}_{\mathrm{t}}^{+}(k) = \mathbf{H}(k) + \mathbf{N}_{\mathrm{t}}(k)\mathbf{S}_{\mathrm{t}}^{+}(k). \tag{8.8}$$

An effective method to reduce the impact of the error caused by the noise term, i.e., the last term in equation (8.8), is to apply averaging with a moving average algorithm over time and/or frequency. For time averaging, typical average window lengths are between 5 and 50, depending on the system configuration. Furthermore, in order to minimize the influence of nonlinearities on channel equalization, a training symbol is usually constructed such that it exhibits a low peak-to-average power ratio (PAPR).

8.3 10 x 122-Gb/s Transmission Experiment with PDM-OFDM

8.3.1 Experimental Setup

The performance of 121.9-Gb/s PDM-OFDM was evaluated in a WDM environment on a long-haul transmission link. The experimental setup is shown in Fig. 8.4. In this experiment, arbitrary waveform generators (AWG) are used at a sampling rate of 10 GSamples/s to emulate the generation of the OFDM basebands. The baseband waveforms that are produced by the AWGs have been calculated off-line and are outputted continuously. The OFDM signal consists of two polarization-multiplexed signals with four multi-band OFDM bands each. The 121.9-Gb/s signal has therefore a total of eight 15.2-Gb/s tributaries. In order to create this multi-band signal, the two OFDM baseband signals are upconverted to four intermediate frequencies (IF), present at 8.7, 14.4, 20.4, and 26.1 GHz. Inset 2 of Fig. 8.4 illustrates the electrical OFDM channel allocation after upconversion. As only two AWGs were available care was taken in this setup that neighboring OFDM bands are generated by different AWGs for decorrelation. Therefore, AWG 1 is used for the first (8.7 GHz) and third OFDM (20.4 GHz) bands and AWG 2 for the second (14.4 GHz) and fourth

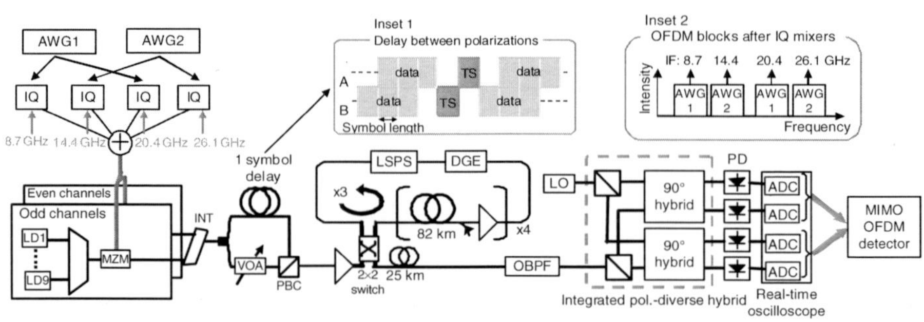

Fig. 8.4 Experimental setup for 10 × 121.9 Gb/s transmission experiment

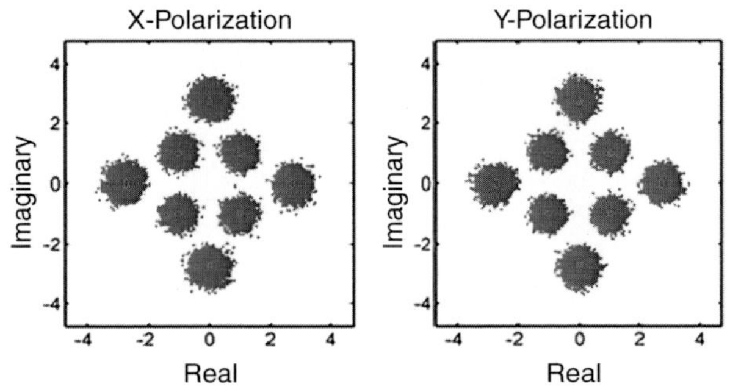

Fig. 8.5 Scatterplot of the first OFDM band of the 121.9-Gb/s PDM-OFDM signal

(26.1 GHz) bands. The FFT size is 1,024, from which 520 subcarriers are effectively used for the transport of data. Although the minimal required cyclic prefix for this configuration in a 1,000-km link is about 1 ns, a larger cyclic prefix of 22 samples (2.2 ns) per OFDM symbol was chosen to relax the synchronization requirements at the receiver after transmission. Together with the cyclic prefix, the OFDM symbol length is 104.6 ns and the OFDM symbol rate 9.6 MHz. A non-rectangular 8-QAM constellation shown in Fig. 8.5 is used for symbol mapping, which provides the maximum symbol distance and consequently the best bit error rate (BER) versus signal-to-noise ratio (SNR) performance [18, 19]. For this WDM experiment, 10 external cavity lasers (ECLs) with an approximate linewidth of 100 kHz are aligned on a 50-GHz ITU grid between 1553.7 and 1557.4 nm. Two parallel Mach–Zehnder modulators are used in this setup for separate modulation of the even and odd channels. The optical modulators are single-ended Mach–Zehnder modulators designed for analog applications (with a high linearity). In order to suppress the carrier of the OFDM signal the modulator is biased in its minimum. After modulation, the even and odd WDM channels are combined using a 50-GHz interleaver. The interleaver is aligned such that the image band of the OFDM signal is rejected. Polarization multiplexing is emulated by splitting the signal with a 3-dB coupler and combining it with a polarization beam combiner. As shown in inset 1 of Fig. 8.4, one arm is delayed by exactly one OFDM symbol (104.6 ns) for decorrelation. The bit rate before coding is 121.9 Gb/s, from which 6% is used for training symbols and 2.15% for the cyclic prefix. Before FEC this results in a data rate of 112.6 Gb/s allowing transport of Ethernet frames with 7% FEC and up to 4% protocol overhead. Figure 8.6 shows the optical spectrum of the 121.9-Gb/s OFDM signal, which clearly indicates the well-defined optical spectrum with a 22.8-GHz bandwidth.

The re-circulating loop consists of four spans of 82-km SSMF without optical dispersion compensation. After every span, amplification is provided by a Raman/EDFA structure with an average on/off Raman gain of $\sim$10 dB. A dynamic gain equalizer (DGE) is used for power equalization and a loop-synchronous polarization scrambler (LSPS) is employed to reduce loop-induced polarization effects. After transmission, 25 km of SSMF is inserted, resulting in a total transmission distance of 1,009 km. At the receiver, the signal is split in two random polarizations and

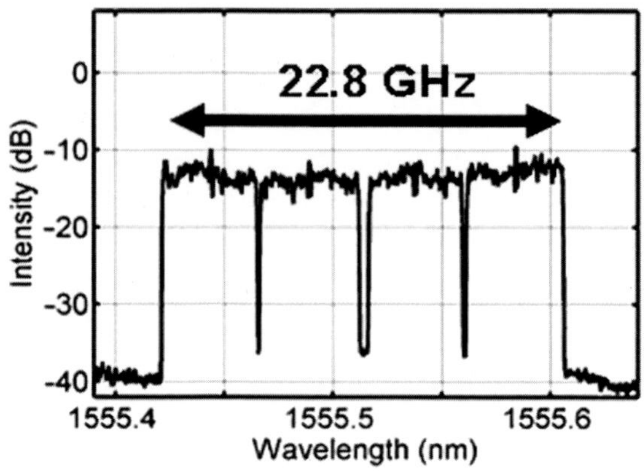

Fig. 8.6 Measured optical spectrum of the 121.9-Gb/s PDM-OFDM signal

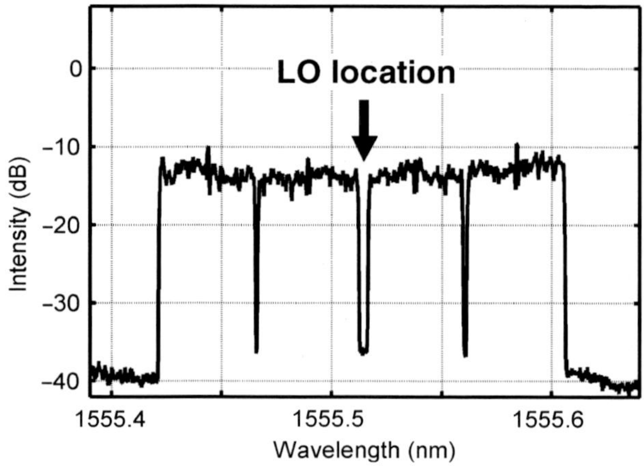

Fig. 8.7 Calculated electrical spectrum of the 121.9-Gb/s signal after coherent detection

is detected with a polarization diverse 90° optical hybrid. An ECL with ∼100-kHz linewidth is used as free running local oscillator (LO) and four single-ended 20-GHz Pin/TIA modules are used for detection. A real-time digital storage oscilloscope is used to sample the four outputs of the optical hybrid. The bandwidth of the oscilloscope is 16 GHz, the sampling frequency is 50 GHz, and the effective number of bits is approximately 5.5 bits. After detection, the data is post-processed off-line. Figure 8.7 shows the computed electrical spectrum after coherent detection. The LO is located in the middle of the OFDM signal, in between the second and third OFDM bands.

Training symbols are periodically inserted into the OFDM signal, so that polarization de-rotation at the receiver can be realized through MIMO processing described in the previous section. In order to distinguish between the training symbols of the PDM tributaries the training symbols should be received one after the other. The delay of the PDM emulator is exactly one OFDM symbol, thus training is realized in this experiment by inserting an empty OFDM symbol before and after the training symbol (see inset 1 of Fig. 8.4). Because of the empty OFDM symbol that is inserted for every training period, the training period consists in the experiment of three symbols instead of two. As a result, the overhead caused by training symbols can be reduced from 6 to 4% when the two polarizations are modulated independently (see for instance [6]) instead of the PDM emulator that is used in this experiment.

In order to compensate for the phase noise of the local oscillator, RF-aided phase noise compensation is implemented [9]. In order to mitigate the influence of amplified spontaneous emission (ASE) noise on the channel estimation, a moving average over 12 training symbols was used for MIMO processing at the receiver. For all reported BER results five sets with each 2.4 million bits have been evaluated, each set with different polarization state settings in the LSPS.

8.3.2 Experimental Results

Figure 8.8 shows the measured back-to-back sensitivities for 60.9 Gb/s (single polarization) and 121.9 Gb/s (PDM). Additionally, the simulated sensitivity for 121.9 Gb/s is shown as well. The required optical signal-to-noise ratio (OSNR) for a BER of 1×10^{-3} is 14 and 17.8 dB/0.1 nm for 60.9 and 121.9 Gb/s, respectively. Apart from the 3-dB expected penalty, a 0.8-dB excess penalty is observed for PDM. This penalty is most likely caused by a less effective phase noise compensation scheme. Because of the one symbol delay in the PDM emulator at the transmitter, there is a 105-ns delay between the RF pilot tones of the two polarizations. At the receiver, the sum of these RF pilot tones is used for phase noise compensation and thus the delay between the polarizations reduces the effectiveness of the

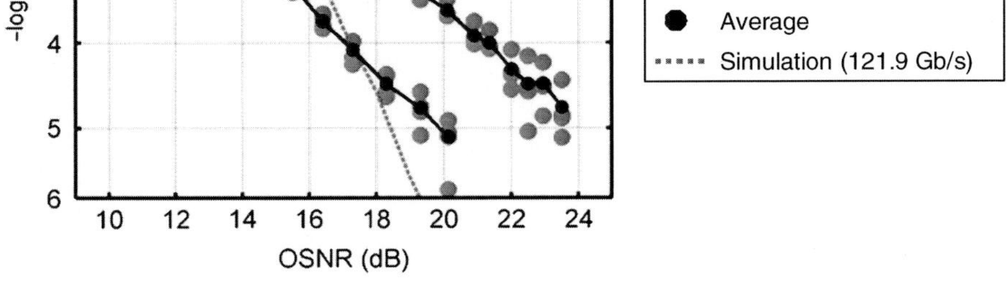

Fig. 8.8 Back-to-Back sensitivity

phase noise compensation scheme. We note that this penalty only applies for our emulated PDM setup and that in a convention PDM setup with a modulator per polarization such a penalty has not been observed [6]. Compared to the simulated sensitivity of 121.9 Gb/s, a penalty of about 2 dB at a BER of 1×10^{-3} is present. For lower BER values an increase in performance penalty between the simulated and experimental curve is observed. This penalty increase is most likely caused by the fact that at lower BER values the residual carrier frequency offset and phase noise after compensation start influencing the system performance [13]. Additionally, the performance of the fourth OFDM band is limited by nonlinearities of the amplifiers and frequency mixer. As a result, a BER floor is observed for this OFDM band, limiting the BER of the 121.9-Gb/s configuration to about 5×10^{-6}.

The fiber launch power optimization for channel 3 (located at 1554.5 nm) after 680-km transmission is shown in Fig. 8.9. The BER performance shows little

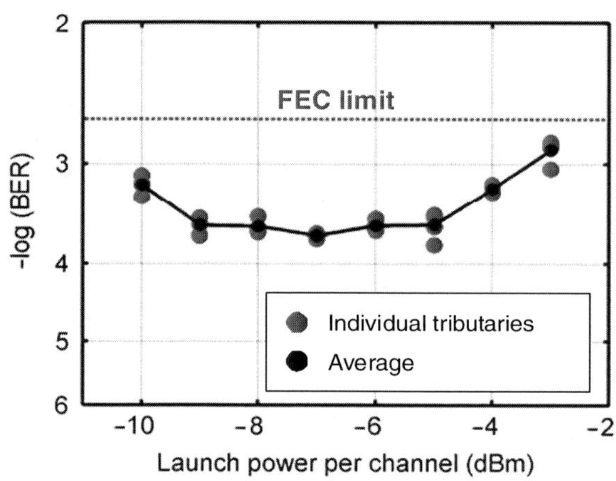

Fig. 8.9 Power excursion after 680-km transmission

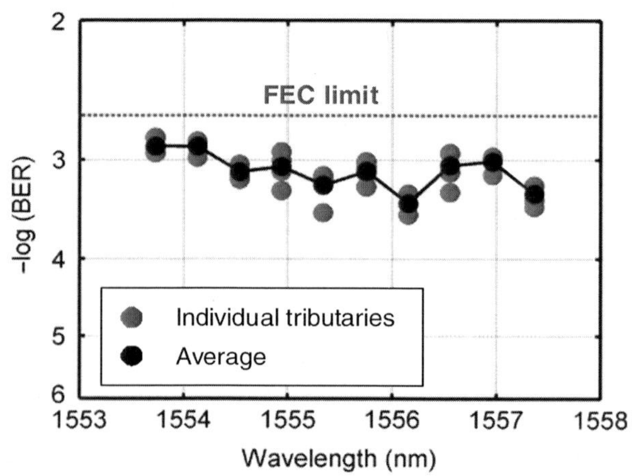

Fig. 8.10 Transmission performance after 1,000-km transmission

variation over a wide launch power range (from −9 to −5 dBm). The nonlinear tolerance could be increased by reducing the constellation size to QPSK, however, this would require a larger electrical bandwidth (which was not available for this experiment). Figure 8.10 shows the performance of the 10 WDM channels after 1,000-km transmission (with −7 dBm/channel launch power). The average OSNR after transmission is ∼21.5 dB and the average BER per channel varies between 3.9×10^{-4} and 1.3×10^{-3}. The channels on the red side of the spectrum slightly outperform the other channels as the noise figure of the EDFA/Raman amplifiers is slightly lower for these wavelengths, which results in a ∼0.4-dB higher OSNR. The obtained BER values are well below the threshold of a concatenated FEC code with 7% overhead (2.3×10^{-3}).

8.4 Conclusion

In this chapter, MIMO processing for polarization de-multiplexing is discussed for polarization division-multiplexed OFDM (PDM-OFDM) transmission systems. Using time-multiplexed training symbols and zero-forcing MIMO processing, the realization 122-Gb/s PDM-OFDM was reported. For MIMO processing itself a negligible OSNR penalty was observed. In a 50-GHz spaced WDM environment, successful transmission is demonstrated of 10 × 121.9 Gb/s over 1,000-km SSMF without any in-line chromatic dispersion compensation.

Acknowledgments The authors would like to thank Dr. T.C.W. Schenk from Philips, Dr. S. Akiba, Dr. M. Suzuki, and Dr. Tanaka from KDDI R&D Laboratories for their support. This work was partly supported by the National Institute of Information and Communications Technology of Japan.

References

1. A.H. Gnauck, G. Charlet, P. Tran, P.J. Winzer, C.R. Doerr, J.C. Centanni, E.C. Burrows, T. Kawanishi, T. Sakamoto, K. Higuma, 25.6-Tb/s C+L-band transmission of polarization-multiplexed RZ-DQPSK signals, in *Proceedings of the Optical Fiber Communications Conference PDP* 19, (2007).
2. H. Masuda, A. Sano, T. Kobayashi, E. Yoshida, Y. Miyamoto, Y. Hibino, K. Hagimoto, T. Yamada, T. Furuta, H. Fukuyama, 20.4-Tb/s (204 × 111 Gb/s) transmission over 240 km using bandwidth-maximized hybrid Raman/EDFAs. in *Proceedings of the Optical Fiber Communications Conference* PDP 20.
3. D.v.d. Borne, N.E. Hecker-Denschlag, G.D. Khoe, H.d. Waardt, PMD-induced transmission penalties in polarization-multiplexed transmission. J Lightwave Technol. **23**(12), 4004–4015 (2005).
4. Y. Han, G. Li, Coherent optical communication using polarization multiple-input-multiple-output. Opt. Express (13), 7527–7534 (2005).
5. C.R.S. Fludger, T. Duthel, D. van den Borne, C. Schulien, E-D. Schmidt, T. Wuth, J. Geyer, E. de Man, G.D. Khoe, H. de Waardt, 10 × 111 Gbit/s, 50 GHz spaced, POLMUX-RZ-DQPSK transmission over 2375 km employing coherent equalisation. J Lightwave Technol. **26**(1), 64–72 (2008).

6. S.L. Jansen, I. Morita, T.C.W. Schenk, H. Tanaka, Long-haul transmission of 16 × 52.5-Gb/s polarization division multiplexed OFDM enabled by MIMO processing. OSA J Opt Network **7**, 173–182 (2008).
7. B.J.C. Schmidt, A.J. Lowery, J. Armstrong, Experimental demonstrations of 20 Gbit/s direct-detection optical OFDM and 12 Gbit/s with a colorless transmitter. in *Proceedings of the Optical Fiber Communications Conference*, Paper PDP 18 (2007).
8. X. Yi, W. Shieh, Y. Ma, Phase noise on coherent optical OFDM systems with 16-QAM and 64-QAM beyond 10 Gb/s. in *Proceedings of the European Conference on Optical Communication*, Tu. 2.5.3 (2007).
9. S.L. Jansen, I. Morita, T.C.W. Schenk, H. Tanaka, 121.9-Gb/s PDM-OFDM Transmission with 2-b/s/Hz Spectral Efficiency over 1,000 km of SSMF. J. Lightwave Technol. **27**(3), 177–188, (2009).
10. T.C.W. Schenk, *RF imperfections in high-rate wireless systems*. (Springer, New York 2008).
11. S.L. Jansen, I. Morita, T.C.W. Schenk, N. Takeda, H. Tanaka, Coherent Optical 25.8-Gb/s OFDM Transmission over 4,160-km SSMF. J Lightwave Technol. **26**, 6–15 (2008).
12. T.M. Schmidl, D.C. Cox, Robust frequency and timing synchronization for OFDM. IEEE Trans. Commun. **45**(12), 1613–1621 (1997).
13. A. van Zelst, T.C.W. Schenk, Implementation of a MIMO OFDM-Based Wireless LAN System. IEEE Trans. Signal Proces. **52**(2), 483–494 (2004).
14. P.W. Wolniansky, G.J. Foschini, G.D. Golden, R.A. Valenzuela, V-BLAST: An architecture for realizing very high date rates over the rich-scattering wireless channel, in *Proceedings of URSI ISSSE*, pp 295–300, 1998.
15. R. van Nee, A. van Zelst, G. Awater, Maximum likelihood decoding in a space division multiplexing system. *Proc. of IEEE VTC* **1**, 6–10 (May 2000).
16. W. Shieh, Maximum-likelihood phase estimation for coherent optical OFDM. in *Proceedings of the European Conference on Optical Communications*, Tu 4.2.2 (2007).
17. I. Barhumi, G. Leus, M. Moonen, Optimal training design for MIMO OFDM systems in mobile wireless channels. Proc. Trans. Sign. Proc. **41**, 1615–1624 (2003).
18. T. Pollet, M. van Bladel, M. Moeneclaey, BER sensitivity of OFDM systems to carrier frequency offset and Wiener phase noise. IEEE Trans. Commun. **43**(2/3/4), 191–193, (1995).
19. A.J. Lowery, L.B. Du, J. Armstrong, Orthogonal frequency division multiplexing for adaptive dispersion compensation in long haul WDM systems, in *Proceedings of the Optical Fiber Communications Conference* 2006, PDP 39.

Chapter 9
No-Guard-Interval Coherent Optical OFDM with Frequency Domain Equalization

Yutaka Miyamoto and Yasuyuki Takatori

Abstract This chapter describes the novel no-guard-interval (GI) coherent optical orthogonal frequency division multiplexing (OFDM) format for high-capacity optical transport network (OTN). Unlike the conventional OFDM configuration, the proposed scheme employs a very small number of OFDM subcarriers, so simple optical analog subcarrier multiplexing can be realized without digital signal processing (DSP) in the OFDM modulation. The scheme also introduces simple OFDM demultiplexing and a digital adaptive time domain or frequency domain equalizer for DSP demodulation in the receiver, without recourse to OFDM overhead (OH) bytes such as GI and training symbols, unlike the conventional OFDM receiver. There is no line rate increase in the OTN channel due to the introduction of these OFDM OH bytes (i.e., the line rate of 111 Gb/s includes 7% OTN overhead, and 103-Gb/s payload is available for bit transparent OTN mapping of 100-GbE signals). The No-GI-OFDM proposal is experimentally tested at the channel rate of 111 Gbit/s as the first step to realizing DWDM long-haul transport at over 10 Tbps.

9.1 Introduction

Internet protocol (IP)-based data now dominates traffic in networks all over the world. In Japan, the broadband access service based on fiber to the home (FTTH) is rapidly growing; the number of FTTH subscribers in Japan exceeded 14 million in DEC 2008, triggering many kinds of latency-sensitive broadband applications based on IP. The continued evolution of the backbone network is required to support this rise in broadband services. The ultra-high-capacity optical transport network

Y. Miyamoto (✉)
NTT Network Innovation Laboratories, NTT Corporation, 1-1 Hikari-no-oka, Yokosuka, Kanagawa, 239-0847 Japan
e-mail: miyamoto.yutaka@lab.ntt.co.jp

M. Nakazawa et al. (eds.), *High Spectral Density Optical Communication Technologies*, Optical and Fiber Communications Reports 6, DOI 10.1007/978-3-642-10419-0_9,

(OTN) is the only network infrastructure that can satisfy such huge traffic demands while ensuring highly reliable operation, administration, and maintenance functionality [1, 2].

The next generation OTN must transport high-speed Ethernet as well as SDH/SONET clients transparently. Standardization on a much higher speed Ethernet interface, 100 GbE, will be finalized by 2010 by the IEEE 802.3 high-speed study group, and International Telecommunication Union – Telecommunication (ITU-T) has standardized the next generation OTN interface "OTU4" that can accommodate and transport 100-Gbps-class signals [3].

The chapter describes the technical challenges facing an advanced modulation format on digital signal processors (DSPs), given the goal of an over 10-Tbps-class OTN that supports 100-Gbps-class client signals. A novel no-guard-interval CO-OFDM format with frequency domain equalization is proposed and shown to offer the transmission performance desired.

9.2 High-Capacity Challenges and Modulation Format Alternatives

It is crucial to enhance the transmission capacity while keeping or enlarging the regenerative repeater spacing of today's OTN, since the available optical bandwidth and the allowable fiber launched power are limited [1]. The long-haul transmission of 100-Gbps channels demands enhanced robustness against chromatic dispersion (CD) and polarization-mode dispersion (PMD) as well as fiber nonlinearity. To mitigate these limitations, high-sensitivity and spectral-efficient coherent modulation formats aided by DSPs, namely digital coherent technologies, are being actively researched. Figure 9.1 shows the advantages of digital coherent technologies in the quest for high-capacity transport [4–17].

The highest reported capacity of 32 Tbps was achieved by using the polarization division multiplexed (PDM) 8-QAM format and the EDFA (C+L-band) at the channel rate of 114 Gbps over 580 km [15]; it is triple the capacity of incoherent OOK formats. The largest capacity distance product of 97 Pb/s km (13.5 Tbps × 7209 km) was realized using the PDM OFDM QPSK format and hybrid EDFA/Raman (extended L-band) amplification [16]; this product is double that possible with incoherent formats.

The choice of modulation format is a very important issue. Figure 9.2 shows the relationship between relative receiver sensitivity and attainable spectral efficiency in the presence of ASE noise. To increase the total system capacity, the high spectral efficiency offered by higher order multilevel modulation formats is very attractive, however, it is necessary to consider the sensitivity decrease implicit with the multilevel formats if there are more than eight levels. The approach of combining a QPSK-based format with other multiplexing schemes such as PDM, subcarrier multiplexing, and orthogonal frequency division multiplexing (OFDM) is another way to achieve high OSNR tolerance [2].

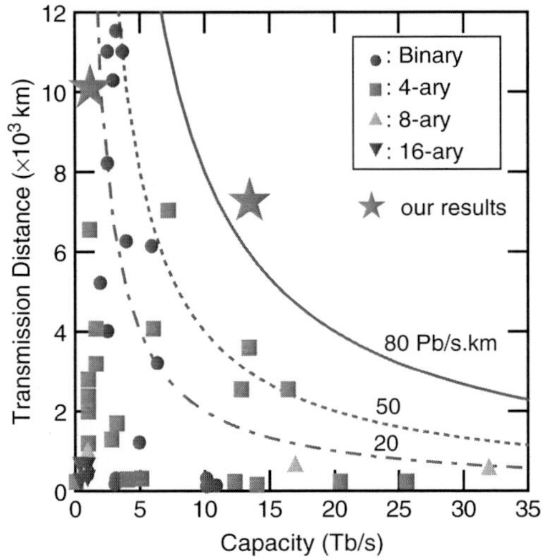

Fig. 9.1 High-capacity studies

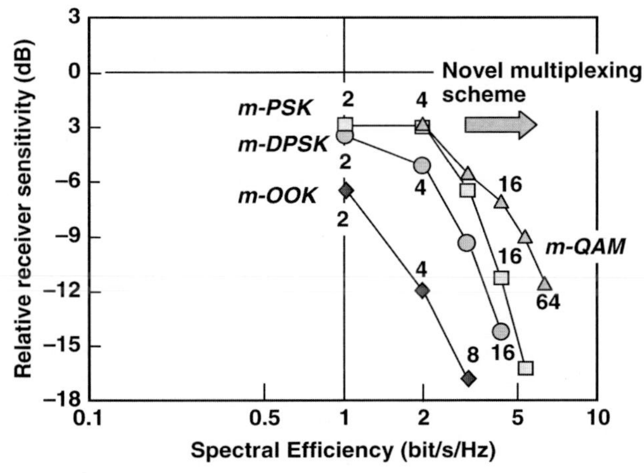

Fig. 9.2 Trade-off in multilevel modulation format performance

9.3 Concept of No-Guard-Interval OFDM

OFDM is very attractive for optical fiber communication because linear waveform distortion can be stably compensated in the frequency domain, and so is being actively studied [7–9, 12–14, 16, 18, 19]. OFDM is well established in radio transmission systems and is generally utilized to transmit large numbers of low-bit-rate orthogonal subcarriers, more than 50. To keep the orthogonal relationship between subcarrier signals, the subcarrier frequency spacing is set to be a multiple of the symbol duration in the frequency domain. Guard interval (GI) and training signal are necessary to adaptively compensate the linear inter-symbol interference due to CD and PMD after transmission. Therefore, some additional line rate increase due to GI is inevitable if CD compensation in excess of 10,000 ps/nm is needed. To handle such OFDM signals, the transmitter uses a DSP running inverse fast Fourier transform (IFFT) and digital-to-analogue converters (DACs), while the receiver uses analogue–to-digital converters (ADCs) and a DSP running FFT.

We proposed a novel OFDM transmission scheme that uses orthogonal frequency components with no GI (No-GI-OFDM: No-Guard-Interval OFDM) [2, 12–14, 16, 18, 19]. Instead of GI, the receiver implements adaptive equalization in either the time or frequency domain to adequately compensate the linear waveform distortion caused by CD and PMD. One feature of No-GI-OFDM is that the line rate is independent of the compensation capability. Any increase in the line rate increase is to be resisted given the speed limits of baseband electronics and the degradation possible in the transmission characteristic. Another benefit is that it uses a very small number of subcarrier signals, less than 10. This reduction in subcarrier number is effective in reducing the peak-to-average power ratio (PAPR), which is useful in suppressing nonlinear signal distortion in the transmission fiber. Yet another benefit is the transmitter configuration; compact modulators can be created using photonic integration and used for OFDM modulation instead of DSPs and DACs. Since optical fiber transmission performance is strongly dependent on the symbol rate due to several Kerr nonlinear effects, this configuration is beneficial in increasing the symbol rate while reducing the need for DSPs and DACs at the transmitter. Figure 9.3 shows the evolution in the basic concept for the two-subcarrier case.

As shown in Fig. 9.3a, the carrier suppressed return-to-zero (CS-RZ) format [20] utilizes just two orthogonal components with the same data signal and the same logic. The two orthogonal beating modes can be easily generated by a Mach–Zender (MZ) modulator (the two-mode generator). The modulator is biased at its null point and is driven by a sinusoidal clock at half the frequency of the symbol rate, R. The two orthogonal beating modes are then modulated with the same data signal by a data modulator. If the data modulation format is NRZ, the CS-RZ-OOK(on–off keying) format is obtained. When the date modulation format is duobinary, the duobinary CS-RZ (DCS-RZ) format, which is equivalent to RZ-AMI (RZ-alternate mark inversion) format, is obtained. [21]. The frequency spacing of the two orthogonal data modes in Fig. 9.3a equals the symbol rate (In the case of a binary format using CS-RZ, the symbol rate equals the line rate). The two orthogonal components carry the same data and are utilized to create a bandwidth-efficient RZ pulse format. Therefore, these RZ-OOK signals are detected by simple direct detection.

As shown in Fig. 9.3b, the continuous phase frequency shift keying (CPFSK) format [22] utilizes two orthogonal NRZ components with the same data signal but opposite logic. An optical mode splitter such as MZ interferometer (MZI) filter is placed between the two-mode generator and the mode-selecting data modulator (Mode SEL.) in order to realize the complementary data modulation of two orthogonal beating modes from the two-mode generator. The two orthogonal modes

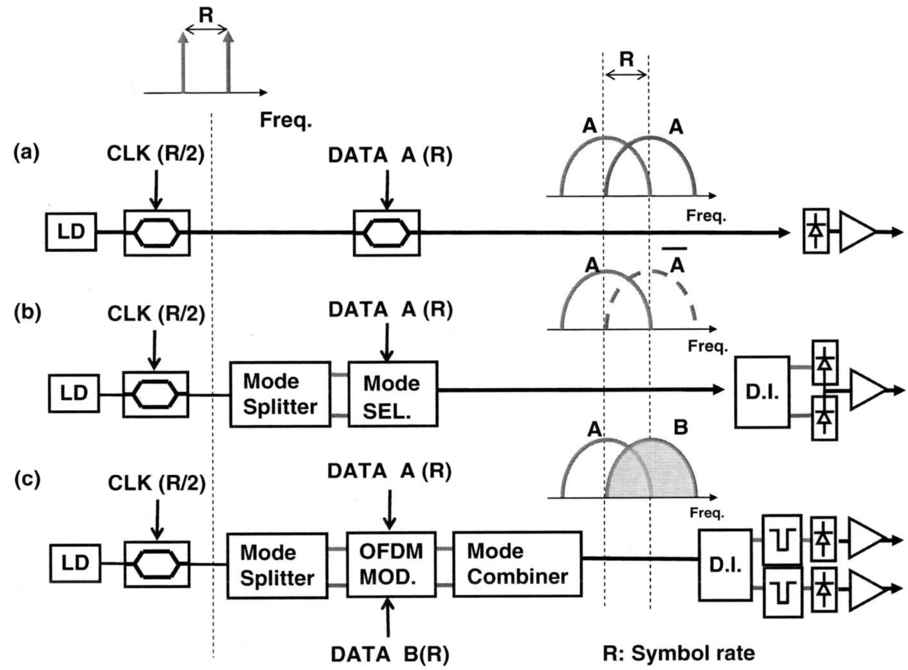

Fig. 9.3 Concept evolution for the two-mode case. (**a**) CS-RZ format. (**b**) CPFSK format. (**c**) No-Guard-Interval OFDM format

are selected according to data A to create the CPFSK data format. In ref. [22], the CPFSK modulator consists of a CS-RZ pulse carver and an asymmetric MZ modulator acting as both a mode-splitting MZI filter and a mode-selecting data modulator. The CPFSK format is a high-sensitivity data format with differential-balanced detection and, its constant amplitude is a favorable feature for suppressing nonlinear effects in ultra-long-haul transmission. The two orthogonal components carry the same data and are utilized to create the high-sensitivity bandwidth-efficient FSK format.

In the case of OFDM with two subcarriers, the OFDM format utilizes two orthogonal components with different data signals and the same symbol rate [2, 12–14, 16, 19] as shown in Fig. 9.3c. An optical mode splitter is inserted between the two-mode generator and the data modulator; the OFDM modulator modulates each orthogonal mode with different data streams A and B. The modulated orthogonal modes are then coherently coupled to generate a two-subcarrier OFDM signal by a mode combiner. To OFDM these two orthogonal subcarriers, the incoherent receiver uses a delay interferometer (D.I.) filter followed by optical sampling gates with the repetition frequency R for optical discrete Fourier transformation (DFT) processing [23]. In ref. [24], the DQPSK format was employed and two 50-Gbit/s DQPSK orthogonal signals were multiplexed to form a 100-Gbit/s two-subcarrier No-GI-OFDM-DQPSK signal. In digital coherent systems, the DFT processing needed for OFDM can be realized using electrical DSPs, which is very attractive for realizing simple receivers that offer high sensitivity as described in the next section.

9.4 PDM No-Guard-Interval CO-OFDM Transmitter and Receiver Configuration

Figure 9.4 shows the proposed configuration of the transmitter (a), the receiver (b), and the DSP block diagram (c) for the No-Guard-Interval Coherent OFDM format with two subcarriers [14]. PDM is applied in order to double the spectral efficiency, over 2 bit/s/Hz, and to reduce the operation speeds of the ADC and DSP at the receiver. In order to realize stable OFDM signal generation, we employ the hybrid integration technology of silica-based planar lightwave circuit (PLC) and $LiNbO_3$ (LN) lightwave circuit [25]. The subcarrier splitter/combiner and data modulators for each subcarrier can be integrated into one module. We have designed and tested two-subcarrier QPSK modulators for PDM 111-Gb/s modulation (13.9 Gbaud).

At the receiver side, a DSP-based polarization-diversity intradyne coherent receiver, which is commonly used in single-carrier coherent reception, is used to detect both in-phase and quadrature components of the received signal for each polarization (Fig. 9.4b). The frequency of the local oscillator (LO) is tuned to the

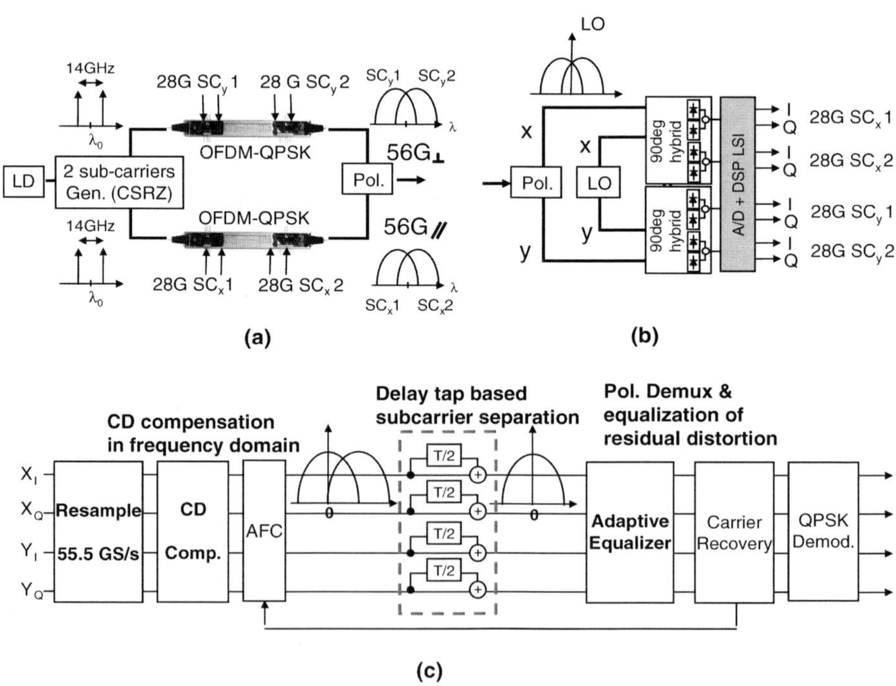

Fig. 9.4 Transmitter and receiver configuration of No-Guard-Interval OFDM

center of the received OFDM signal to reduce the required bandwidth of the electronic devices.

A detailed block diagram of the DSP is shown in Fig. 9.4c, which is similar to that for the single-carrier intradyne receiver except for the subcarrier separation part. First, CD is compensated by fixed-tap linear filters, since frequency domain equalization greatly reduces the DSP complexity [26–28]. Each subcarrier is then demultiplexed as follows: after shifting each subcarrier frequency to zero, the data sequence of each subcarrier, d_n, is obtained by DFT:

$$d_n = \frac{1}{N}\sum_{k=0}^{N-1} S\left(\frac{kT}{N}\right)\exp\left(j\frac{2\pi}{N}nk\right), \tag{9.1}$$

where N is the number of subcarriers, $S(t)$ the received signal, and T the symbol interval. In the two-subcarrier case, this is simply realized by a delay tap with a half symbol interval. Polarization demultiplexing and signal equalization are then performed by adaptive filters consisting of four complex-valued finite impulse response (FIR) filters arranged in a butterfly structure. In our transmission experiments, we used three-symbol length FIR filters optimized by the constant modulus algorithm (CMA) and least mean square (LMS) algorithm. Carrier recovery is performed by using the "Viterbi-and-Viterbi" carrier phase estimation (CPE) method, and finally the binary data are recovered by demodulating the equalized signal.

Figure 9.5 shows a complexity comparison of the time domain equalizer (TDE) and the frequency domain equalizer (FDE) without GI in the receiver [26, 28]. Since the total CD value exceeds 10,000 ps/nm (i.e., this requires more than 100 taps), FDE has a clear advantage over TDE in reducing DSP complexity.

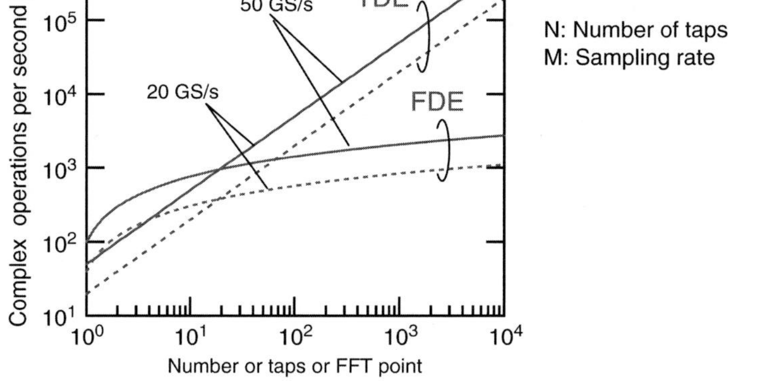

Fig. 9.5 Complexity comparison of equalizer configuration (TDE vs. FDE)

9.5 111 Gbps No-Guard-Interval OFDM Transmitter and Receiver Performance

The required bandwidth of the OFDM signal is theoretically the same as that of a single-carrier signal. Since the transmitter of the typical optical transmission system does not use a steep digital rolloff filter, the proposed No-GI-OFDM with only two subcarriers offers a spectrally efficient signal compared to the single-carrier signal as shown in Fig. 9.6 [13, 14]. The bandwidth of 42 GHz is narrow enough for 50-GHz spacing WDM.

The back-to-back OSNR tolerance of 111-Gb/s two-subcarrier CO-OFDM signals was verified [13] as shown in Fig. 9.7a. The OSNR tolerance of the 111-Gb/s single-polarization CS-RZ-DQPSK format was compared to that of differential direct detection (DD). CO-OFDM exhibits about 3 dB better OSNR tolerance thanks to its coherent detection, and the required OSNR at the bit error rate (BER) of 1×10^{-3} ($Q = 9.8$ dB) was 15.8 dB, which almost matches that of single-carrier RZ-DQPSK with a digital coherent receiver [11].

Figure 9.7b shows the tolerance of 111-Gb/s two-subcarrier CO-OFDM (circles) to differential group-delay (DGD) [14]. In this experiment, 12-tap (3-symbol interval) adaptive FIR filters were used to equalize the PMD-induced waveform distortion. The measured DGD tolerance of the single-polarization 111-Gb/s CS-RZ-DQPSK format is also shown. The allowable DGD of No-GI CO-OFDM, defined as that yielding 1-dB Q-penalty, was about 70 ps, which corresponds to about 1 symbol interval. This allowable DGD is about 8 times larger than that permitted by the 111-Gb/s CS-RZ-DQPSK format, and about twice that possible with 10-Gb/s NRZ-OOK.

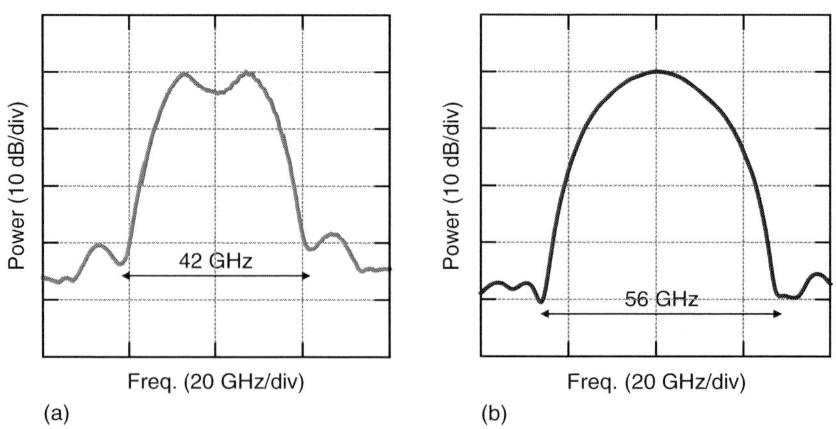

Fig. 9.6 111 Gbit/s Optical spectra. (**a**) No-GI CO-OFDM. (**b**) single-carrier DP-QPSK

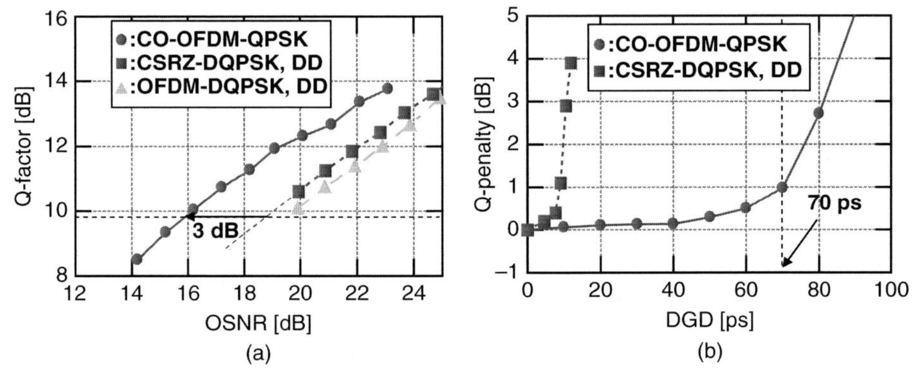

Fig. 9.7 Transmitter and receiver performance of 111 Gbps PDM No-GI-OFDM QPSK format. (**a**) OSNR tolerance. (**b**) PMD tolerance

9.6 13.5-Tbps WDM Transmission Using 111-Gbps PDM No-Guard-Interval OFDM QPSK Format

We have tested the ultra-long-haul WDM transmission performance of the PDM No-Guard-Interval OFDM QPSK format [16]. At the transmitter, 135-CW optical carriers (1563.05–1619.62 nm) for 135-channel transmission with 50-GHz spacing were generated by laser diode light sources.

The transmission line consisted of a 240.3-km re-circulating loop. The loop consisted of three 80.1-km spans of low-loss and low-nonlinearity pure silica core fiber (PSCF) [29], three inline P-EDFAs and gain equalizers (GEQs) [30], two P-EDFAs for loop-loss compensation, and a loop-synchronous polarization scrambler (LSPS). The loss coefficient of the PSCF and the loss of the 80.1-km span were 0.160 dB/km and 13.5 dB at 1570 nm, respectively, and the effective area was 110 μm^2 (average value over the three spans). The CD ranged from 21.1 to 24.3 ps/nm/km for the wavelengths from 1563 to 1620 nm; PMD was 0.078 ps/$\sqrt{}$km. A backward-pumped DRA with a second-order pumping scheme [16] in the extended L-band, 1563 to 1620 nm, was used to improve the OSNRs of the transmitted signals. The signal power launched into the PSCF was about −4 dBm/ch for the 135-channel transmission.

Figure 9.8a and b shows the measured Q-factors and optical output WDM spectra of a 13.5 Tbit/s–7,209 km experiment and a 1 Tbit/s–10,093 km experiment, respectively. The Q-factors of all channels were better than the Q-limit of 9.1 dB (dashed lines in Figs. 9.8a and b) which yields BERs below 1×10^{-12} with the use of ITU-T G.975.1 enhanced FEC. The total CD compensated in the receiver is as high as 239,000 ps/nm after 10,093-km transmission. The Q-margin will be improved to 0.7 dB with the use of recent commercial 10-Gb/s FEC techniques with 7% overhead [17]. A recent novel nonlinear compensation scheme has the potential to further extend the transmission distance to over 10,000 km [31].

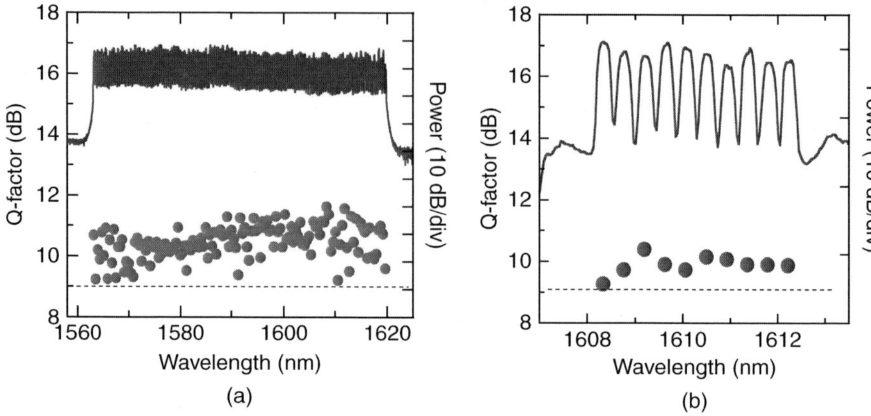

Fig. 9.8 111-Gbps No-GI-C0-OFDM channel long-haul WDM transmission performance. (**a**) 13.5 Tbps – 7209-km transmission. (**b**) 1 Tbps – 10,093-km transmission

9.7 Conclusions

The recent progress in high-capacity optical transport based on DSP-aided coherent modulation formats was reviewed. We proposed the novel spectrally efficient No-Guard-Interval CO-OFDM format for high-capacity OTN. It offers greatly enhanced PMD tolerance and CD tolerance at the channel rate of 111 Gbit/s thanks to novel OFDM modulation/demodulation schemes and frequency domain equalization in the receiver with very low DSP requirements. We demonstrated over 10 Tbps capacity at over 7,209 km with spectral efficiencies higher than 2 bit/s/Hz using the No-GI-Interval CO-OFDM format.

Acknowledgments The authors thank Kazuo Hagimoto, Yoshinori Hibino, Shinji Matsuoka, Kazuyasu Okada and Masato Mizoguchi for their continuous encouragement. The authors also thank Eiji Yoshida, Akihide Sano, Hiroji Masuda, Etsushi Yamazaki, Takauki Kobayashi, Eiichi Yamada, Riichi Kudo, Koichi Ishihara, Masato Tomizawa, Shigeki Aisawa, and Takuya Ohara for fruitful discussions and materials for this chapter. The authors thank Akimasa Kaneko, Shinji Mino, and Takashi Yamada for the PLC-LN modulator fabrication. This work is supported in part by the National Institute of Information and Communications Technology (NICT) of Japan under "Universal Link Project."

References

1. K. Hagimoto, Core Network. NTT R&D Review 3, (2005).
2. Y. Miyamoto, A. Sano, H. Masuda, E. Yoshida, S. Aisawa, Enhancing the capacity beyond terabit per second for transparent optical transport network. Proc. ECOC, 10.5.1 (2007).
3. T. Ohara, O. Ishida, Standardization activities for optical transport network. NTT Tech. Rev. **7**(3), (2009).

4. R. Noe, Phase noise-tolerant synchronous QPSK/BPSK baseband-type intradyne receiver concept with feed forward carrier recovery. J. Lightwave Technol. **23**(2), 802–807 (2005).
5. S. Tsukamoto, D.-S. Ly-Gagnon, K. Katoh, K. Kikuchi, Coherent demodulation of 40-Gbit/s polarization-multiplexed QPSK signal with 16-GHz spacing after 200-km transmission. Proc OFC/NFOEC2005, PDP29, (2005).
6. S.J. Savory, A.D. Stewart, S. Wood, G. Gavioli, M.G. Taylor, R.i. Killey, P. Bayvel, Digital equalisation of 40 Gbit/s per wavelength transmission over 2480 km of standard fibre without optical dispersion compensation. Proc. of ECOC2006, Th2.5.5 (2006).
7. A.J Lowery, L. Du, J. Armstrong, Orthogonal frequency division multiplexing for adaptive dispersion compensation in long haul WDM systems. OFC2006, PDP39 (2006).
8. W. Shieh, C. Athaudage, Coherent optical orthogonal frequency division multiplexing. Electron. Lett. **42**, 587–589, May 11 (2006).
9. S.L. Jansen, I. Morita, N. Takeda, H. Tanaka, 20-Gb/s OFDM Transmission over 4,160-km SSMF enabled by RF-pilot tone phase noise compensation. Proc.OFC/NFOEC2007, PDP15 (2007).
10. C. Laperle, B Villeneuve, Z. Zhang, D. McGhan, H. Sun, M. O'Sullivan, Wavelength division multiplexing (WDM) and polarization mode dispersion (PMD) performance of a coherent 40 Gbit/s dual-polarization quadrature phase shift keying (DP-QPSK) transceiver. OFC/NFOEC2007, PDP16 (2007).
11. C.R.S. Fludger, T. Duthel, D. van den Borne, C. Schmidt, T. Wuth, E. de Man, G. D. Khoe, H. de Waardt, 10 × 111 Gbit/s, 50 GHz spaced, POLMUX-RZ-DQPSK transmission over 2375 km employing coherent equalisation. Proc. OFC/NFOEC2007 PDP22 (2007).
12. E. Yamada, A. Sano, H. Masuda, T. Kobyashi, E. Yoshida, Y. Miyamoto, Y. Hibino, K. Ishida, Y. Takatori, K. Okada, K. Hagimoto, T. Yamada, H. Yamazaki, Novel no-guard-interval PDM CO-OFDM transmission in 4.1 Tb/s(50 × 88.8-Gb/s) DWDM Link over 800 km SMF including 50-GHz spaced ROADM Nodes. In Proceedings of the Optical Fiber Communication Conf./National Fiber Optic Engineers' Conf. OFC/NFOEC2008, PDP8 (2008).
13. E. Yamada, E. Yamada, A. Sano, H. Masuda, E. Yamazaki, T. Kobayashi, E. Yoshida, K. Yonenaga, Y. Miyamoto, K. Ishihara, Y. Takatori, T. Yamada and H. Yamazaki, "1 Tbit/s (111 Gbit/s/ch × 10 ch no-guard-interval CO-OFDM transmission over 2100 km DSF. Electron. Lett. **44**(24) 1417–1418 (2008).
14. A. Sano, E. Yamada H. Masuda, E.Yamazaki, T. Kobayashi, E. Yoshida, Y. Miyamoto, S. Matsuoka, R. Kudo, K.Ishihara, Y. Takatori, M. Misoguchi, K. Okada, K. Hagimoto, H. Yamazaki, S.Kamei, H. Ishii, 13. 4-Tb/s (134 × 111-Gb/s/ch) No-guard-interval coherent OFDM transmission over 3600 km of SMF with 19-ps average PMD. Proc. European Conf. on Optical Communications, ECOC2008, Th3E1 (2008).
15. X. Zhou, J. Yu, M.-F, Huang, Y. Shao, T. Wang, P. Magill, M. Cvijetic, L. Nelson, M. Birk, G. Zhang, S.Y. Ten, H. B. Matthew, S. K. Mishra, 32 Tb/s(320 × 114 Gbps) PDM-RZ-8QAM transmission over 580 km of SMF-28 ultra-low-loss fiber. OFC/NFOEC2009, PDPB4 (2009).
16. H. Masuda, E. Yamazaki, A. Sano, T. Yoshimatsu, T. Kobayashi, E. Yoshida, Y. Miyamoto, S. Matsuoka, Y. Takatori, M. Mizoguchi, K. Okada, K. Hagimoto, T. Yamada, S. Kamei, 13.5-Tb/s (135 × 111-Gb/s/ch) No-guard-interval coherent OFDM transmission over 6,248 km using SNR maximized second-order DRA in extended L-band. Proc. Optical Fiber Communication Conf./National Fiber Optic Engineer's Conf. OFC/NFOEC2009, PDPB5, *(2009)*.
17. G. Charlet, M. Saisi, P. Tran, M Bertolini, H. Mardoyan, J. Renaudier, O.B. Pardo, S. Bigo, 72 × 100 Gb/s Transmission over transoceanic distance, using large effective area fiber, hybrid Raman-Eribium amplification and coherent detection. OFC/NFOEC2009, PDPB6, (2009).
18. T. Kobayashi, A. Sano, E. Yamada, Y. Miyamoto, H. Takara, A. Takada, Electro-optically subcarrier multiplexed 110 Gb/s OFDM signal transmission over 80 km SMF without dispersion compensation. Electron. Lett. **44**(3) 225–226 (2008).
19. A. Sano, H. Masuda, E. Yoshida, T. Kobayashi, E. Yamada, Y. Miyamoto, F. Inuzuka, Y. Hibino, Y. Takatori, K. Hagimoto, T. Yamada, Y. Sakamaki, 30 × 100-Gb/s all-optical OFDM transmission over 1300 km SMF with 10 ROADM nodes. Proc. European Conf. on Optical Communications ECOC2007, PD1.7 (2007).

20. Y. Miyamoto, A. Hirano, K. Yonenaga, A. Sano, H. Toba, K. Murata, O. Mitomi, 320-Gbit/s (8 × 40 Gbit/s) WDM transmission over 367-km with 120-km repeater spacing using carrier-suppressed return-to-zero format. Electron. Lett. **35**(23) 2041–2042, November (1999).
21. Y. Miyamoto, K. Yonenaga, A. Hirano, H. Toba, K. Murata, H. Miyazawa, Duobinary carrier-suppressed return-to-zero format and its application to 100 GHz-spaced 8 × 43-Gbit/s DWDM unrepeated transmission over 163 km. Electron. Lett. **37**(23), 1395–1396, (2001).
22. Y. Miyamoto, S. Kuwahara, T. Yamada, S. Suzuki, High-speed CPFSK WDM signal transmission using PLC-LN hybrid asymmetric MZ modulator. Proc. OFC2005 OTuL2, Anahiem (2005).
23. H. Sanjoh, E. Yamada, Y. Yoshikuni, Optical orthogonal frequency division multiplexing using frequency/time domain filtering for high spectral efficiency up to 1 bit/s/Hz', Paper ThD1, OFC2002.
24. A. Sano, H. Masuda, E. Yoshida, T. Kobayashi, E. Yamada, Y. Miyamoto, F. Inuzuka, Y. Hibino, Y. Takatori, K. Hagimoto, T. Yamada, Y. Sakamaki, 30 × 100-Gb/s all-optical OFDM transmission over 1300 km SMF with 10 ROADM nodes. Proc. European Conf. on Optical Communications (ECOC2007), PD1.7 (2007).
25. T. Yamada, Y. Sakamaki, T. Shibata A. Kaneko, A. Sano, Y. Miyamoto, Compact 111-Gbit/s Integrated RZDQPSK modulator using hybrid assembly technique with silica-based PLCs and LiNbO3 Devices. OFC2008, OThC3 (2008).
26. K. Ishihara, T. Kobayashi, R. Kudo, Y. Takatori, A. Sano, E. Yamada, H. Masuda, M. Matsui, M. Mizoguchi, Y. Miyamoto, Frequency-domain equalisation without guard interval for optical transmission systems, Electron. Lett. **44**(25), 1480–1482, December (2008).
27. B. Spinnler, F.N. Hauska, M Kuschnerov, Adaptive equalizer complexity in coherent optical receivers. ECOC2008, We.2.E.4 (2008).
28. R. Kudo, T. Kobayashi, K. Ishihara, Y. Takatori, A. Sano, E. Yamada, H. Masuda, Y. Miyamoto, PMD compensation in optical coherent single carrier transmission using frequency-domain equalization. Electron. Lett. **45**(2), 124–125 (2009).
29. K. Nagayama et al., Ultra-low-loss (0.1484 dB/km) pure silica core fibre and extension of transmission distance. Electron. Lett, **38**(20), 1168–1169 (2002).
30. H. Masuda, Y. Miyamoto, Low-noise extended L-band phosphorus co-doped silicate EDFA consisting of novel two-stage gain-flattened gain blocks. Electron. Lett. **44**(8), 1082–1083 (2008).
31. E. Yamazaki, H. Masuda, A. Sano, T. Yoshimatsu,T. Kobayashi, E. Yoshida, Y. Miyamoto, R. Kudo, K. Ishihara, M. Matsui, Y. Takatori, Multi-staged nonlinear compensation in coherent receiver for 12 015 km WDM transmission of 10-ch x 111 Gbit/s no-guard-interval co-OFDM. Electron. Lett. **45**(13), 695–697 (2009).

Chapter 10
QPSK-Based Transmission System: Trade-Offs Between Linear and Nonlinear Impairments

Takeshi Hoshida and Jens C. Rasmussen

Abstract QPSK-based modulation formats are promising for spectrally efficient DWDM long-haul and metro systems because of their higher tolerance to linear and nonlinear impairments. Several variants of QPSK-based format implementations including non-coherent RZ-DQPSK, coherent SP-QPSK, and coherent DP-QPSK will be introduced and discussed from a viewpoint of trade-off between nonlinear and linear tolerances.

10.1 Introduction

Most of the contemporary long-haul and metro photonic networks based on reconfigurable optical add-drop multiplexing (ROADM) and/or wavelength cross-connect (WXC) are deployed with 100-GHz channel spacing and with 10-Gb/s channels. The emerging traffic demand necessitates the capacity upgrading of such networks by means of higher spectral density, i.e., denser channel spacing, e.g., 50-GHz, and increased channel capacity, e.g., 40 Gb/s, 100 Gb/s and beyond. It is at the same time desired to maintain the same transmission reach as the current networks while achieving such a high spectral density.

Spectrally efficient modulation formats, in particular the ones based on quadrature phase shift keying (QPSK), are among the most promising enablers of such an ultimate goal. The reason why QPSK is promising in particular can be understood from the theoretical trade-off relationship between spectral efficiency (symbol rate) and required signal-to-noise ratio (SNR) to achieve a certain bit error ratio (Fig. 10.1). While the theoretical difference between the binary PSK (BPSK) and the QPSK in required SNR is minimum, the spectral efficiency increase beyond 2 bit/symbol starts to impair the required SNR and thus the transmission reach rather rapidly.

T. Hoshida and J.C. Rasmussen (✉)
Fujitsu Laboratories Limited, 1-1 Kamikodanaka 4-Chome, Nakahara-ku, Kawasaki 211-8588
e-mail: hoshida@jp.fujitsu.com

M. Nakazawa et al. (eds.), *High Spectral Density Optical Communication Technologies*, Optical and Fiber Communications Reports 6, DOI 10.1007/978-3-642-10419-0_10,

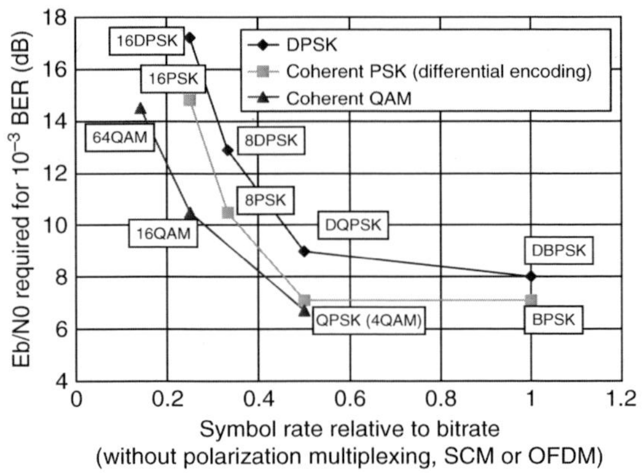

Fig. 10.1 Comparison of multi-level modulation formats

10.2 Options of QPSK-Based Transceiver Implementation

There are many variations of transceiver implementation employing QPSK-based modulation formats as shown in Fig. 10.2. Additional multiplexing schemes such as polarization multiplexing and/or multiple subcarriers, i.e., subcarrier multiplexing (SCM) or orthogonal frequency division multiplexing (OFDM), can be employed to achieve even higher spectral efficiencies, without sacrificing the required optical SNR (OSNR) performance in principle. Optional RZ pulse carving can be used in the transmitter. Receiver can be configured either with direct detection or with coherent detection to enable further optimization depending on the target applications.

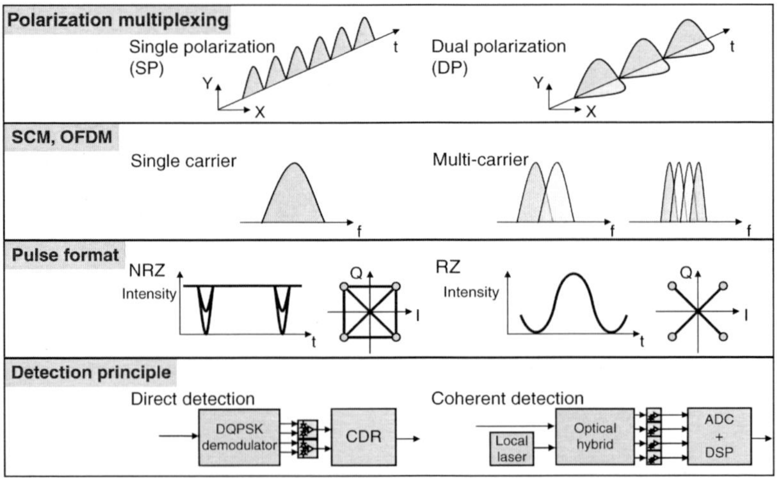

Fig. 10.2 Options of QPSK-based modulation formats

Table 10.1 Variants of QPSK-based modulation format implementation format and their relevance to transmission performance

		Relevant to						
				Linear distortions			Nonlinear distortions	
	Options	OSNR tolerance	Symbol rate	CD	PMD	PDL	Intra-channel	Cross-channel
Polarization multiplexing	SP or DP	Primarily no	Yes	Yes	Yes	Yes	Yes	Yes
SCM OFDM	Single carrier or multicarrier	Primarily no	Yes	Yes	Yes	Primarily no	Yes	Yes
Pulse format	NRZ or RZ	Yes	No	A little	A little	Primarily no	Yes	A little
Detection principle	Coherent or direct	Yes	No	Very much	Very much	Yes	Yes	Yes

Table 10.1 summarizes how those implementation options might affect different aspects of transmission performance. Reduction of symbol rate brought by the use of dual-polarization (DP) instead of single polarization (SP) and/or by the use of multiple subcarriers influences every aspect related to waveform distortion, i.e., linear and nonlinear distortions. It should be noted here that the use of DP also affects the tolerance to polarization mode dispersion (PDL) as it will be discussed later. Additional RZ pulse carving to the QPSK signal can improve the OSNR tolerance when the receiver electrical filter is not optimal and intra-channel nonlinear distortion. At the same time it also impacts chromatic dispersion (CD) tolerance, polarization mode dispersion (PMD) tolerance, and cross-channel nonlinear effects such as cross-phase modulation (XPM). The choice of detection principle strongly impacts every aspects of the performance.

In the following discussion some specific investigations on the above aspects are introduced; Sect. 10.3 discusses the PDL on DP signals; Sect. 10.4 discusses how differently SP and DP signals are affected by the cross-phase modulation induced by the co-propagating 10-Gbit/s on–off keying (OOK) signals; Sect. 10.5 compares nonlinear distortion tolerances between direct and coherent detection receivers.

10.3 DP Signal Impairment Due to PDL

PDL can be a serious limiting factor of transmission reach in particular when the number of concatenated in-line node equipment, which is the major source of PDL, is large. Some example is described in ITU-T G.680 Appendix I, where maximum PDL value as high as 9.2 dB with the outage probability of 4.2×10^{-5} is referred to [1]. In this section, PDL-induced penalties in 43-Gb/s DP signal transmission are discussed, focusing on the problems specific to DP signals.

Figure 10.3 shows two modes of DP signal degradation due to PDL: (a) power inequality and (b) polarization mixing between two polarization tributaries. In (a) the first mode, PDL causes a significant SNR degradation due to variation of optical power levels between orthogonal polarization components. The worst-case scenario occurs when the two polarization tributaries are aligned with lower-loss and higher-loss axes of the PDL, respectively. This SNR degradation cannot be mitigated by any possible receiver implementations. Moreover, PDL-induced unequal signal power levels at the receiver side may cause even further BER degradation due to the influences of electrical noise. The second degradation mode, polarization mixing, becomes the most evident when PDL is aligned at 45° with respect to the two polarization axes, which causes overlapping of two polarization components. Some types of receivers are known to mitigate polarization mixing to some extent when it is not very significant [2].

In order to discuss SNR degradation of DP signals two models are considered: with lumped PDL and with distributed PDL (Fig. 10.4). In distributed PDL model, concatenation of 25 PDL elements with equal losses followed by amplifiers that

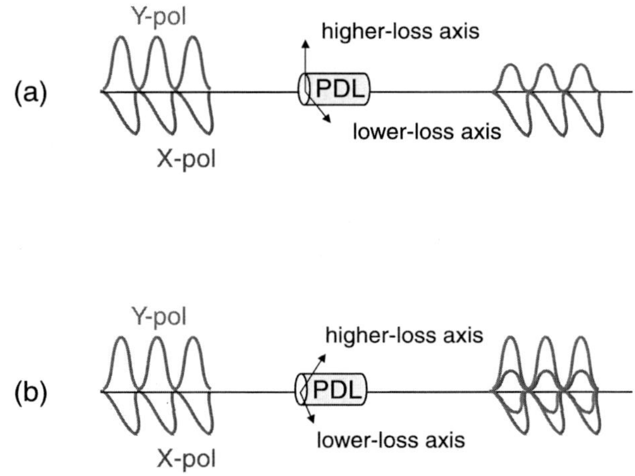

Fig. 10.3 DP signal degradation due to PDL (**a**) power inequality between polarization tributaries and (**b**) polarization mixing

generate ASE was assumed. Here, the PDL value of each concatenated element (PDLi) was varied from 0.02 to 0.24 dB and their axes were aligned with each other to provide a total PDL $= 25 \times$ PDLi, ranging from 0.5 to 6 dB, and the worst-case condition was assumed where one polarization tributary (Y) is aligned with higher-loss axis of PDL and the other polarization tributary (X) is aligned with the lower-loss axis. Q-penalty was calculated from the received BER values of X- and Y-polarization tributaries (Fig. 10.4). The effective Q-factors were calculated based on the average of BER values for both polarization components. In this case, Y-polarization channel suffers from BER degradation while X-polarization channel enjoys slight BER improvement due to signal power increase. As a whole, effective Q-factor suffers large degradation due to PDL. Even though the calculation was performed on a DP-RZ-DQPSK system, similar performance should be observed for

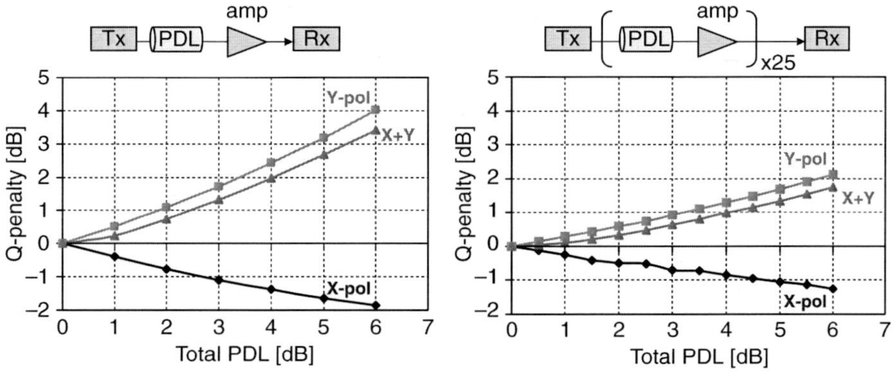

Fig. 10.4 Simulated PDL penalty on DP signals at OSNR of 16 dB (**a**) Lumped PDL model and (**b**) distributed PDL model

any DP systems. It should be noted here that the difference in lumped and distributed ASE addition is large. Lumped model (Fig. 10.4b) may be more useful in specifying and testing the PDL tolerance of transceiver although such a model hardly represents real-life transmission link situations. The distributed model (Fig. 10.4b) may seem to represent actual system situations better than the lumped model, but it is not generally accurate because PDL penalty depends on the specific link parameters such as loss variation among the spans along a link and thus can deviate from either case.

From the above discussions it can be understood that DP system design, compared to the SP system design, will require more careful treatment of PDL in the link budgeting.

10.4 Impact of Cross-Phase Modulation on SP- and DP-RZ-DQPSK Signals

One concern on QPSK-based system is the impact of cross-phase modulation (XPM). The phase deviation induced by XPM gives a direct impairment to the PSK signals, which is different from the situation with OOK systems. Moreover, phase separation between symbols that are smaller in QPSK than in BPSK should result in higher susceptibility to XPM.

XPM impact is in particular significant when conventional OOK signals, e.g., 10-Gb/s NRZ channels, are co-propagating in the same fiber, when the fiber dispersion coefficient is not large enough to induce sufficient walk-off between wavelength channels [3]. When discussing the impact on DP signals, two mechanisms on how XPM could impair the signal should be distinguished: nonlinear phase noise and nonlinear polarization crosstalk (Fig. 10.5). The former effect, existing also with the SP systems, has been studied in the previous works [4]. On the other hand, the latter effect that is unique to the DP case is based on the fast random polarization modulation induced by XPM.

In order to compare the significance of the former and of the latter, a numerical simulation was performed with 450-km transmission of 43-Gb/s SP- and DP-RZ-DQPSK signals with neighboring 11.1-Gb/s signals having 50-GHz channel spacing. The relative polarization angle between the 43-Gb/s SP- or DP-RZ-DQPSK and 11.1-Gb/s NRZ channels were changed from 0° to 90° and other simulation parameters were as follows: fiber chromatic dispersion coefficient of 3.83 ps/nm/km, fiber dispersion slope of 0.084 ps/nm^2/km, link distance of 75 km$\times$6 span, fiber input power of -1 dBm/ch, and channel spacing was 50 GHz. As it can be seen in Fig. 10.6, the Q-penalty for DP is larger than that for SP case consistently, and the penalty for DP-RZ-DQPSK becomes worst when the 11.1-Gb/s NRZ neighbors were aligned with one of the two polarization tributaries. This result suggests that the XPM-induced polarization crosstalk is not as significant as the XPM-induced phase crosstalk at least under the specific system condition assumed.

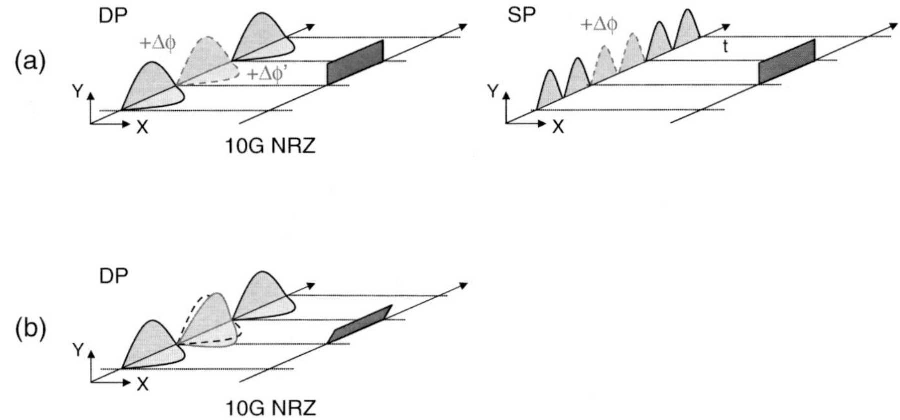

Fig. 10.5 XPM-induced penalty under 10-Gb/s NRZ co-propagation (**a**) XPM-induced phase noise and (**b**) XPM-induced polarization crosstalk

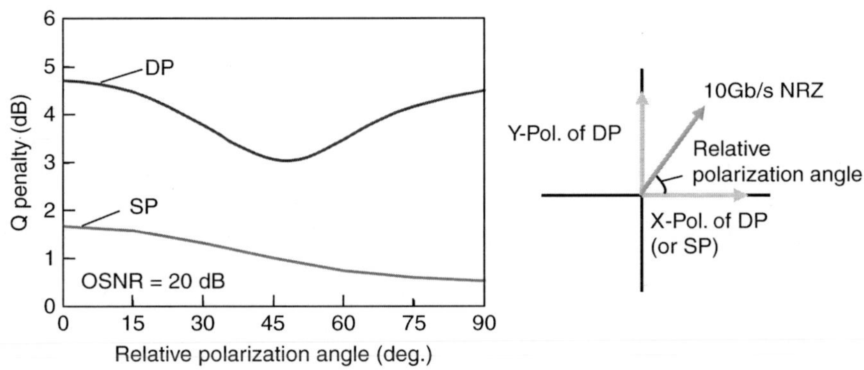

Fig. 10.6 Simulated XPM-induced penalty under 10-Gb/s NRZ co-propagation

In the following, experimental comparison between SP- and DP-RZ-DQPSK employing direct-detection receivers is reported, focusing on XPM-induced impairment in a transmission over NZ-DSF with co-propagating 11.1-Gb/s NRZ channels (Fig. 10.7). Eighty DFB laser diodes were prepared as the laser sources in C-band (1530.33–1561.83 nm), with 50-GHz channel spacing on the ITU-T grid. The 5 of the 80 at 1531.90, 1538.19, 1546.12, 1554.13, and 1558.98 nm were modulated in RZ-DQPSK format at a symbol rate of either 21.5 Gbaud (for SP-RZ-DQPSK) or 10.7 Gbaud (for DP-RZ-DQPSK). The bit sequence was arranged to be a pseudo random bit sequence (PRBS) 15 after decoding in the receiver and was used to drive the I- and Q- branches of a DQPSK modulator. RZ pulse carver made these signals into RZ pulse train having 50% duty. The inter-pulse delay between polarization tributaries within one-bit period was half of a symbol period (~46 ps) by default in order to minimize the impact of intra-channel nonlinear penalty [5]. The transmitted waveforms are shown in Fig. 10.7 for SP and DP (time-interleaved), respectively.

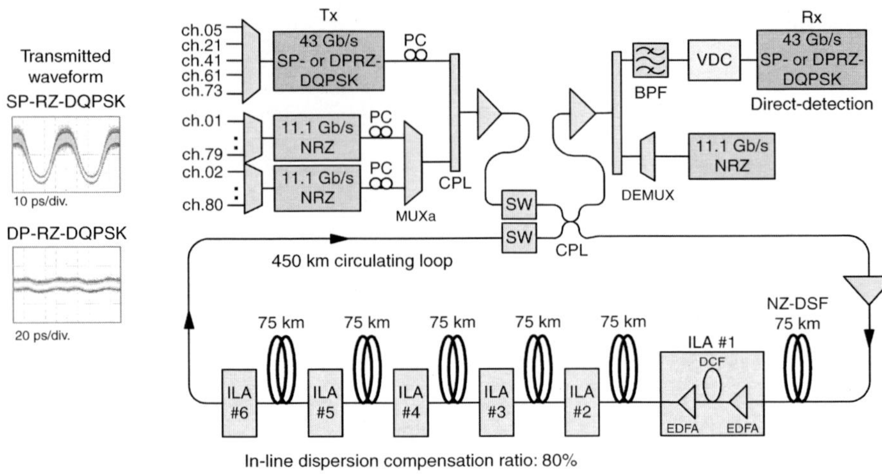

Fig. 10.7 Setup for transmission experiment

The 11.1-Gb/s NRZ modulation on the other 75 channels were turned on and off to quantify the impact of XPM, where the odd- and even-number channels were modulated separately using asynchronous PRBS 23. The relative polarization angle between 43-Gb/s SP-RZ-DQPSK (or the X-polarization tributary of DP-RZ-DQPSK) and 11.1-Gb/s NRZ channels was adjusted to 0° at the ingress point to the first fiber span using the polarization controllers to assess the worst-case condition. The transmission line was a 450-km long circulating loop that consisted of six spans of 75-km NZ-DSF, having a span loss of 16.1 dB. The 80% of the total chromatic dispersion due to NZ-DSF transmission line was compensated by DCFs [6]. At the receiver side, a virtually imaged phased array (VIPA) variable dispersion compensator (VDC) was inserted in the 43 Gb/s receiver to compensate for the residual dispersion for each channel. Figure 10.8 shows the measured Q-values for the channel 41 (1546.12 nm) as a function of transmission distance with and without

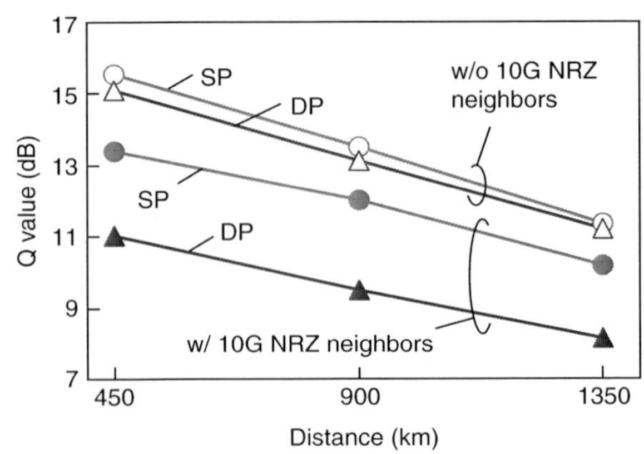

Fig. 10.8 Comparison of transmission performance of 43-Gb/s SP- and DP-RZ-DQPSK

the 11.1 Gb/s NRZ signal co-propagation. The fiber input power was −1 dBm/ch. The OSNRs after 450, 900, and 1,350 km were 19.7, 16.9, and 14.4 dB, respectively. The Q-values for SP- and DP-RZ-DQPSK were almost similar within the measurement accuracy (∼0.2 dB) regardless of distance when there were no NRZ neighbors. When the NRZ modulation was turned on, however, the Q-value for DP became about 2 dB worse than that for the SP case. Considering that the impact of polarization mixing is less than that of XPM phase noise in this setup, the different XPM tolerance observed for SP and DP signals are mainly ascribed to the difference in the baud-rate [7].

10.5 XPM Tolerance Comparison between Direct and Coherent Detection Receivers

The XPM tolerance is also dependent on receiver implementation, i.e., direct and coherent detection, because the two detection methods rely on different phase detection principles that have different sensitivity to the phase noise.

In this section, transmission performance of 43-Gb/s single-polarization RZ-DQPSK signals with coherent and direct-detection receivers is compared experimentally. In addition a direct measurement experiment on the XPM phase noise by using a continuous wave (CW) probe light is introduced.

The experimental setup was basically the same as the one shown in Fig. 10.7. The 5 channels of the 80 were modulated with SP-RZ-DQPSK format at a symbol rate of 21.5 Gbaud, the laser sources for which were temperature-controlled DFB lasers with a typical linewidth less than 10 MHz. Residual dispersion at the receiver input was compensated either by using a VIPA VDC in the case of the 43-Gb/s direct-detection (DD) receiver (Rx) or by means of digital processing in the case of the 43-Gb/s coherent receiver. The coherent-Rx consisted of a 90° optical hybrid and two pairs of balanced photodiodes, a local oscillator laser with a nominal linewidth of 100 kHz, and a polarization controller for polarization alignment. The frequency mismatch between the local oscillator and the signal carrier was maintained to be within about 300 MHz. Electrical waveforms were then sampled at a rate of 40 GSa/s for 40 μs durations using a digital sampling oscilloscope with an electrical bandwidth of 15 GHz. Then offline post-processing was performed on a personal computer. The post-processing algorithm included resampling, equalization, carrier recovery, and error-counting modules. The resampling routine resampled the data from 40 GSa/s to 43 GSa/s to ensure that there are two samples per symbol. Equalization was performed by a 19 tap, T/2-spaced FIR filter. A novel carrier recovery stage whose details are reported elsewhere [8, 9] then corrected for the frequency and phase offsets between the local oscillator and the source laser. Differential decoding was applied after the symbol decision to account for possible phase-slips. Errors are then counted for the 1.7 million symbols to evaluate the effective Q-factor.

First, effective Q-factors after 0, 450, and 900 km transmission with a fiber input power of −4 dBm/ch were evaluated at a received OSNR of 13 dB (Fig. 10.9). For

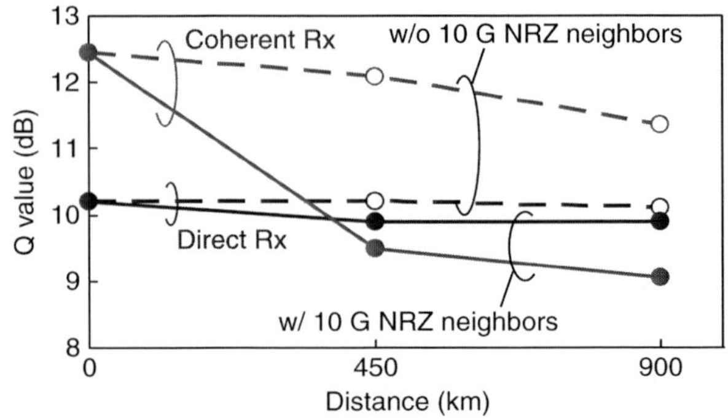

Fig. 10.9 Comparison of direct detection and coherent detection of 43-Gb/s SP-RZ-DQPSK signals

coherent-Rx, the aperture width for averaging in the carrier recovery was optimized for each data point [10, 11]. At 0 km, the Q-factor for coherent-Rx is 2.3 dB better than for DD-Rx as was anticipated from theory. Thanks to this merit, the coherent-Rx outperformed the DD-Rx on that transmission system without co-propagating 11.1-Gb/s NRZ signals although it was specifically optimized for the DD-Rx [6]. With co-propagating 11.1-Gb/s NRZ signals, however, the coherent-Rx showed degrading performance while the DD-Rx did not show a significant degradation, and as a result the coherent-Rx performed worse than the DD-Rx and for transmission distances of 450 km and 900 km in spite of the back-to-back Q-merit of 2.3 dB. The above results strongly suggest that the coherent-Rx is by far more sensitive to the XPM phase noise. In order to verify this, the channel at 1546.12 nm (channel 41) was replaced by a CW probe to detect the phase noise with and without co-propagating 11.1 Gb/s NRZ signals. After 900-km transmission, the CW light was received at an OSNR of 13 dB with the coherent-Rx front-end and I- and Q-components were recorded with a sample rate of 20 GSa/s. In the post-processing of the signal two types of carrier phase detector were employed to discuss the difference between DD-Rx and coherent-Rx: the first one, DD-Rx-like phase detector, included a delayed detection circuit while the second one, coherent-Rx-like phase detector, lacked such a block, Coh-Rx-like phase detector. Figure 10.10 shows the CW signal constellations right after the carrier phase detectors. When the 10-G NRZ neighbors were absent, no significant difference was observed between the constellations after the two-phase detectors. When the 10-G OOK neighbors co-propagated with the probe CW light, however, the coherent-Rx-like phase detector suffered much wider phase deviation than the DD-Rx-like phase detector. In order to obtain further insight the phase noise spectra were plotted as shown in Fig. 10.11, while the XPM phase noise had a wide spectrum that would overlap with the QPSK modulation spectrum, DD-Rx-like phase detector successfully attenuates it in the entire frequency range regardless of the existence of 10-G NRZ neighbors.

While the above results clearly indicate that coherent-Rx is more vulnerable to the XPM phase noise, it is also implied that a coherent receiver can include DD-Rx-like phase detector to mitigate its impact and can realize similar performance as DD-Rx at least.

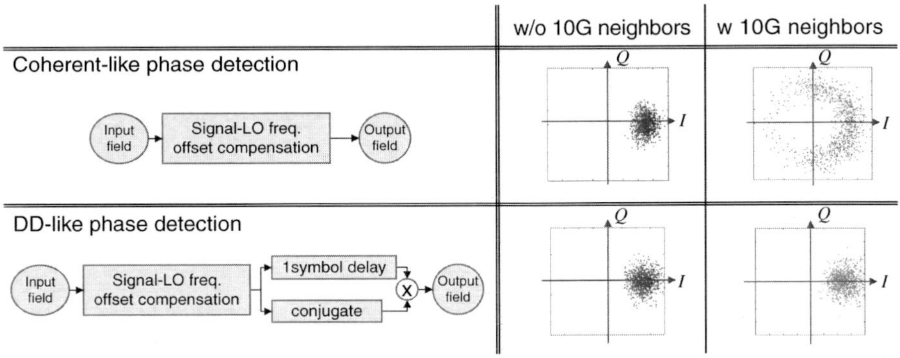

Fig. 10.10 Measured XPM phase noise induced by 900-km transmission by using a CW probe

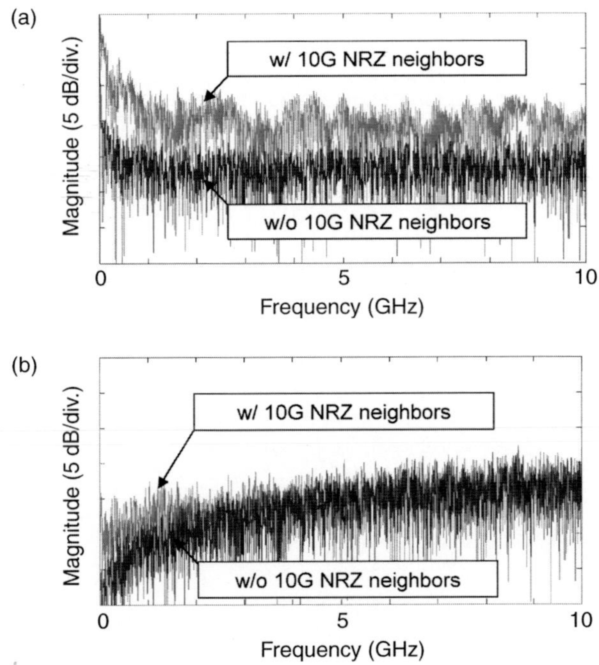

Fig. 10.11 XPM phase noise spectra (**a**) after coherent-Rx-like phase detection and (**b**) after DD-Rx-like phase detection

10.6 Summary

QPSK-based modulation formats are attractive from a viewpoint of OSNR vs. spectral efficiency trade-off. Its various implementation options can offer different balance among cost, size, and performance of transceivers. In addition, there seems to be a trade-off in terms of different aspects of transmission impairments through the results introduced in the above: DP formats are more tolerant to linear distortions than SP formats thanks to its lower symbol rates, while it can be less tolerant to

PDL and XPM; coherent receivers can deliver significant improvement to linear distortion tolerance as well as OSNR tolerance while it can be more vulnerable to XPM impairments. Thus optimum choice should depend on system conditions such as bit-rate, transmission fiber types, fiber input power, dispersion map, and necessity of co-propagation with OOK channels.

References

1. ITU-T Recommendation G.680, Appendix II.3.2, July (2007).
2. C. Laperle et al., Wavelength division multiplexing (WDM) and polarization mode dispersion (PMD) performance of a coherent 40Gbit/s dual-polarization quadrature phase shift keying (DP-QPSK) transceiver. In OFC2007, PDP16, March 2007.
3. C. Xie et al., Impact of inter-channel nonlinearities on 10Gbaud NRZ-DQPSK WDM transmission over Raman amplified NZDSF spans. In ECOC2007, paper 10.4.3, September 2007.
4. G. Charlet et al., Nonlinear interactions between 10 Gb/s NRZ channels and 40 Gb/s channels with RZ-DQPSK or PSBT format over low dispersion fiber. In ECOC2006, paper Mo3.2.6, September 2006.
5. D. van den Borne et al., 1.6-b/s/Hz spectrally efficient transmission over 1700 km of SSMF using 40 $\times$ 85.6-Gb/s POLMUX-RZ-DQPSK. J. Lightwave Technol. **25**(1), 222–232, January (2007).
6. K. Nakamura et al., Dispersion map suitable for hybrid 10 Gb/s NRZ and 40 Gb/s RZ-DQPSK transmission over 50 GHz-spaced network with low dispersion fibre. In ECOC 2007, paper 10.4.4, September 2007.
7. O. Vassilieva et al., Symbol rate dependency of XPM-induced phase noise penalty on QPSK-based modulation formats. In ECOC2008, We1E4, September 2008.
8. L. Li et al., Wide-range, accurate and simple digital frequency offset compensator for optical coherent receivers. In OFC/NFOEC2008, OWT4, March 2008.
9. Z. Tao et al., Multiplier-free phase recovery for optical coherent receivers. In OFC/NFOEC2008, OWT2, March 2008.
10. D. van den Borne et al., Carrier phase estimation of coherent equalization of 43-Gb/s POLMUXNRZ-DQPSK transmission with 10.7-Gb/s NRZ neighbors, In ECOC 2007, paper We7.2.3, September 2007.
11. T. Tanimura et al., Non-linearity tolerance of direct detection and coherent receivers for 43-Gb/s RZ-DQPSK signals with co-propagating 11.1-Gb/s NRZ signals over NZ-DSF. In OFC/NFOEC2008, OTuM4, March 2008.

Chapter 11
Real-Time Digital Coherent QPSK Transmission Technologies

Timo Pfau and Reinhold Noé

Abstract The high data rates in optical communication generate stringent constraints on algorithms and technologies suitable for real-time coherent optical receivers based on digital signal processing (DSP). In this chapter both the algorithmic and technological requirements for real-time DSP-based receivers are summarized and illustrated by examples. In addition, the evolution of real-time coherent QPSK transmission from the first prototypical implementation to the first commercial product is reviewed.

11.1 Introduction

The high data rates in optical communication of 43 Gb/s, 112 Gb/s, or even above generate stringent constraints on algorithms and technologies suitable for real-time coherent optical receivers. The essential receiver components (Fig. 11.1) are a polarization-diversity optical front end with optical 90° hybrids, O/E conversion, analog-to-digital converters (ADC), and a digital signal processing unit (DSPU). In this chapter, we summarize both algorithmic and technological requirements for real-time coherent receivers and present results from QPSK transmission experiments.

11.2 Algorithmic Requirements

Coherent receivers are widely used in radio and wired communication. Researchers in these areas already developed powerful algorithms, e.g., to track the carrier phase or to compensate for distortions. The reasons why these algorithms cannot simply be

---

T. Pfau (✉)
University of Paderborn, EIM-E, Optical Communication and High-Frequency Engineering, Warburger Str. 100, D-33098 Paderborn, Germany. Current Affiliation: Bell Laboratories, Alcatel-Lucent, 600-700 Mountain Ave, Murray Hill, NJ 07974, USA.
e-mail: timo.pfau@alcatel-lucent.com

M. Nakazawa et al. (eds.), *High Spectral Density Optical Communication Technologies,* Optical and Fiber Communications Reports 6, DOI 10.1007/978-3-642-10419-0_11,

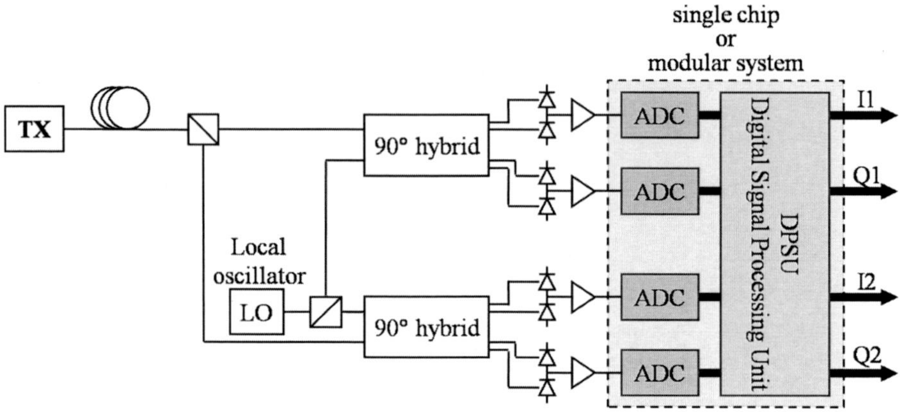

Fig. 11.1 Coherent optical receiver structure

applied in coherent optical receivers are the stringent requirements imposed by the high data rates in optical communication. This section summarizes these requirements.

11.3 Feasibility of Parallel Processing

The DSPU cannot operate directly at the sampling clock frequency of the analog-to-digital converter, which is in general 10 GHz or higher, but requires demultiplexing to process the data in m parallel modules at clock frequencies below 1 GHz as shown in Fig. 11.2 [1]. This allows automated generation of the layout, which is indispensable due to the complexity of usually more than 1 million required logic gates for the DSPU.

A comparison between the sampling clock frequency and the admissible clock frequency for the DSPU shows that at least $m = 16$ parallel modules are required. Algorithms for real-time applications should therefore theoretically allow parallel processing with an unlimited number of parallel channels. In order to fulfill this requirement the calculations within one module must not depend on the result of the same calculation in another parallel module.

A good example for the feasibility of parallel processing is the comparison of finite (FIR) and infinite (IIR) impulse response filters. Figure 11.3 depicts their structures in both serial and parallel systems. It can be seen that it is easily possible to parallelize an FIR filter. Although the output signal

$$y_k = \alpha_0 x_k + \alpha_1 x_{k-1} + \alpha_2 x_{k-2} \tag{11.1}$$

depends on information provided by several parallel modules it does not depend on the result of the same calculation performed in these modules. In contrast, it is a big challenge to realize the parallel structure shown for the IIR filter, because the result

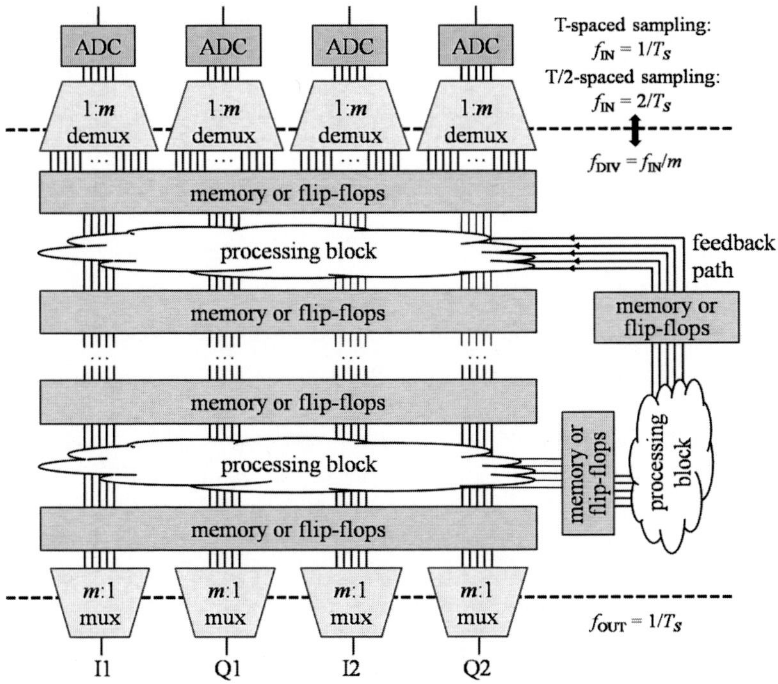

Fig. 11.2 Interfacing between ADCs and DSPU and internal structure of the DSPU

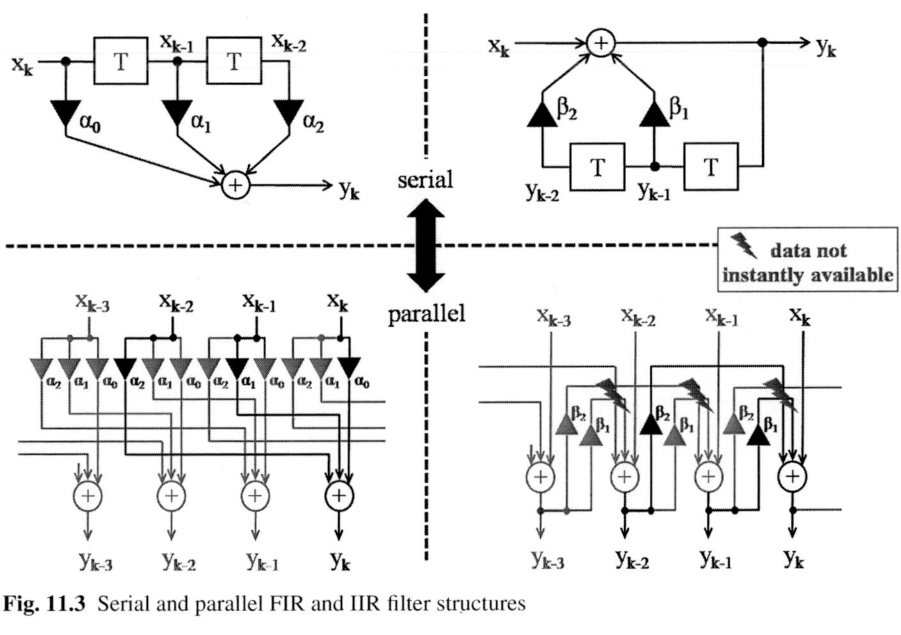

Fig. 11.3 Serial and parallel FIR and IIR filter structures

$$y_k = x_k + \beta_1 y_{k-1} + \beta_2 y_{k-2} \tag{11.2}$$

depends on results calculated at the same time in other parallel modules. A very low clock frequency would be needed to allow all calculations to be executed within one clock cycle. But this would result in an even higher number of required parallel modules, hence creating a vicious circle.

11.4 Hardware Efficiency

Another important constraint for real-time coherent receiver algorithms is that they must allow for a hardware-efficient implementation [1]. This requirement also originates from the parallel processing in the DSPU. Since most of the required algorithm blocks have to be implemented m times within the DSPU, computationally intensive algorithms require a huge amount of chip area and therefore increase power consumption and cost. The algorithms considered for a chip implementation should therefore not only be evaluated by performance, but also by hardware efficiency.

One way to increase hardware efficiency is to use signal transformations, e.g., FFT/IFFT, log-function or transformation of in-phase and quadrature signal pairs into polar coordinates [2]. Although the transformation itself requires additional hardware effort, this is often beneficial because subsequent calculations are simplified. For example, an FFT can reduce a convolution to multiplications, or the log-function or polar coordinates allow to replace multiplications by summations.

The use of lookup tables (LUT) is another effective way to reduce the required hardware. Coordinate transformations and nonlinear functions are the main candidates for a LUT implementation, but it can be sometimes even beneficial to replace a multiplication by a LUT, especially if the required precision is low.

This directly brings us to the last but most important way to increase efficiency: the optimization of the required precision. In microprocessors and computers it is common to use standard precisions which are usually 32 or 64 bit for floating point variables and 8–64 bit for integer variables. This is useful to allow the efficient compilation of software. However, in an application-specific hardware design each additional bit increases the required number of gates and hence chip area, power consumption, and cost. Therefore all calculations should be optimized to just the precision that is necessary to achieve the required accuracy.

11.5 Tolerance Against Feedback Delays

The last important algorithmic consideration for real-time applications is the tolerable feedback delay [3]. In simulation or offline processing feedback delays of one symbol are easy to achieve, but this is impossible in a real-time system designed for multi-Gb/s operation. The reasons are the parallel processing and massive pipelining, which are required to cope with the high data rates. Pipelining means that only fractions of the whole algorithm are processed within one clock cycle, and the intermediate results are stored in buffers (e.g., memory or flip-flops) as shown in Fig. 11.2. Therefore it can take easily 100 symbol durations until a received symbol has an impact on the feedback signal. Algorithms for polarization control and CD/PMD compensation usually employ integral controllers with time constants in the microsecond range to update their tap coefficients. There the additional delay due to pipelining, which is in the nanosecond range, can be neglected. But the feedback delay can have a severe impact on the performance of algorithms that require an instantaneous feedback, e.g., decision-directed carrier recovery, which is often used in offline signal processing for higher order QAM [4, 5].

Figure 11.4 shows the general structure of such a decision-directed carrier and data recovery module with an optimum feedback delay $\Delta = 1$ [6]. The input signal of the system is given by

$$\underline{Y}_k = \underline{X}_k e^{j\psi_k} + \underline{n}_k \tag{11.3}$$

The phase estimator uses the estimated carrier phase $\hat{\psi}_{k-1}$ to derotate the input symbol $\underline{Y}_k$, and the result is fed into a decision device ($[\ ]_\mathrm{D}$ denotes the output of the decision device).

$$\underline{\hat{X}}_k = \left[\underline{Y'}_k\right]_D = \left[\underline{Y}_k e^{-j\hat{\psi}_{k-1}}\right]_D \tag{11.4}$$

The carrier phase estimate $\hat{\psi}_k$ is calculated with

$$\hat{\psi}_k = \sum_{i=0}^{N_\mathrm{CR}} w_i \left(\arg\left\{\underline{Y}_{k-i}\right\} - \arg\left\{\underline{\hat{X}}_{k-i}\right\}\right), \sum_{i=0}^{N_\mathrm{CR}} w_i = 1, \tag{11.5}$$

where w_i are weighting coefficients (e.g., Wiener filter coefficients) and N_CR is the FIR filter length. If only preceding symbols are used as inputs to the FIR filter, a feedback delay of $\Delta = 1$ becomes theoretically possible.

However, as explained above, in a practical implementation it is impossible to achieve $\Delta = 1$. Figure 11.5 shows the structure of a decision-directed carrier recovery in a parallel and pipelined implementation. It can be seen that due to

Fig. 11.4 Decision-directed carrier recovery with $\Delta = 1$

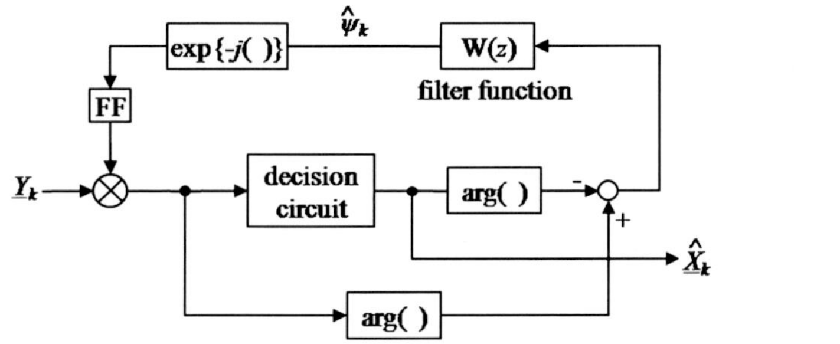

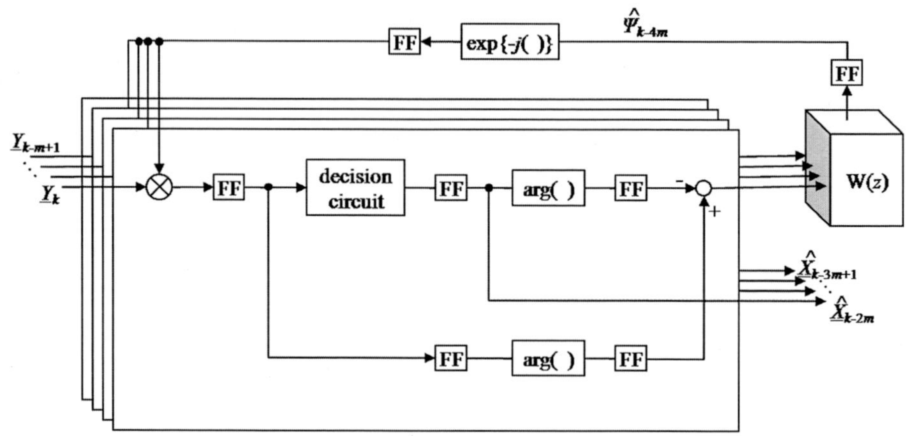

Fig. 11.5 Decision-directed carrier recovery in a realistic receiver with parallel and pipelined signal processing

the parallel processing of p consecutive samples the feedback delay between two symbols within one module is equal to the degree of parallelism p. Even if the latest phase estimate from all parallel outputs is fed back into each module as shown in Fig. 11.5, the average delay amounts to $(p + 1)/2$. Taking also the number of pipeline stages l into account [represented by the flip-flops (FF) in Fig. 11.5], which means that it takes l clock cycles until an input sample has an impact on the feedback value, the total average feedback delay becomes as follows:

$$\Delta = (l - 1) \cdot p + \frac{p + 1}{2}. \tag{11.6}$$

Note that the minimum number of pipeline stages is 1, because this is the minimum number of clock cycles required in a digital feedback system [see Eq. (11.4)].

To determine the effect of Δ on the phase noise tolerance of the receiver, let $\hat{\psi}_{k-\Delta}$ be a perfect estimate of $\psi_{k-\Delta}$, i.e., $\hat{\psi}_{k-\Delta} = \psi_{k-\Delta}$ and $\underline{Y}_k$ is only corrupted by phase noise. With Eq. (11.4), $\underline{Y}'_k$ can be written as

$$\underline{Y}'_k = \underline{Y}_k \exp\{j(\psi_k - \psi_{k-\Delta})\} = \underline{Y}_k \exp\{j\psi'_k\}. \tag{11.7}$$

Phase noise can be modeled as a Wiener process with $\psi_k = \sum_{i=-\infty}^{k} f_i$, where the f_i's are independent and identically distributed random Gaussian variables with zero mean and variance $\sigma_f^2 = 2\pi\,(\Delta f \cdot T_S)$, Δf is the 3 dB linewidth of the laser, T_S is the symbol duration [7]. Therefore ψ'_k is given by

$$\psi'_k = \psi_k - \psi_{k-\Delta} = \sum_{i=k-\Delta}^{k} f_i \tag{11.8}$$

ψ_k' is a random Gaussian variable with zero mean and variance

$$\sigma_\psi^2 = \Delta \cdot \sigma_\Delta^2 = 2\pi \left(\Delta \cdot \Delta f \cdot T_S\right) \tag{11.9}$$

Hence the standard deviation of the phase noise increases by the factor $\sqrt{\Delta}$.

In a practical implementation, assuming, e.g., parallel processing with $p = 32$ and $l = 5$ pipeline stages, the average feedback delay is $\Delta = 144.5$ samples. This shows that a decision-directed carrier recovery is fairly unfeasible because the phase noise tolerance is reduced by a factor of $\sim$12. Consequently any feedback loop must be avoided in the carrier recovery process, especially for higher order QAM constellations with their inherently smaller phase noise tolerance.

11.6 Technological Requirements

There are two major options, how to design the chipset for a real-time digital coherent receiver. ADCs and DSPU can either be integrated in a single chip to ease interfacing and reduce the footprint [8] or in a modular approach for optimized performance, where ADCs and DSPU can be developed in different technologies for maximum bandwidth and high integration, respectively [9]. Both approaches have their significant advantages. The integration of analog-to-digital converters with the DSPU of course reduces the footprint compared to a multi-chip solution and eases the interfacing, because the attachment of the ADCs to the DSPU can be supported by tools. However, as both the ADCs and the DSPU must be designed in the same technology, a compromise has to be found between best analog performance and bandwidth and integration with the highest density and low power consumption. Here lies the advantage of the multi-chip solution. As the different ICs can be designed in different technologies optimal for the respective tasks, the performance of the overall system can be improved. Additionally this approach also gives the opportunity to use, e.g., ADCs provided by a third party and just develop the DSPU, which reduces development cost and time to market. But this comes at the price that significant effort has to be spent for the design of the chip interface. Considering that each of the four ADCs has a 5–6 bit resolution and that the data is eventually already demultiplexed in the ADC ICs, at least 20 and up to more than 100 multi-Gb/s synchronous connections are required between the ICs, which is not a trivial design problem.

Another upcoming option for the design of real-time digital coherent receivers is the application of field-programmable gate arrays. This approach is closely related to the multi-chip design, only the application-specific ASIC is replaced by an FPGA. Due to their flexibility and increasing performance – today's FPGA already offers multiple serial inputs with up to 10 Gb/s data rate and almost one million logic gates – they become an interesting alternative to full-custom ASICs. But although the FPGA performance increases tremendously they do not yet meet the requirements for coherent digital 40 Gb/s or 100 Gb/s receivers. Therefore a complete chip design will still be necessary in the near future to realize commercial systems. But FPGAs are indispensable anyway – for prototyping and design verification at lower symbol rates.

11.7 Real-time Implementations of Digital Coherent QPSK Receivers

The first real-time implementation of a digital coherent QPSK receiver was published in July 2006 by T. Pfau et al. [10]. An FPGA and commercially available ADCs were used to coherently recover a 800 Mb/s QPSK-modulated optical signal. The experimental setup is shown in Fig. 11.6.

The results from this first experiment showed that the implementation of a real-time digital coherent QPSK receiver was feasible and that at 10 Gbaud it will tolerate the linewidth of standard DFB lasers (Fig. 11.7). At a sum linewidth times symbol rate product $\Delta f \cdot T = 7.5 \times 10^{-5}$ using the ECL, error-free transmission was

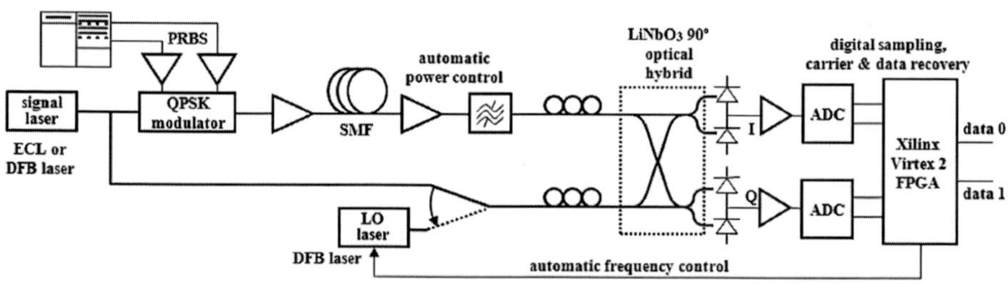

Fig. 11.6 Real-time synchronous QPSK transmission setup

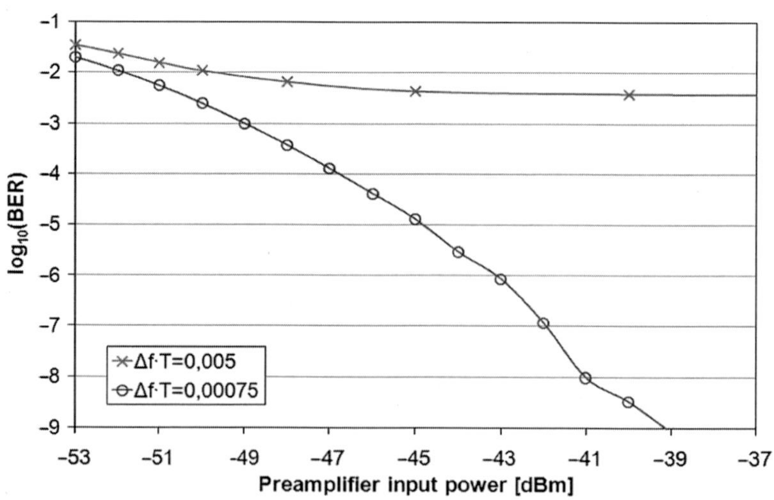

Fig. 11.7 BER vs. preamplifier input power for different sum linewidth times symbol rate products

achieved. Even if the DFB lasers were employed, which raised $\Delta f \cdot T$ to 5×10^{-3}, the signal could still be recovered, albeit with a bit error rate floor at 3.5×10^{-3}.

In the following experiments the data rates were successively increased [11, 12]. However, until then only 2 bit/symbol were transmitted and the polarizations had to be aligned manually. This changed when in July 2007 the University of Paderborn published the results of a polarization-multiplexed QPSK (PM-QPSK) transmission with real-time coherent digital receiver and automatic polarization tracking [13]. These first receiver implementations were all based on FPGAs and commercially available ADCs, as this approach does not require an expensive and time-consuming IC design. But therefore the achieved data rates with FPGAs and commercially available ADCs are limited to $\leq$10 Gb/s until today [14].

That the development of an ASIC is possible was demonstrated by Nortel, who published the first results of a real-time coherent QPSK receiver based on a specifically developed ASIC also in July 2007 [8]. Due to possible sampling rates of 20 Gsample/s of the chip, this was the first implementation running at the standard data rate of 40 Gb/s. Later the data rate was even pushed to 46 Gb/s [15]. With twofold oversampling, polarization control, and PMD/CD compensation, this chip is today available in a commercial product and employed in the field. Figure 11.8 summarizes all mentioned milestones until the commercialization of the first coherent digital QPSK transceiver.

From the scientific perspective the development and study of real-time coherent transmission systems are essential. It allows to investigate effects that could be covered by simulations or offline experiments only with enormous effort, if at all. These are, e.g., dynamic effects, long-term behavior, or start-up issues. Additionally the algorithms employed in the DSPU prove to be applicable for real-time operation and all limiting effects such as various quantization effects are taken into account.

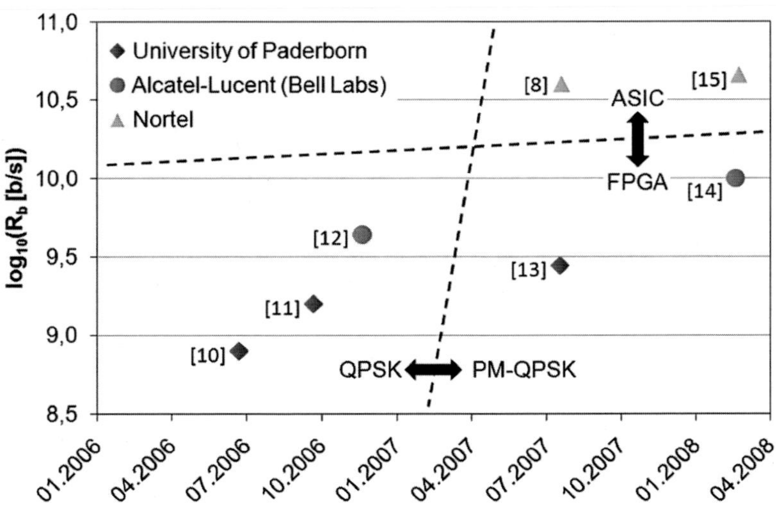

Fig. 11.8 Milestones in the development of real-time digital coherent QPSK receivers and the achieved bit rates R_b

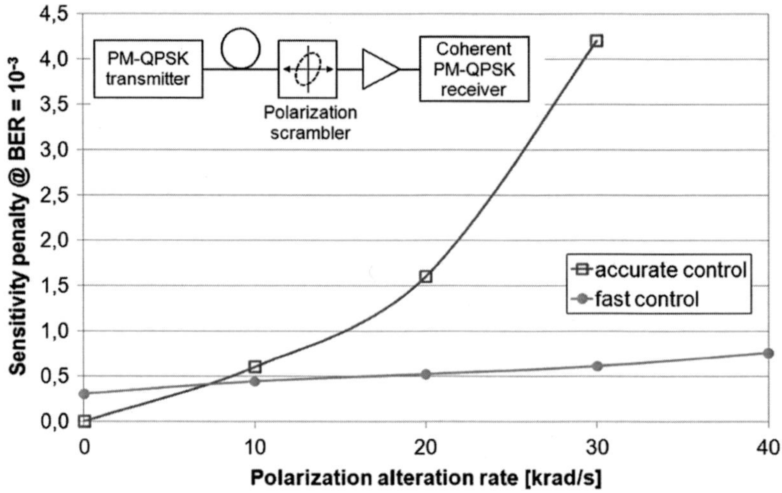

Fig. 11.9 Sensitivity penalties of a 2.8 Gb/s coherent QPSK receiver due to fast polarization changes

Figure 11.9 gives an example for the investigation of such dynamic effects [16]. It shows the sensitivity penalties of a 2.8 Gb/s coherent polarization-multiplexed QPSK receiver against polarization fluctuations with different alteration rates (given in rad/s on the Poincaré sphere). The results show that depending on the selection of either a fast or an accurate but slower polarization controller polarization changes with an alteration rate up to 40 krad/s can be compensated with less than 1 dB sensitivity penalty.

References

1. R. Noé, PLL-free synchronous QPSK polarization multiplex/diversity receiver concept with digital I&Q baseband processing. IEEE Photon. Technol. Lett. **17**(4), 887–889, April (2005).
2. J. Sitch, Implementation aspects of high-speed DSP for transmitter and receiver signal processing. Proc. IEEE/LEOS Summer Topicals'07. July 23–25, 2007, Ma4.3.
3. F. Derr, Coherent optical QPSK intradyne system: concept and digital receiver realization. IEEE J. Lightwave Technol. **10**(9), 1290–1296 (1992).
4. Y. Mori et al., Transmission of 40-Gbit/s 16-QAM signal over 100-km standard single-mode fiber using digital coherent optical receiver. Proc. ECOC'08. September 21–25, 2008, Tu.1.E.4.
5. P. Winzer, A. Gnauck, 112-Gb/s polarization-multiplexed 16-QAM on a 25-GHz WDM Grid. Proc. ECOC'08. September 21–25, 2008, Th.3.E.5.
6. E. Ip, M. Kahn, Feedforward carrier recovery for coherent optical communications. IEEE J. Lightwave Technol. **25**(9), 2675–2692, September (2007).
7. F. Munier et al., Estimation of phase noise for QPSK modulation over AWGN channels. Proc. GigaHertz'03 Symposium. November 4–5, 2003.
8. K. Roberts, Electronic dispersion compensation beyond 10 Gb/s. Proc. IEEE/LEOS Summer Topicals'07. July 23–25, 2007, Ma2.3.

9. T. Pfau et al., Towards real-time implementation of coherent optical communications. Proc. OFC/NFOEC'09. 22–26. March 2009, OThJ4.
10. T. Pfau et al., Real-time synchronous QPSK transmission with standard DFB lasers and digital I&Q Receiver. Proc. OAA/COTA'06. June 28–30, 2006, CThC5.
11. T. Pfau et al., 1.6 Gbit/s real-time synchronous QPSK transmission with standard DFB lasers. Proc. ECOC'06. September 24–28, 2006, Mo4.2.6.
12. A. Leven et al., Real-time implementation of 4.4 Gbit/s QPSK intradyne receiver using field programmable gate array. Electron. Lett. **42**(24), 1421–1422, November 23 (2006).
13. T. Pfau et al., PDL-tolerant real-time polarization-multiplexed QPSK transmission with digital coherent polarization diversity receiver. Proc. IEEE/LEOS Summer Topicals'07. July 23–25, 2007, Ma3.3.
14. A. Leven, N. Kaneda, Y. Chen, A real-time CMA-based 10 Gb/s polarization demultiplexing coherent receiver implemented in an FPGA. Proc. OFC/NFOEC'08. February 24–28, 2008, OTuO2.
15. L. Nelson et al., Performance of a 46-Gbps dual-polarization QPSK transceiver in a High-PMD Fiber. Proc. OFC/NFOEC'08. February 24–28, 2008, PDP9.
16. M. El-Darawy et al., Realtime 40 krad/s polarization tracking with 6 dB PDL in Digital synchronous polarization-multiplexed QPSK receiver. Proc. ECOC'08. September 21–25, 2008, We.3.E.4.

Chapter 12
Challenge for Full Control of Polarization in Optical Communication Systems

Shiro Ryu

Abstract Countermeasures against the receiver sensitivity degradation due to the state of polarization change in single-mode fibers in coherent optical communication systems are reviewed from early efforts in 1980s to the latest research results. It is pointed out that the polarization-diversity schemes are considered to be the most promising solutions among various countermeasures for the future applications owing to its applicability to the systems with analogue as well as digital signal processing.

12.1 Introduction

Almost 30 years have passed since the first experiments of coherent optical communication systems [1]. It is essential for coherent heterodyne/homodyne detection that the state of polarization (SOP) of the signal light coincides with that of the local oscillator light. However, it is well known that the SOP of the signal light fluctuates as it propagates along the fibers due to external disturbances such as environmental temperature change, vibration, and stress applied to the fibers. This situation causes the receiver sensitivity fluctuation in coherent systems. A lot of efforts in the field have been devoted to solve the above-mentioned problem.

In this chapter, first we briefly look at actual polarization fluctuation in installed optical cables. Then, we review the research and development works regarding the countermeasures against the polarization fluctuation problem in coherent systems, i.e., polarization-maintaining fibers, polarization control schemes, polarization-diversity schemes, and polarization scrambling schemes from the first proposal to the most up-to-date applications including the digital coherent systems with a digital

S. Ryu (✉)
Network Division, SoftBank Telecom Corporation, 1-9-1, Higashi-shimbashi, Minato-ku, Tokyo 105-7316, Japan
e-mail: shiro.ryu@tm.softbank.co.jp

M. Nakazawa et al. (eds.), *High Spectral Density Optical Communication Technologies*, Optical and Fiber Communications Reports 6, DOI 10.1007/978-3-642-10419-0_12,

signal processing (DSP)-based receiver. Finally, it is concluded that the polarization-diversity schemes are the most promising for the coherent systems in the future photonic network.

12.2 Polarization Fluctuation in Single-Mode Fibers

In single-mode optical fibers, an HE_{11} mode, a propagation mode, degenerates into two orthogonally polarized modes, i.e., HE_{11x} and HE_{11y} modes due to core deformation in manufacturing process or by external disturbances such as stress to the fibers [2]. Figure 12.1 shows the two modes in single-mode fibers.

In general, there is the phase velocity difference between the two modes and that is called birefringence. The two modes also have different group velocities. The polarization-mode dispersion (PMD) is expressed as the differential group-delay time between the two modes. The PMD causes waveform distortion as the signal light propagates along the fibers. In installed single-mode fibers, the SOP of the signal light changes frequently due to the variation of the external disturbances.

Figure 12.2 shows a basic configuration of the coherent optical communication systems [3]. In coherent systems, the signal light from the transmitter is heterodyne or homodyne detected by the local oscillator light in a receiver. It is essential that the SOP of the signal and local oscillator light should coincide at a surface of a photodetector in a receiver, whereas the SOP of the signal light fluctuates due to the reasons above. This situation causes severe receiver sensitivity degradation in coherent systems.

Then, we briefly look at the SOP fluctuation speed. Figure 12.3 shows a power spectrum of the SOP fluctuation after a polarizer when a 45-km-long submarine cable was installed in the 6000-m-deep sea [4]. Figure 12.3 indicates that the maximum SOP fluctuation speed is about 50 Hz due to the fast mechanical vibration of the cables during installation. Since the measurement was carried out, it had been considered that the SOP fluctuation speed in terrestrial optical cables must be slower than that.

However, the recently measured SOP fluctuation in terrestrial cables has shown that the maximum fluctuation speed due to fiber movement and/or vibration

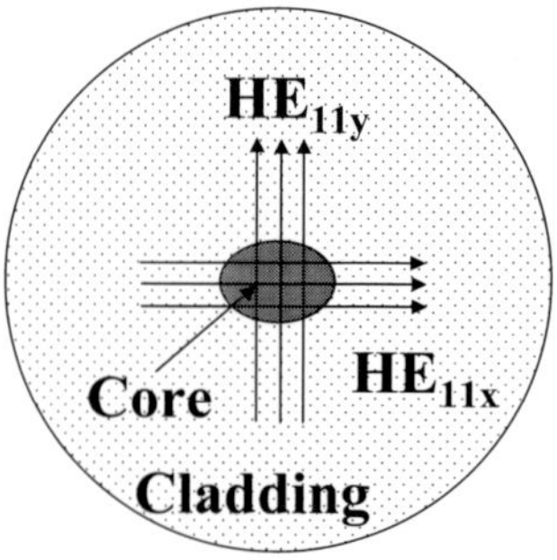

Fig. 12.1 HE_{11x} and HE_{11y} modes in single-mode fibers

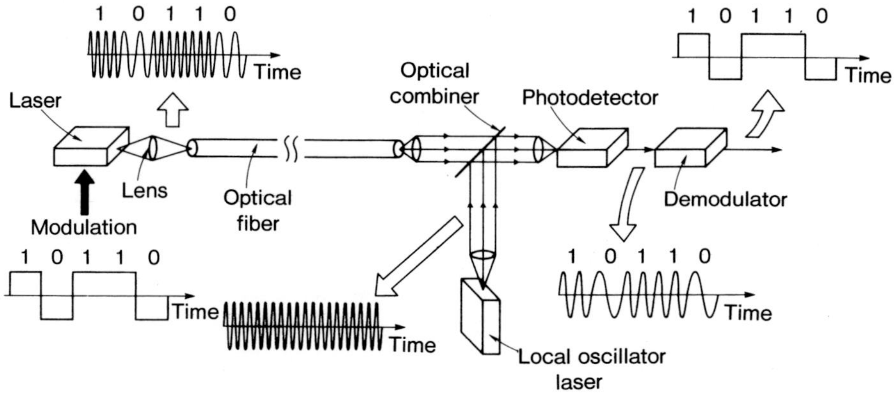

Fig. 12.2 Basic configuration of coherent optical communication systems. Reproduced with permission from Ryu [3] © 1995 by Artech House, Inc.

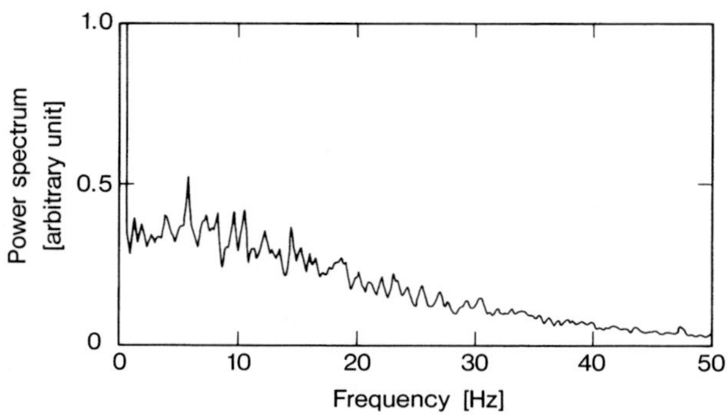

Fig. 12.3 Power spectrum of SOP fluctuation in submarine cable installation. Reprinted with permission from Ryu et al. [4] © 1998, IET

becomes the order of 10 kHz [5]. This discovery has made an impact on the coherent systems since the systems should cope with higher SOP fluctuation speed than the previously measured one for the submarine cable installation.

12.3 Solutions for SOP Fluctuation

In this section, the solutions for receiver sensitivity degradation in coherent systems are reviewed from the early proposals to the most up-to-date systems.

12.3.1 Polarization-Maintaining Fiber

A polarization-maintaining fiber (PMF) is able to preserve the linear SOP along the whole transmission lines by introducing anisotropy and decreasing mode coupling between orthogonal modes [2]. Various types of PMFs have been proposed so far including elliptical-core fibers, stress-induced birefringent fibers, and side-pit fibers [2].

However, the PMFs have not been used for long-haul transmission lines due to the following two reasons: The first reason is that the polarization extinction ratio deteriorates after long-haul transmission and this affects the transmission performances; the second reason is that the loss of the PMFs is relatively high as compared with that of conventional single-mode fibers.

For these reasons, the PMFs have not been used as the transmission lines for coherent optical communication systems, but they have found other commercial applications, e.g., the interconnection of optical devices for which the polarization-maintaining capabilities are required.

12.3.2 Polarization Control Scheme

A polarization control scheme is another solution for the SOP fluctuation problem. If the SOP of the signal light is adaptively controlled by a polarization controller to be a linear polarization with a direction that is the same as a local oscillator, the maximum heterodyne/homodyne detection efficiency can be achieved.

A polarization control scheme has been investigated from early 1980s, and in early experiments a polarization controller made with a $LiNbO_3$ electrooptical crystal with a half-wave voltage of 66 V at a wavelength of 1.62 μm was used to compensate for the SOP change in 1-km-long single-mode fibers [6]. In the experiments, pressure was intentionally applied to the fibers by fiber touch to artificially generate the SOP fluctuation in the fibers.

A polarization controller with a $LiNbO_3$ electrooptical crystal has been used for many years. Latest research results have reported an endless polarization controller which can track very fast polarization fluctuation at 9 krad/s on the Poincare sphere by the aid of a field programmable gate array (FPGA)-based controller [7].

Recently, polarization-division multiplexing (PDM) technology, in which orthogonally polarized signals are used for doubling the transmission capacity, is attracting much attention due to rapid growth of data traffic demand. The SOP of the PDM signals also fluctuates due to external disturbances applied to the fibers. Hence, it is necessary to adaptively control the SOP of the PDM signals so that original PDM signals with x- and y-polarization (horizontal and vertical polarization) can be recovered. A polarization control scheme can also be applied for this purpose.

We recently succeeded in a field trial of 160 Gbit/s signal transmission using the PDM signals using a polarization control scheme [8]. Figure 12.4 shows a configuration of a field trial [8]. The 160 Gbit/s wavelength-division multiplexing (WDM) signals were generated by using return-to-zero differential quadrature phase-shift keying (RZ-DQPSK) modulation and PDM technology. The symbol rate of the signals was 40 GBaud. The WDM transmitter was located at *Yokohama Laboratory*.

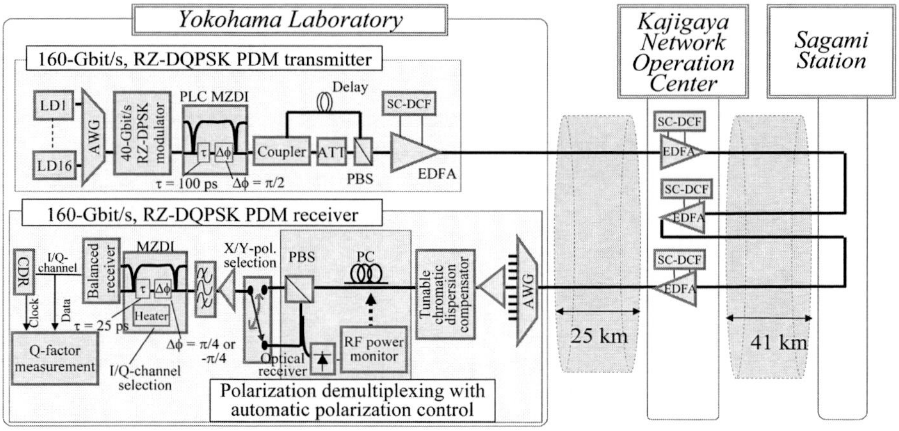

Fig. 12.4 Configuration of a field trial. Reprinted with permission from Yagi et al. [8] © 2008, Optical Society of America

The signals from a WDM transmitter were launched into four spans of installed single-mode fiber cables with a total length of 214 km. The erbium-doped fiber amplifiers (EDFAs) were used to compensate for the loss of each span.

Then, we describe each part more in details. The transmitter has 16 DFB lasers whose wavelengths range from 1534.3 to 1558.2 nm with a channel spacing of 200 GHz. The WDM signals are RZ-DPSK modulated by a phase modulator. To generate a DQPSK signal the RZ-DPSK signals are launched into a planer lightwave circuit (PLC)-based Mach–Zehnder delay interferometer (MZDI) with a four-bit delay (100 ps) and π/2 phase shift [9]. Then, the PDM signals are generated by an optical circuit with a polarizing beam splitter (PBS) as shown in Fig. 12.4.

At a receiver, the chromatic dispersion (CD) is compensated by a tunable CD compensator. Then, the signal launched into a polarization control and demultiplexing optical circuit that automatically controls the SOP of the PDM signals into initial orthogonally polarized signals. The RZ-DQPSK signals corresponding to each SOP is demodulated by a MZDI with one-bit delay, resulting in demodulated I- and Q-channel signals.

Then we carried out a field trial through 214-km-long fibers. Figure 12.5 shows the variation of Q-factor for the received four channels when no automatic polarization control was applied [8]. As is seen from Fig. 12.5, a Q-factor had been worse than the forward error correction (FEC) limit of 9.1 dB for almost all time of the measurement due to the SOP fluctuation in the installed fibers. Then, a Q-factor was measured with automatic polarization control. Figure 12.6 shows the results [8]. As we can see from Fig. 12.6, a Q-factor had been kept better than the FEC limit for more than 120 hours (5 days). From the results of the experiments above, we can say that an automatic polarization control scheme is efficient for the PDM systems that suffer from the receiver sensitivity degradation due to the SOP fluctuation in installed optical fiber cables.

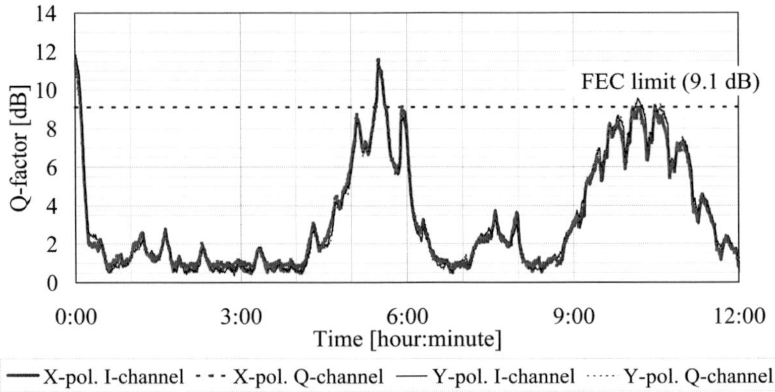

Fig. 12.5 Variation of Q-factor with time without automatic polarization control. Reprinted with permission from Yagi et al. [8] © 2008, Optical Society of America

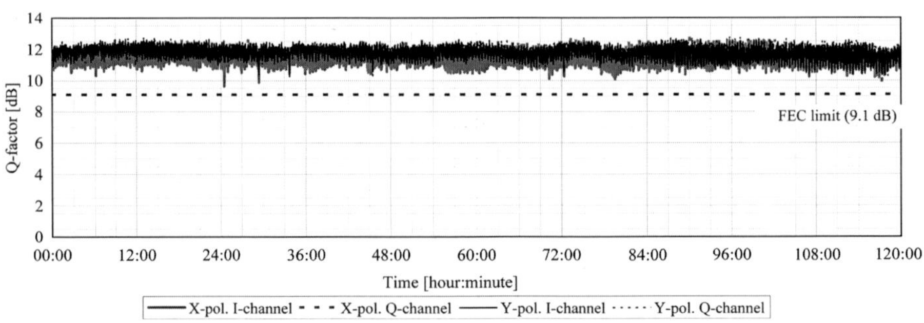

Fig. 12.6 Variation of Q-factor with time with automatic polarization control. Reprinted with permission from Yagi et al. [8] © 2008, Optical Society of America

12.3.3 Polarization-Diversity Scheme

A polarization-diversity scheme was first proposed in 1983 [10]. Figure 12.7 shows a basic configuration of a polarization-diversity receiver [10]. In a polarization-diversity scheme, the received signal light is launched into a polarization beam splitter such as a Wollaston prism which splits the input light into orthogonally polarized components. The local oscillator light is also divided into orthogonally polarized light. Then, signal and local oscillator light are introduced into separate photodetectors prepared for x- and y-polarization. Hence, perfect match of the SOP for signal and local oscillator light is achieved at each photodetector.

Intermediate frequency (IF) signals for each polarization should be combined using automatic phase control circuit in Fig. 12.7 since relative phase of the IF signals fluctuates as a result of the SOP fluctuation in the fibers.

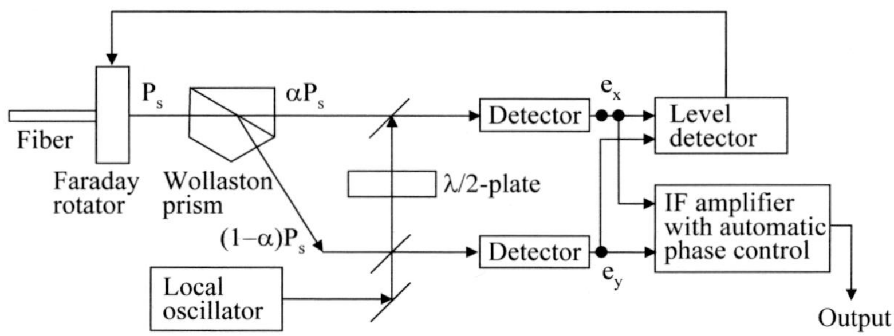

Fig. 12.7 Polarization-diversity receiver. Source: Okoshi et al. [10] Copyright © 1983 IEICE

Signal combining can also be done in the baseband domain. If the heterodyne detected signals for x- and y-polarized light are demodulated separately and recombined again, phase control is not necessary since the demodulated signals have no longer phase information. The first sea trial of coherent optical communication systems was successfully carried out in 1988 using a polarization-diversity scheme with baseband combining [4].

Figure 12.8 shows a setup for the first sea trial [4]. In the trial, 45-km-long submarine cables were installed by a cable ship. The signal with frequency-shift keying (FSK) modulation at 560 Mbit/s was launched into the cable. The fibers in the cables are loop backed at an end box of the cables, so that the signal was transmitted through 90 km and received by a polarization-diversity receiver installed in the

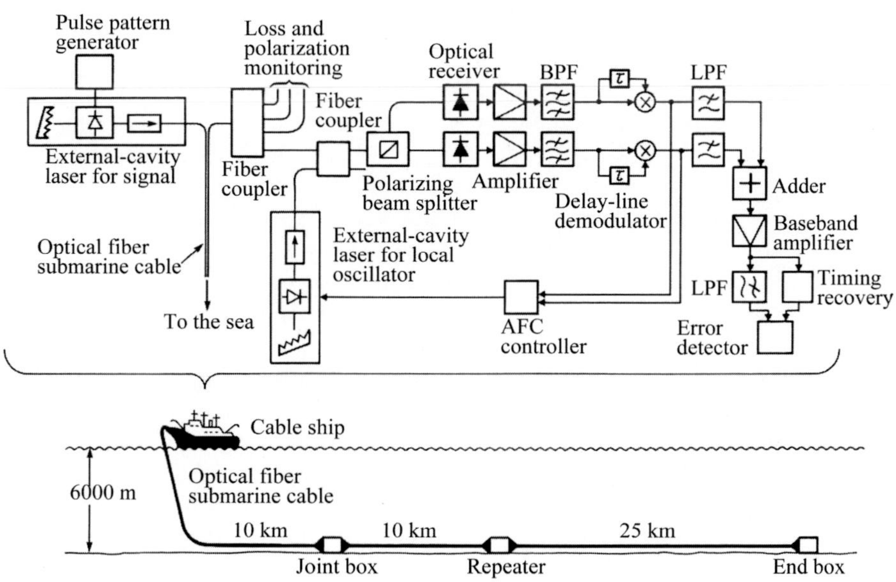

Fig. 12.8 Configuration of first sea trial of coherent systems. Reprinted with permission from Ryu et al. [4] © 1988, IET

Fig. 12.9 Variation of bit error rate with polarization-diversity operation on/off. Reprinted with permission from Ryu et al. [4] © 1988, IET

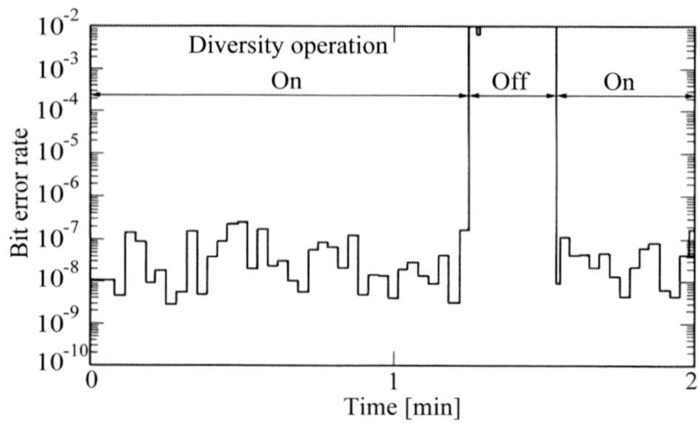

cable ship. In the polarization-diversity receiver, baseband combining is used and the bit error rate (BER) characteristics were measured with time. Figure 12.9 shows measurement results of the variation of the BER with time with a polarization-diversity operation on/off [4]. As it was already described in Sect. 12.2, fast polarization fluctuation with a speed up to 50 Hz occurred in the cables, and the BER could never be measured without a polarization-diversity operation. However, the BER had been kept to be stable with a polarization-diversity operation. The results indicate that a polarization-diversity scheme is quite promising for the SOP fluctuation problem in the fibers.

In 1987, a polarization-diversity receiver incorporating phase diversity was proposed. In a phase-diversity receiver [11], heterodyne detection with low IF frequency (offset frequency) is done for in-phase and quadrature parts of the signals, so baseband combining after demodulation recovers the data information correctly. We can easily combine phase- and polarization-diversity functions [11].

After about 20 years, coherent optical communication systems have come into a new stage, i.e., optical phase estimation is now possible by using DSP technology [12]. A receiver configuration incorporating phase and polarization diversities with DSP is shown in [12]. In the receiver, maximum ratio combining is used for digitized signals before phase estimation. Polarization insensitive operation of 10-Gbit/s binary PSK system was confirmed using the receiver [12].

We can say that a receiver incorporating phase and polarization diversities are quite effective for old analogue coherent systems as well as digital coherent systems with DSP.

A polarization-diversity configuration can also be applied to a coherent DSP receiver for the PDM systems. We can adopt the same configuration as a polarization-diversity receiver for coherent systems. In a receiving part, some combination of initial x- and y-components of the PDM signals are obtained at both receivers, so the polarization control is necessary to recover the initial linear SOP. For this purpose, constant modulus algorithm (CMA) that is a digital control algorithm for the x- and y-components of the received signals can be applied [13]. The CMA can be considered as an active polarization control in an electrical domain.

Figure 12.10 summarizes the evolution of the polarization-diversity schemes. As was already discussed, a receiver incorporating polarization and phase diversities

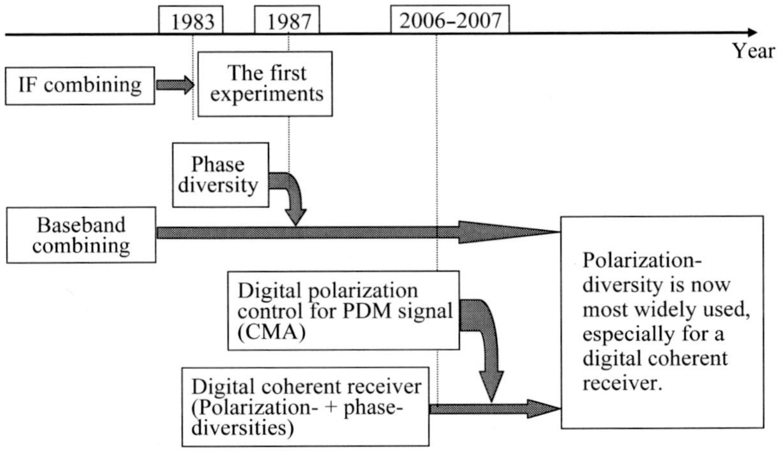

Fig. 12.10 Evolution of polarization-diversity schemes

with digital signal processing is now widely used for a digital coherent receiver. Moreover, the scheme can also be applied to the PDM signals by adding digital polarization control (CMA) in a receiver.

12.3.4 Polarization Scrambling Scheme

A polarization scrambling scheme was proposed in 1987 [14]. In early experiments in [14], the SOP of the data signals with a rate of 20 Mbit/s was scrambled at a rate of 80 Mbit/s. If a rate of the polarization scrambling is sufficiently higher than a data rate, the SOP of the signal and local oscillator light perfectly matches for almost half a bit period. Hence, coherent detection is possible by paying 3-dB receiver sensitivity penalty in the scheme [14]. However, a polarization scrambling scheme had not been studied actively since a scrambling speed should be sufficiently higher than a data rate, which is unrealistic requirement in commercial systems.

It was also found that polarization scrambling at a speed that corresponds to the duration of one FEC frame was quite effective for mitigation of the system penalty caused by the PMD in long-haul submarine cable applications [15].

As mentioned above, polarization scrambling schemes are not suitable for improving receiver sensitivity degradation in coherent systems, but they found other applications, e.g., in submarine cable systems.

12.4 Evolutions of Technology for SOP Fluctuation Problem

Figure 12.11 summarizes how technology has evolved for SOP fluctuation problem. As was previously discussed, the PMF and polarization scrambling methods have not been adopted as the solutions for the problem, but they found other applications. Polarization control schemes have been used and they are now also being applied

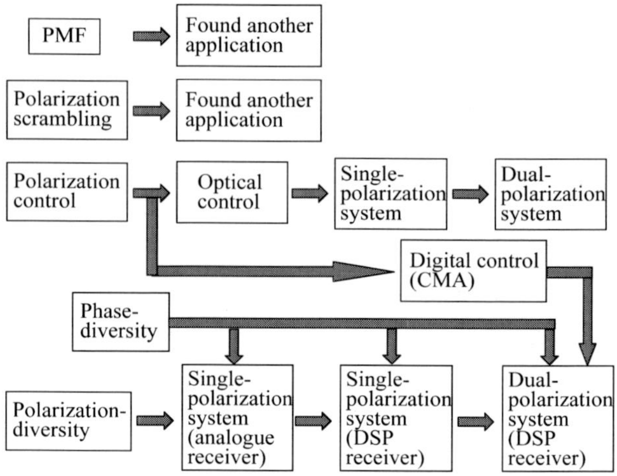

Fig. 12.11 Evolution of various schemes for SOP fluctuation problem

to the PDM systems. However, the polarization control schemes require active control subsystems for controlling the SOP, which may impose additional cost on the systems.

Polarization-diversity schemes have been used from old analog coherent systems to the up-to-date digital coherent systems with a DSP receiver, and they can also be applied to the PDM systems with digital polarization control in software domain.

Among the schemes, polarization-diversity schemes are the most widely used in research fields of coherent optical communication systems. The author believes that the polarization-diversity schemes will evolve into the most practical solutions for the future optical communication systems.

References

1. T. Okoshi, K. Kikuchi, Frequency stabilization of semiconductor lasers for heterodyne-type optical communication systems. Electron. Lett. **16**(5), 179–181 (1980).
2. T. Okoshi, Single-polarization single-mode optical fibers. IEEE J Quant Electron. **QE-17**(6), 879–884 (1981).
3. S. Ryu, *Coherent lightwave communication systems*, (Artech House, Boston/London, 1995).
4. S. Ryu, S. Yamamoto, Y. Namihira, K. Mochizuki, H. Wakabayashi, First sea trial of FSK heterodyne optical transmission system using polarisation diversity. Electron. Lett. **24**(7), 399–400, March (1988).
5. P.M. Krummrich, E.-D. Schmidt, W. Weiershausen, A. Mattheus, Field trial results on statistics of fast polarization changes in long haul WDM transmission systems. OFC/NFOEC, OThT6, (2005).
6. Y. Kidoh, Y. Suematsu, K. Furuya, Polarization control on output of single-mode optical fibers. IEEE J. Quant. Electron. QE-17(6), 991–994 (1981).
7. B. Koch, A. Hidayat, H. Zhang, V. Mirvoda, M. Lichtinger, D. Sandel, R. Noe, Optical Endless polarization stabilization at 9 krad/s with FPGA-based controller. IEEE Photon. Technol. Lett. **20**(12), 961–963 (2008).

8. M. Yagi, S. Satomi, S. Ryu, Field trial of 160-Gbit/s, polarization-division multiplexed RZ-DQPSK transmission system using automatic polarization control. OFC/NFOEC, OThT7, (2008).
9. B. Milivojevic, A.F. Abas, A. Hidayat, S. Bhandare, D. Sandel, R. Noe, M. Guy, M. Lapointe, 160 Gbit/s, 1.6 bit/s/Hz RZ-DQPSK polarization-multiplexed transmission over 230 km fiber with TDC. ECOC'04, We1.5.5, (2004).
10. T. Okoshi, S. Ryu, K. Kikuchi, Polarization-diversity receiver for heterodyne/coherent optical fiber communications. IOOC'83, 30C3–2.
11. T. Okoshi, Y.H. Cheng, Four-port homodyne receiver for optical fibre communications comprising phase and polarisation diversities. Electron. Lett. **23**(8), 377–378 (1987).
12. S. Tsukamoto, Y. Ishikawa, K. Kikuchi, Optical homodyne receiver comprising phase and polarization diversities with digital signal processing. ECOC2006, Mo4.2.1.
13. S.J. Savory, G. Gavioli, R.I. Killey, P. Bayvel, Transmission of 42.8 Gbit/s polarization multiplexed NRZ-QPSK over 6400km of standard fiber with no optical dispersion compensation. OFC/NFOEC 2007, OTuA1.
14. T.G. Hodkinson, R.A. Harmon, D.W. Smith, Polarisation-insensitive heterodyne detection using polarisation scrambling. Electron. Lett. **23**(10), 513–514 (1987).
15. C.R. Davidson, H. Zhang, Y. Cai, L. Liu, J. –X. Cai, A. Pilipetskii, M. Nissov, N.S. Bergano, Direct measure of system margin enhancement by polarization scrambling. OFC2004, WE1.

Part III
Opto-electronics Devices

Chapter 13
Semiconductor Lasers for High-Density Optical Communication Systems

Hiroyuki Uenohara

Abstract Semiconductor lasers have several advantages compared to other types of lasers (a fiber ring laser, a solid-state laser, a gas laser), such as compactness, integrability to other components (a modulator, a power monitor photodiode, a wavelength tuning component). In addition, a monolithic tunable laser can be fabricated in the semiconductor by combining wavelength tuning regions and a phase control region together with an active region. That is why a monolithic integrated semiconductor tunable laser is very attractive not only for a compact light source for large-capacity, optical communication systems but also for economical backup resource. However, there are several technical issues to achieve high-quality semiconductor tunable lasers applicable to high-density optical communication systems due to the severe requirements of narrow spectral linewidth, wide tunability, and high wavelength stability.

In this chapter, several technical methodologies for improving spectral linewidth and wavelength tunability are overviewed.

13.1 Requirements to Spectral Linewidth

For realization of large-capacity and long-haul WDM optical transmission systems, significant distortion by both chromatic and polarization mode dispersion should be overcome. For that solution, modulation scheme with narrow signal bandwidth is the recent trends. For example, multi-level modulation format, such as DQPSK and M-QAM, and multi-carrier modulation, such as OFDM, have been intensively investigated. For these modulation schemes, high-sensitive receiver is necessary, and therefore, transmission signal is detected with digital coherent detection technique. Therefore, even for a tunable semiconductor laser, narrow spectral linewidth

H. Uenohara (✉)
Precision and Intelligence Laboratory, Tokyo Institute of Technology, 4259 Nagatsuta, Midori-ku, Yokohama 226-8503, Japan
e-mail: uenohara.h.aa@m.titech.ac.jp

M. Nakazawa et al. (eds.), *High Spectral Density Optical Communication Technologies*, Optical and Fiber Communications Reports 6, DOI 10.1007/978-3-642-10419-0_13,

and frequency stability should be necessary for suppressing the phase noise and the fluctuation of intermediate frequency.

The simulation results of power penalty as a function of spectral linewidth per laser per data rate have been reported by Seimetz at OFC2008 [1]. In these simulations, phase compensation and decoding in the digital signal processors are assumed. As can be seen in this figure, the required spectral linewidth depends on the modulation scheme. In case of data rate of 40 Gbps, more than 1 MHz linewidth is sufficient for QPSK and 8 PSK. But for increasing multi-level up to square 16 QAM, less than 100 kHz linewidth is required.

Research of laser diodes with narrow spectral linewidth has been conducted for coherent optical communication systems from late 1980s to early 1990s. The main research reports are summarized in Table 13.1 [2–7]. In the LD itself, the narrowest data of 3.6 kHz was reported in the corrugation pitch-modulated DFB laser with strained MQW active layer in 1993 [2]. In the DBR laser, 40 kHz linewidth was reported in 1996 [3]. By using electrical and optical feedback schemes, the linewidth was drastically reduced to 7 Hz [7]. But to avoid the complicated architecture and to achieve the compact local oscillators for WDM systems, LDs without external components are desirable. Figure 13.1 indicates the development of the reported data of spectral linewidth [2, 3, 8–14]. Several types of active layers and structures such as distributed feedback (DFB) and distributed Bragg reflector (DBR) are compared in this figure. There is a little difference between DFB and DBR structures, and spectral linewidth is reduced by introducing MQW and strained layer structure into the active layer. The main technical points for reducing spectral linewidth will be reviewed in the following sections.

Table 13.1 Reports of semiconductor lasers with narrow spectral linewidth

	Schemes	Mechanism	Features ● : Pros, ○ : Cons	$\Delta\nu$	Ref.
LD	Strained MQW DFB CPM DFB	α parameter ↘ spatial hole burning ↘	● no mode hop	3.6kHz	2
LD	MQW DBR	α parameter ↘	● small spatial hole burning → $\Delta\nu$ ↘ ○ mode hop	40kHz	3
LD	External Cavity DFB	Q ↗	○ mode instability by phase mismatch	900kHz	4
Elec./Opt. Control	Electrical Feedback	Freq. noise detect → Bias current control	● High sensitivity ● Narrower than Quantum noise	250Hz	5
Elec./Opt. Control	Optical Feedback	Q ↗	● FP resonator : wide bandwidth	10kHz	6
Elec./Opt. Control	Elec./Opt. Feedback	Opt. + Elec. feedback	○ complicated	7Hz	7

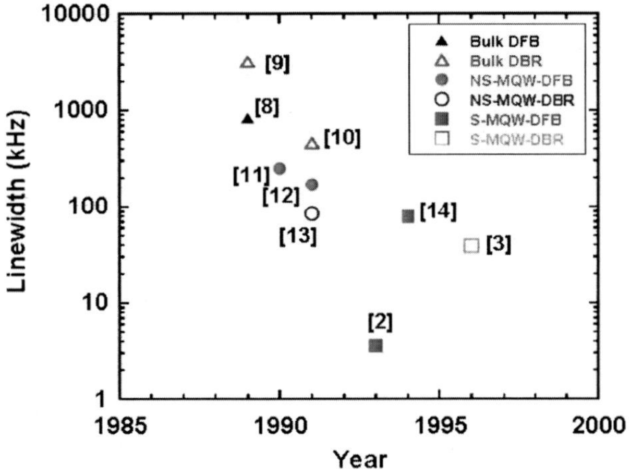

Fig. 13.1 Reported spectral linewidth vs. reported year

13.2 Spectral Linewidth of Semiconductor Lasers

The linewidth is expressed by modified Schawlow–Towns equation [Eq. (13.1)] [15].

$$\Delta\nu = \frac{v_g^2 \cdot h\nu \cdot n_{sp} \cdot \alpha_{th} \cdot (\alpha_{loss} + \alpha_{th})}{8\pi P_o}(1+\alpha^2), \tag{13.1}$$

where v_g is group velocity of light, h is Plank's constant, ν is frequency of light, n_{sp} is spontaneous emission factor, α_{th} is threshold loss, α_{loss} is laser cavity loss, α is linewidth enhancement factor (or α-parameter), and P_o is output optical power, respectively. As can be seen in Eq. (13.1), it is proportional to threshold gain, squared α-parameter, and inversely proportional to optical power. So, it can be controlled by designing these parameters.

Now, several kinds of spectrum narrowing scheme are explained briefly.

(1) α-parameter control

First, reducing α-parameter is effective. α-parameter is defined as the ratio of refractive index change against carrier density to the differential gain coefficient as shown in Eq. (13.2):

$$\alpha = -\frac{4\pi}{\lambda}\frac{\frac{\partial n}{\partial N}}{\frac{\partial g}{\partial N}}. \tag{13.2}$$

For increasing differential gain, introduction of strained MQW active layers is successful [2, 3, 11–14]. α-Parameter is reduced about half, and the resultant linewidth becomes about one-fourth compared to bulk active layers. In case of

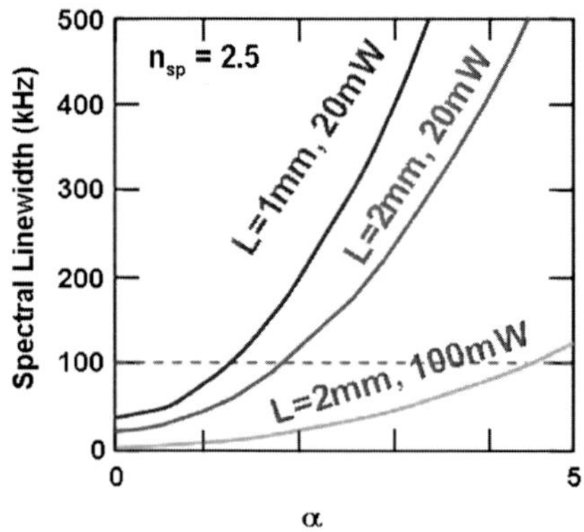

Fig. 13.2 Simulation results of spectral linewidth as a function of α-parameter

decreasing the refractive index change, it can be down to zero in quantum dots, so the maturity of quantum dot structure is expected [16, 17].

The simulation of spectral linewidth as a function of α-parameter is indicated in Fig. 13.2. In the calculations, wavelength of 1.55 μm and n_{sp} of 2.5 were assumed. Cavity length of 1 and 2 mm, and output power of 20 and 100 mW were used as parameters. If output power is set at 20 mW, α-parameter of less than 2 is required to get linewidth of less than 100 kHz, so the selection of active layer material is still important.

(2) Reduction of threshold gain

Next, reducing threshold gain is also one of the most effective approaches for narrow spectral linewidth. The threshold gain α_{th} is expressed as:

$$\alpha_{th} = \alpha_{in} + \frac{1}{L} \ln\left(\frac{1}{R_{eff}}\right) \tag{13.3}$$

and

$$R_{eff} = \tanh^2(\kappa L), \tag{13.4}$$

where α_{in} is internal loss of the laser cavity, L is cavity length, R_{eff} is effective mirror reflectivity in case of DFB or DBR lasers (lossless case, Fresnel reflectivity of the cleaved facet in case of a Fabry–Perot laser), and κ is a coupling coefficient of the grating. From Eqs. (13.1) and (13.3), we can see that spectral linewidth decreases with large κL or long cavity structure due to the reduction of cavity loss. Simulation results of the spectral linewidth as a function of cavity length are shown in Fig. 13.3. In simulation, α_{in} of $10\,\mathrm{cm}^{-1}$, R_{eff} of 0.3, n_{sp} of 2.5, and wavelength of 1.55 μm were assumed. As can be seen in this figure, longer than 700-μm cavity is necessary for achieving linewidth of less than 100 kHz. But too large κL for DFB lasers causes the rebroadening by spatial hole burning, so careful design is needed. Some examples will be shown in the next section [2, 11, 13].

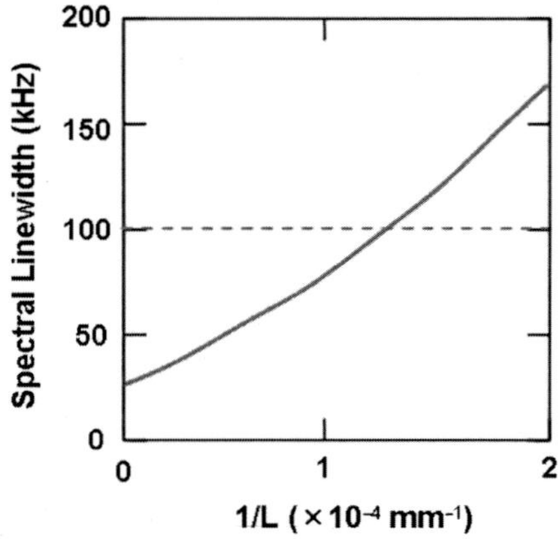

Fig. 13.3 Simulation results of spectral linewidth as a function of inverse cavity length

(3) Suppression of linewidth rebroadening

In Fig. 13.4, spectral linewidth as a function of reciprocal output power is indicated schematically. Ideally, it decreases with increasing output power, and it approaches to zero according to Eq. (13.1). But in reality, it cannot be down to zero because there are mainly two factors for avoiding spectral narrowing; one of the causes is spectral rebroadening, by which it increases with increasing output power unexpectedly as shown in Fig. 13.5 [11]. So, we have to care about linewidth rebroadening. In the DFB lasers, reduction of spatial hole burning is essential. In the quarter wavelength-shifted DFB lasers, the photon density along the cavity shows the peak intensity at the phase-shifted region as shown in Fig. 13.6. Under the high-output power condition, the carrier density along the cavity becomes non-uniform due to the strong stimulated emission, and this causes linewidth rebroadening. Then, special structure for flattening the photon density such as corrugation pitch modulation [2] or multi-electrode structures [18] should be needed. The spontaneous emission recombination in the separate confinement heterostructure (SCH) or barrier layers is found to be another cause of linewidth rebroadening [19].

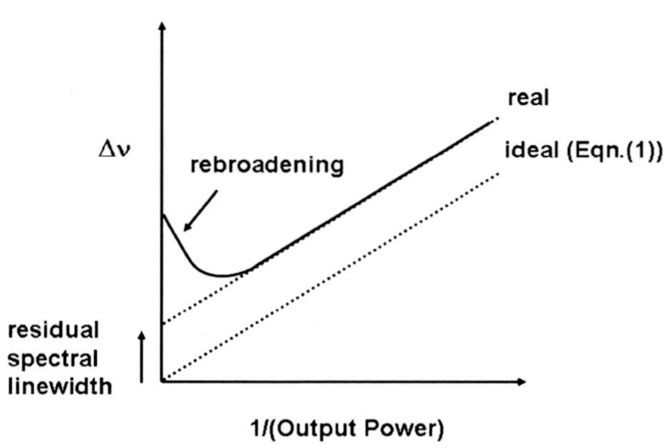

Fig. 13.4 Schematic image of spectral linewidth as a function of inverse output power

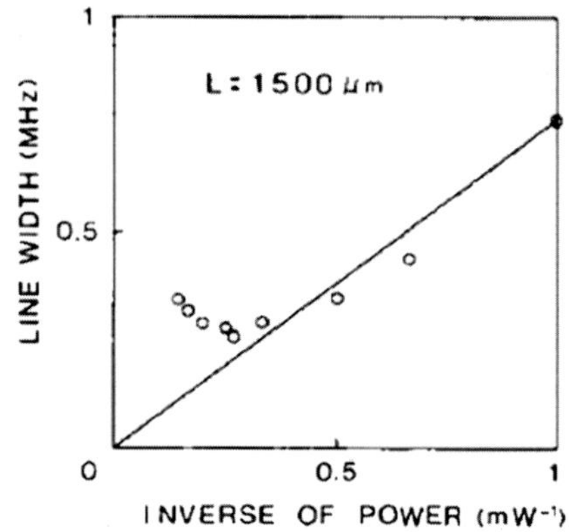

Fig. 13.5 Experimental results of spectral linewidth as a function of inverse output power (after M. Kitamura et al. [9])

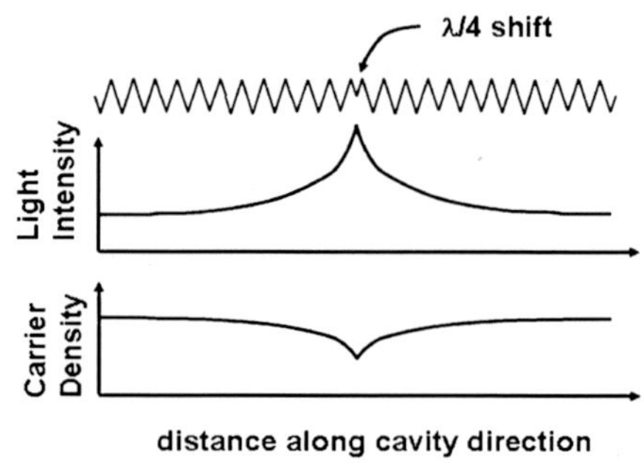

Fig. 13.6 Schematic image of grating structure (*top*), internal photon density (*middle*), and carrier density (*bottom*) along the cavity of a quarter wavelength-shifted DFB laser

(4) Residual spectral linewidth

Another one is the residual spectral linewidth, and it is defined as the linewidth offset derived from the interpolation of the linear region as shown in Fig. 13.4. The reason for this is considered to be $1/f$ noise [20], and solution for this is still being under investigation.

13.3 Reports of Narrow Spectral Linewidth Semiconductor Lasers

Now, I will show you some reports of the narrow spectral linewidth semiconductor lasers.

13.3.1 DFB and DBR Lasers

As already mentioned in Table 13.1, there are mainly two types of single longitudinal mode semiconductor lasers with narrow spectral linewidth. One is a DFB laser, and some spectral narrowing structures are shown in Fig. 13.6. A DFB laser has peri-

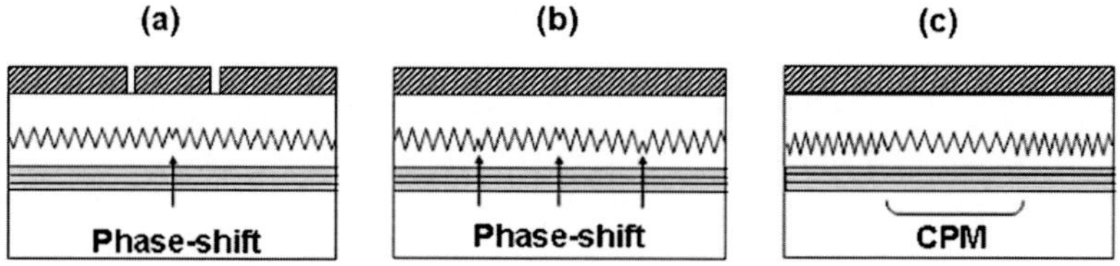

Fig. 13.7 Several types of DFB lasers for flattening the internal photon density. (**a**) Multi-electrode. (**b**) Multi-phase-shift. (**c**) Corrugation pitch modulation

odical grating structure in the waveguide layer on or below the MQW active layer with the grating pitch $\Lambda = m \cdot (\lambda_{\text{Bragg}}/2n_{\text{eff}})$, where m is an integer (m=1 in usual), λ_{Bragg} is Bragg wavelength, and n_{eff} is effective refractive index. Figure 13.7a–c are a multi-electrode structures [18], a multi-phase-shift structure [8], and a corrugation pitch modulation structure [2], respectively. In each structure, photon density becomes flatter than the conventional one electrode, one quarter wavelength shifted, and uniform grating DFB laser by non-uniform current injection, multi-peaking of the photon density, or phase mismatch in the CPM structure. Figure 13.8 indicates the corrugation pitch-modulated (so-called CPM) DFB laser [12]. To flatten the photon density along the cavity, the grating pitch at the center region is slightly longer than that at the rest of the region. The active layer consists of four periods of 1% compressively strained InGaAs and lattice-matched InGaAsP MQW structure. The cavity length is 1200 μm. Then, several schemes such as strained quantum well for reducing α-parameter, long cavity, and grating structure for flattening photon density along the cavity are introduced in this laser. Calculation of the light intensity as a function of the position in the cavity is indicated in Fig. 13.9a. With

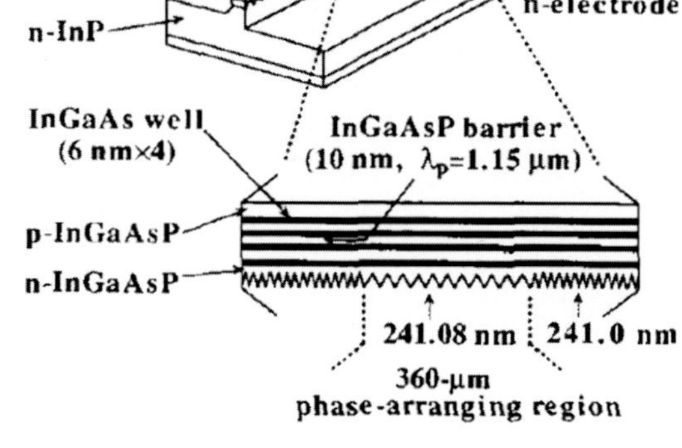

Fig. 13.8 Schematic structure of a corrugation pitch-modulated (CPM) DFB laser (after M. Okai et al. [12])

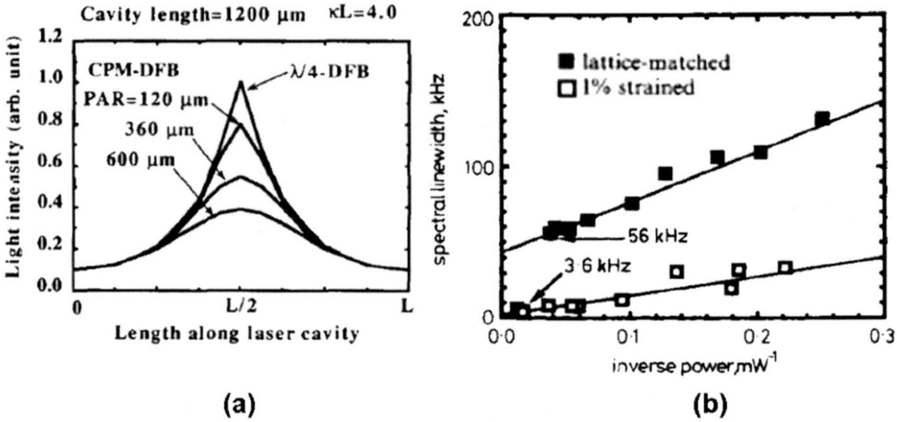

Fig. 13.9 Simulation results of internal photon density along the cavity (**a**) and experimental results of spectral linewidth as a function of inverse output power (**b**) (after M. Okai et al. [2, 12])

increasing the length of the pitch modulation region, the peak intensity decreases and the uniformity of the light intensity increases. In case of the pitch modulation length of 360 μm, the peak intensity becomes about half of that of the uniform pitch DFB laser. The spectral linewidth as a function of inverse power is shown in Fig. 13.9b. The record narrow linewidth of 3.6 kHz was achieved at the output power of 50 mW. Note that the noise from the power supply and the ambient temperature fluctuation broaden the linewidth, and battery and vacuum chamber were used in this experiment.

Another candidate is a DBR laser, and typical structure is indicated in Fig. 13.10. It has grating structures in the waveguide layer like in a DFB laser, but they are used as wavelength-selective reflection mirrors. Then, they are formed at the front and the rear sides of the laser cavity, and the active layer is formed between them. In case of the DBR laser, it is generally said that spectral linewidth can be narrowed compared to the DFB laser because the photon density along the cavity is intrinsically flat. The DBR laser with spectral linewidth of 40 kHz has been reported [3]. This laser consists of etched grating structure on one side, and the cleaved facet on the other side (Fig. 13.11a). The active layer is composed of compressively strained InGaAs and GaAs MQW structure. The cavity length is totally 620 μm. By reducing

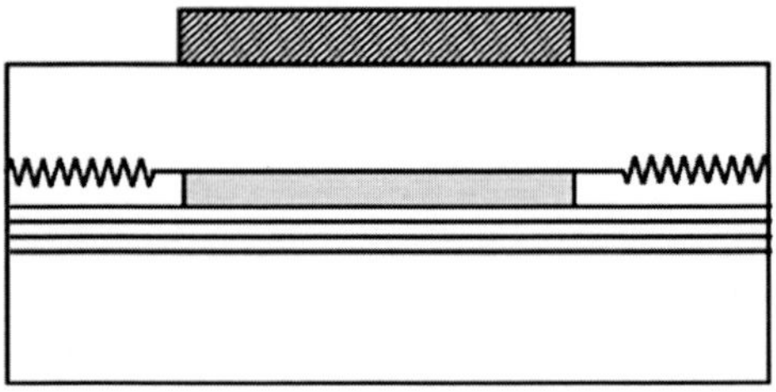

Fig. 13.10 Schematic structure of a DBR laser

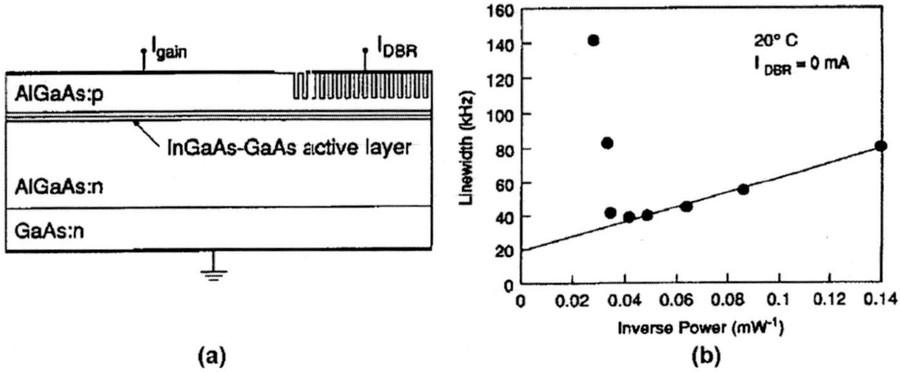

Fig. 13.11 Schematic structure of an asymmetric cladding DBR laser (**a**) and spectral linewidth as a function of inverse output power (**b**) (after G.M. Smith et al. [3])

the surface and substrate mode losses, ultra-narrow spectral linewidth was achieved successfully at the output power of 25 mW (Fig. 13.11b).

13.3.2 External Cavity Semiconductor Lasers

When the external cavity structure is introduced, the longer cavity length than the semiconductor laser itself works for reducing the cavity loss and the threshold gain, and therefore, narrow linewidth can be achieved. An external cavity laser consisting of semiconductor-based gain chip and a PLC-based Bragg grating is produced [21]. The specification of spectral linewidth is 10 kHz. Good temperature stability of the lasing wavelength can be obtained due to the smaller thermal coefficient of refractive index in silica than semiconductors.

13.4 Comparison between Several types of Narrow Spectral Linewidth Semiconductor Lasers

Then, what type of laser is suitable, DFB, DBR, and external cavity lasers? By all structures, narrow spectral linewidth of 10–40 kHz can be achieved. But in case of the DFB laser, no mode-hop operation can be obtained, and this point is the advantage of the DFB laser over the DBR laser.

On the other hand, if we desire to obtain wide tunability in small footprint, the DBR laser is superior to other types of lasers. And it is generally said that the DBR laser is superior to the DFB laser due to the flat light distribution along the cavity.

As for the external cavity laser, the cavity loss can be small due to long cavity. And flexible filter design can be introduced.

Then, we should still make efforts to achieve wide tunable, narrow spectral linewidth lasers.

13.5 Reported Technologies and Issues of Tunable Semiconductor Lasers

Next, the possibility of tunable semiconductor lasers [22, 23] with narrow spectral linewidth and high frequency stability will be considered. Moreover, stable lasing wavelength is important to get stable tracking in the coherent detection as well as narrow spectral linewidth. Previous to these considerations, several schemes for tunable lasers are reviewed. Many types of tunable lasers have been intensively studied such as a multi-electrode DFB [18, 24, 25] (Fig. 13.12a), a tunable twin guide (TTG) DFB laser with a tuning layer in the grown layers [26] (Fig. 13.12b), a multi-electrode DBR laser [27] (Fig. 13.12c), a vertical cavity surface emitting laser (VCSEL) with cantilever mirror [28, 29] (Fig. 13.12d), an external cavity lasers with a movable grating [30] (Fig. 13.12e), and so on. In principle, lasing wavelength is tuned by current injection or temperature control for DFB and DBR lasers. Refractive index of the grating structure or a tuning layer is changed, and then the Bragg wavelength is also varied. But in conventional structure, refractive index change of the semiconductor (GaAs-based material for 0.85 μm range, and InP-based material for 1.3 and 1.55 μm ranges) is smaller than 1%, and therefore, the lasing wavelength (Bragg wavelength) change is also less than 1%. For the reported cases, a tuning range was 7.2 nm for a multi-electrode DFB laser [24], 8 nm for a TTG DFB laser [26], and 5.8 nm for a multi-electrode DBR laser [27].

Then, to get wide tunability within the whole C-band and/or L-band, special structure should be introduced, for example, arrayed DFB lasers with several wavelengths [31, 32], multi-electrode structure such as a super structure grating (SSG), a sampled grating (SG), a digitally sampled grating (DS) DBR laser [33–35], and the

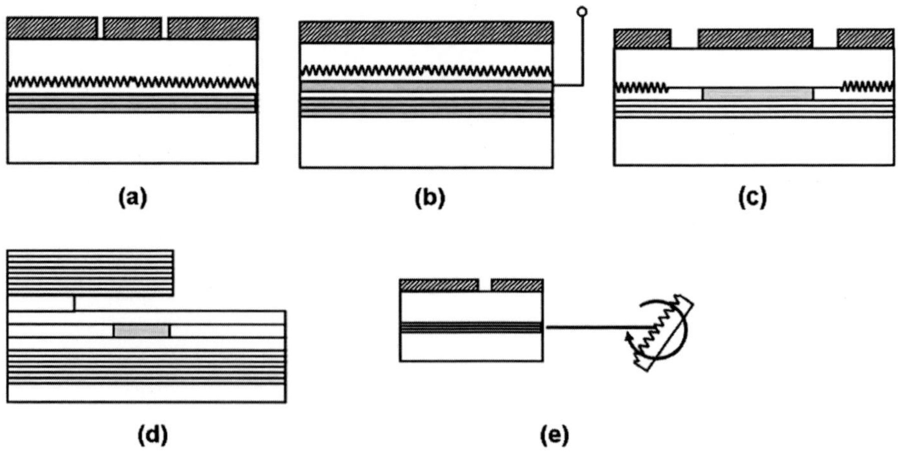

Fig. 13.12 Schematic structure of several types of semiconductor tunable lasers. (**a**) A multi-electrode DFB laser. (**b**) A tunable twin guide DFB laser. (**c**) A multi-electrode DBR laser. (**d**) A tunable vertical cavity surface emitting laser (VCSEL). (**e**) An external cavity laser with a semiconductor optical amplifier

Table 13.2 Reports of several types of tunable lasers (tuning range and spectral linewidth)

Structure	Mechanism	Tuning range (nm)	Tuning speed	$\Delta\nu$	Ref.
DFB-LD array	Target λ select+ n change by T	37 nm by 12 DFBs	~ ms	<4 MHz	28
SSG/SG-DBR	Bernier effect+ n change by I or T	40 nm ($\Delta\nu = 400$ kHz) 140 nm (Q.C.)	~ ns	400 kHz	29
External cavity laser (SOA+ tunable filter)	Bernier effect+ n change by I or T	132 nm	~ μs	120 kHz	34

integration structure using an SOA and a wavelength-selective filter such as a ladder filter, ring resonators, a movable grating with an MEMS, or a liquid crystal tunable filter [36–42]. This section focuses on those structures. Table 13.2 summarizes both spectral linewidth and wavelength tuning range of several types of tunable lasers. And Fig. 13.13 illustrates those data. As can bee seen in the figure, an external cavity laser has narrower spectral linewidth and wider tuning range than a DBR laser. But the tuning speed of the DBR laser is faster in principle than the external cavity laser with an MEMS or a liquid crystal tunable mirror because of the faster response of semiconductor. But note that current injection or temperature control should be used when tuning, and then, carrier plasma effect and thermal drift are dominant for wavelength drift, and these effects should be suppressed. In case of a tunable VCSEL with cantilever mirror, the position of cantilever is controlled by an electrostatic force or bending with different thermal expansion between a top layer and bottom layer of the DBR mirror. Large change of the spacing between the active layer and the top mirror can be obtained, then wide tunability covering the whole C-band has been achieved [28, 29]. However, the volume of the active layer is much smaller than that of the waveguide-type lasers, and then the spectral linewidth reported is from several megahertz to several tens of megahertz [43–45].

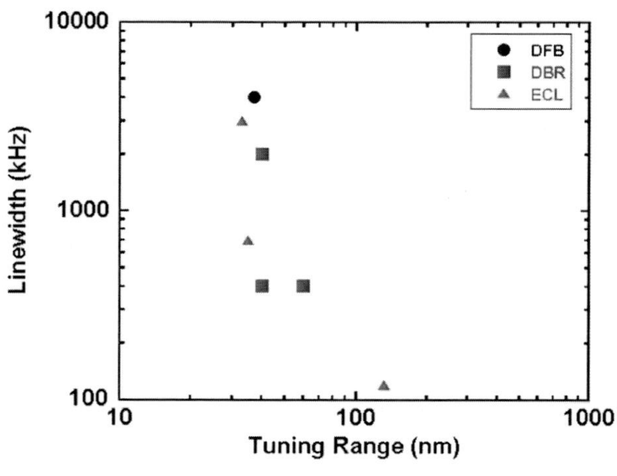

Fig. 13.13 Reported data of spectral linewidth and tuning range

13.5.1 DFB Laser Array (Wavelength Selectable Laser)

In a DFB laser, wavelength tuning by thermal control is limited by the thermal coefficient of refractive index (or lasing wavelength), the increase of threshold current, and the decrease of output power. The thermal coefficient of the lasing wavelength is about 0.08 nm/K, and therefore, tuning range is around 3 nm with 40° temperature change. So, in the tuning range of the whole C-band (30 nm wavelength range), more than 10 DFB lasers are needed in a single tunable laser. A typical monolithic wavelength tunable laser (or this type of laser is often called a wavelength selectable laser) is shown in Fig. 13.14a. Ten to twelve DFB lasers with different lasing wavelength and the same spacing are fabricated on the same wafer, and each output is combined into a single waveguide by using a multi-mode interference (MMI) coupler. To compensate the coupling loss, a booster SOA is also integrated on the same chip. Spacing of 25 GHz, 180 ch (37 nm tuning range), 50 mW output power at room temperature, and linewidth of <4 MHz have been reported [32].

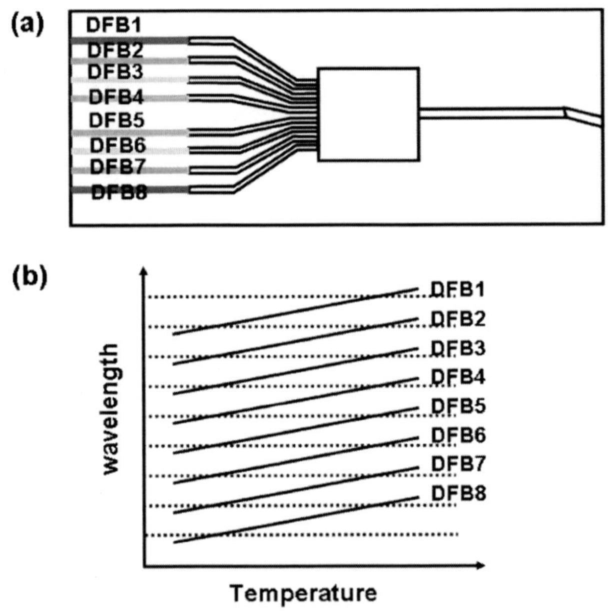

Fig. 13.14 Schematic structure of a DFB laser array (**a**) and concepts of wide wavelength tuning (**b**)

13.5.2 Super Structure Grating or Sampled Grating DBR Laser

In a DBR laser, wider tuning range than a DFB laser can be achieved. Schematic structure of a three-electrode DBR laser is shown in Fig. 13.15a. It consists of three regions: an active region, a phase control region, and a wavelength tuning region. Grating structures are formed in the wavelength tuning region. In case of the uniform pitch grating structure, the Bragg wavelength is shifted toward shorter or longer

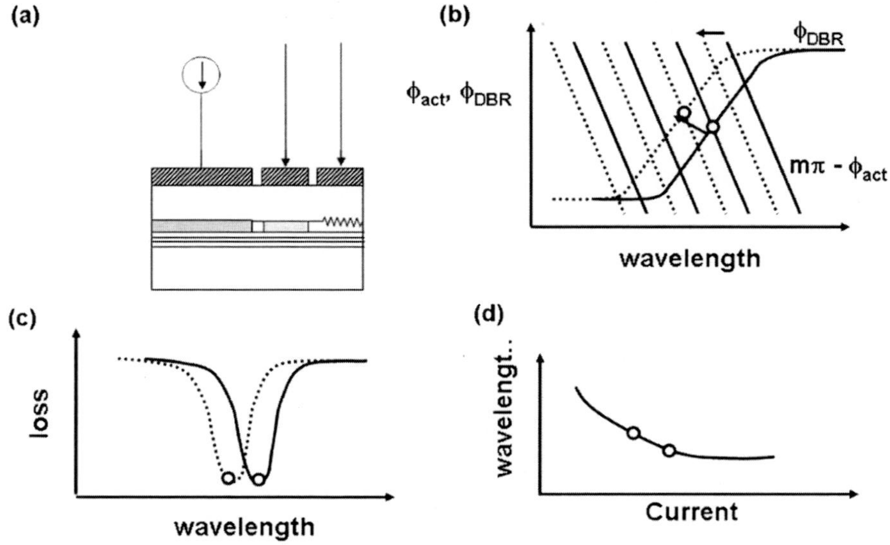

Fig. 13.15 Operation principle of a multi-electrode tunable DBR laser. (**a**) Cross-sectional view. (**b**) Optical phase as a function of wavelength. (**c**) Optical loss by the DBR region. (**d**) Lasing wavelength as a function of tuning current

wavelength side by increasing or decreasing injection current into the wavelength tuning region due to carrier plasma effect. Or the same effect can be achieved by decreasing or increasing device temperature due to the thermal change of refractive index. Consider the optical phase at the interface between the active region and the wavelength tuning region. Assume that the optical phase in the active region is ϕ_{act} and that in the wavelength tuning region is ϕ_{DBR}; the phase matching condition for lasing is expressed by Eq. (13.5) as

$$2\phi_{act} + 2\phi_{DBR} = 2m\pi, \tag{13.5}$$

where m is an integer. The change of ϕ_{act} (Fabry–Perot mode) and $m\pi - \phi_{DBR}$ (DBR mirror) as a function of wavelength is drawn in Fig. 13.15b. Before tuning and after tuning are shown by the solid lines and the broken lines, respectively. Under initial condition before tuning (solid lines), the lasing wavelength is obtained at the cross point between two curves closer to the wavelength with minimum optical loss (Fig. 13.15c). When the Bragg wavelength is varied toward shorter wavelength, the minimum optical loss point is also shifted to shorter wavelength. If the wavelength after tuning is closer to the adjacent mode, the lasing wavelength is changed to the next mode, and this is called mode hopping. But if we shift the Fabry–Perot mode toward shorter wavelength by injecting current into the phase control region (broken lines), the lasing wavelength keeps to the same mode as shown in Fig. 13.15d. Therefore, very wide continuous wavelength shift can be achieved.

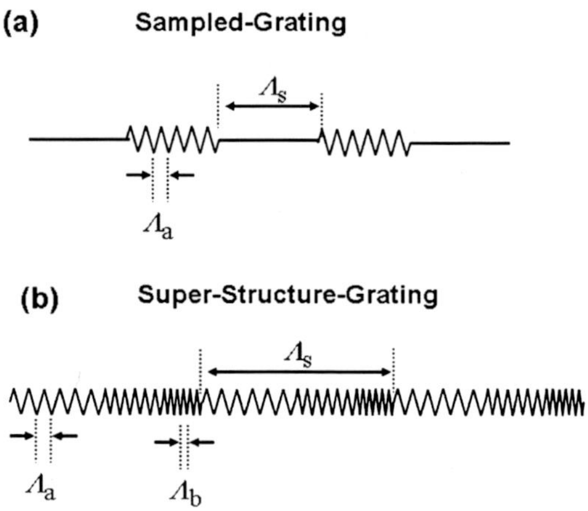

Fig. 13.16 Structure of a sampled grating (**a**) and a super structure grating (**b**)

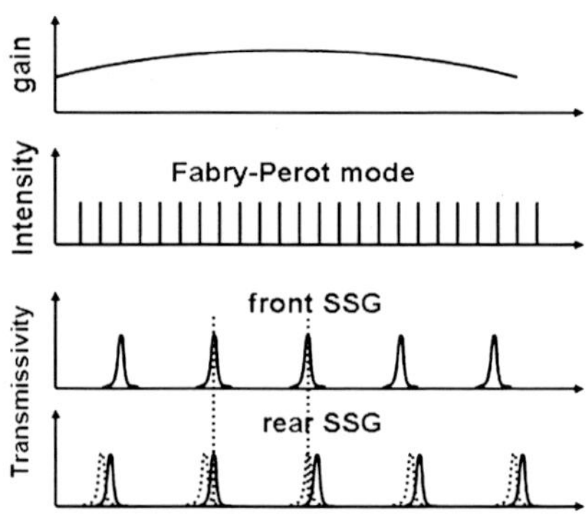

Fig. 13.17 Operation principle of an SG or SSG DBR laser

To extend the tuning range, an SG or an SSG is introduced. The SG structure consists of several gratings with a period of Λ_a and a spacing of Λ_s (Fig. 13.16a). As for an SSG-DBR laser, the grating has chirped with the periods from Λ_a to Λ_b in one period of Λ_s (Fig. 13.16b). It has periodical reflectivity as indicated in Fig. 13.17. But the front and rear wavelength tuning regions have slightly different periods, and the period of the reflection peaks is also different. Then, the lasing wavelength is mode-hopped to the adjacent mode from solid to broken lines. Because of Vernier effect, very wide tuning is possible with small current change. The spectral linewidth of $<$400 kHz and the tuning range of 40 nm has been reported for an SSG-DBR laser [33], but the tuning range can be widened up to 140 nm under quasi-continuous tuning mode [46].

13.5.3 External Cavity Laser

Other type of semiconductor-based lasers for wide tuning is an external cavity laser. The basic structure is shown in Fig. 13.18a. It is composed of a gain medium (semiconductor optical amplifier is one of the most promising candidates), a tunable band

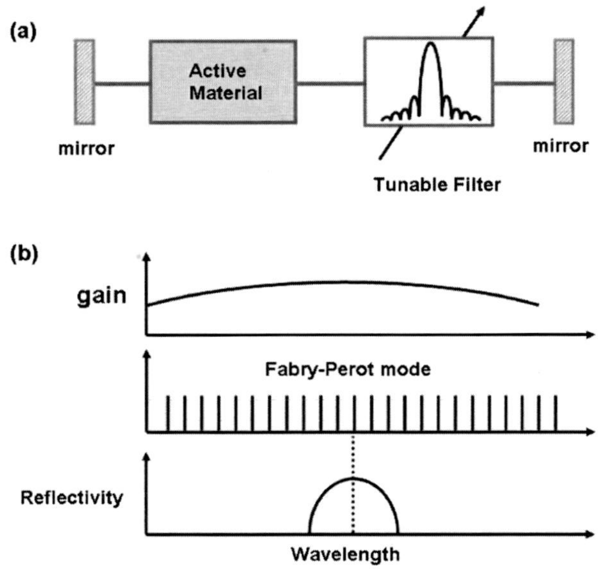

Fig. 13.18 Structure of an external cavity tunable laser (**a**) and its operation principle (**b**)

pass filter, and mirrors on both facets. Gain, Fabry–Perot modes, and the transmission spectrum of the band pass filter are indicated in Fig. 13.18b. Lasing wavelength should be within the optical gain bandwidth, and it is selected at one of the Fabry–Perot modes closer to the wavelength of the minimum optical loss defined by the band pass filter. Compared to a DFB and a DBR laser, the gain bandwidth is wider for the external cavity laser because current injection level can be higher by the use of an SOA. For a tunable wavelength filter, several kinds have been demonstrated such as ring resonators [37, 39, 42], a ladder filter [36], and so on. And as for the structure, hybrid integration of an SOA and a planar lightwave circuit (PLC)-type waveguide filter [39, 42](an example is shown in Fig. 13.19 [39]), and all semiconductor integration [36, 37] have been achieved. An acousto-optic tunable filter (AOTF) also performs superior characteristics. Wide tuning region of 132 nm and a linewidth of 120 kHz are demonstrated [38].

13.5.4 Dynamic Wavelength Drift and its Suppression

At wavelength switching operation, wavelength drift occurs, and this gives some impacts on the coherent detection. There are several mechanisms of wavelength drift. In case of the current injection-type tunable laser such as an SSG-DBR laser, refractive index change due to carrier plasma effect causes blue shift when current increases, and red shift when current decreases with the same speed as carrier density change. Refractive index is also varied with thermal heating, and red shift occurs when current increases, and blue shift occurs when current decreases with the

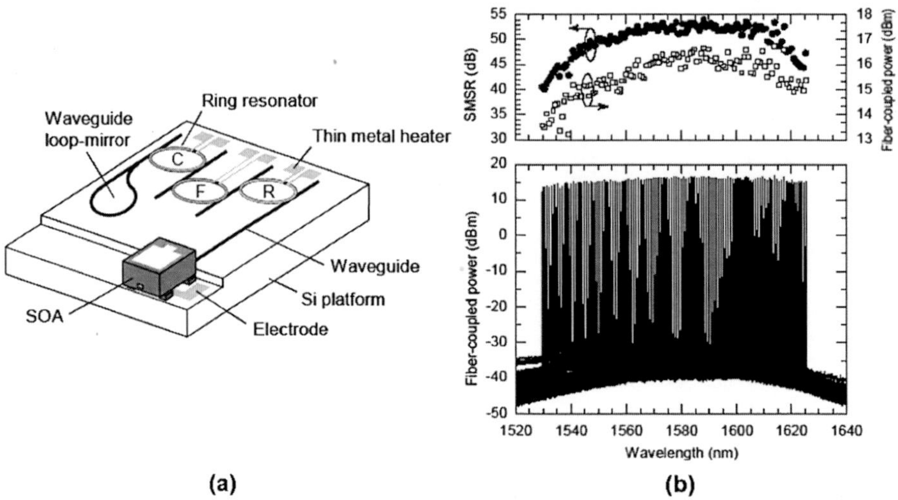

Fig. 13.19 Structure of an external-cavity tunable laser with PLC-based ring resonators (**a**) and SMSR and output power (**b**) (after R. Todt et al. [39])

thermal constant in the order of milliseconds. In each case, the resultant wavelength drift is about several gigahertz to several tens of gigahertz. To stabilize the lasing wavelength when tuning, scheme for stabilizing inner temperature is one of the candidates. In case of a multi-electrode DBR laser (SSG-DBR-LD), thermal drift compensator is introduced (Fig. 13.20) [47]. It is located adjacent to an SSG-DBR region. When tuning, injection current into the tuning region is changed. And the injection current into the thermal drift compensator is also changed to keep the total injection power to be constant. By this method, wavelength drift could be decreased to less than 1 GHz and mode-hop-free operation was achieved, although the increase of the total power is the trade-off.

In case of a DFB laser array, wavelength stabilization is achieved by monitoring the transmission signal output from a Fabry–Perot etalon, and the detected signal is the feedback to the device temperature. The schematic structure of an optical transmitter module is illustrated in Fig. 13.21a. Time transient of wavelength is shown in Fig. 13.21b, and less than 1 pm (125 MHz) wavelength fluctuation has been achieved [48].

13.5.5 Dependence of Wavelength Tuning Scheme on Spectral Linewidth

How to tune the lasing wavelength is important in terms of keeping the spectral linewidth narrow. The comparison of the spectral linewidth between electrical and thermal tuning has been reported [33]. The experimental results are shown in

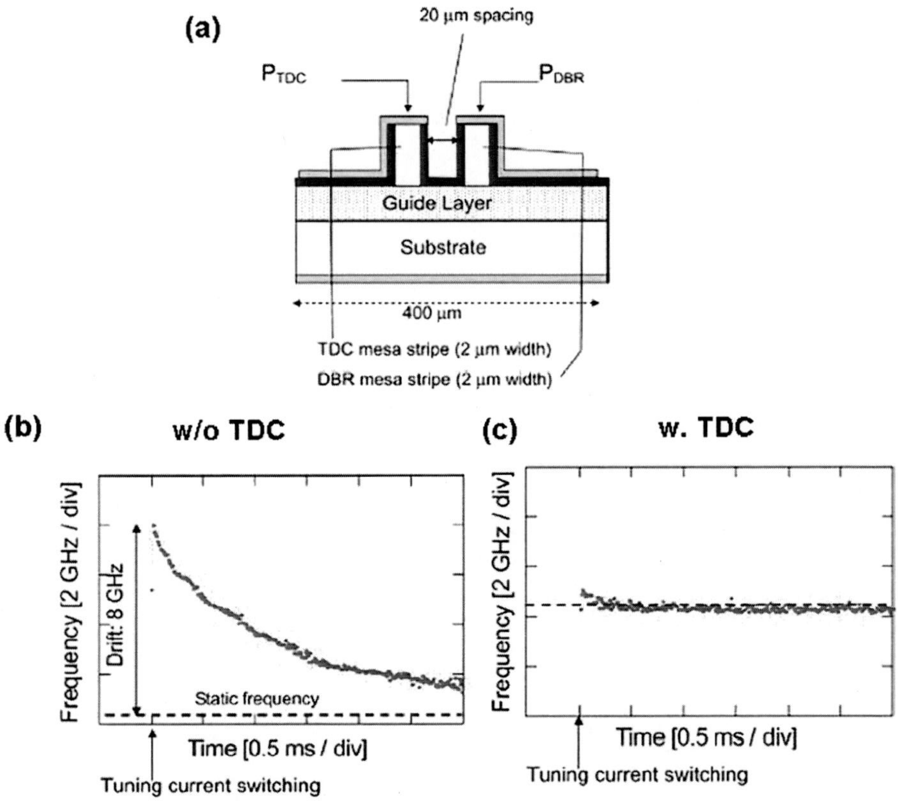

Fig. 13.20 Cross-sectional view of an SSG-DBR laser with thermal drift compensators (TDCs) (**a**), wavelength drift without TDC (**b**), and wavelength drift with TDC (**c**) (after N. Fujiwara et al. [46])

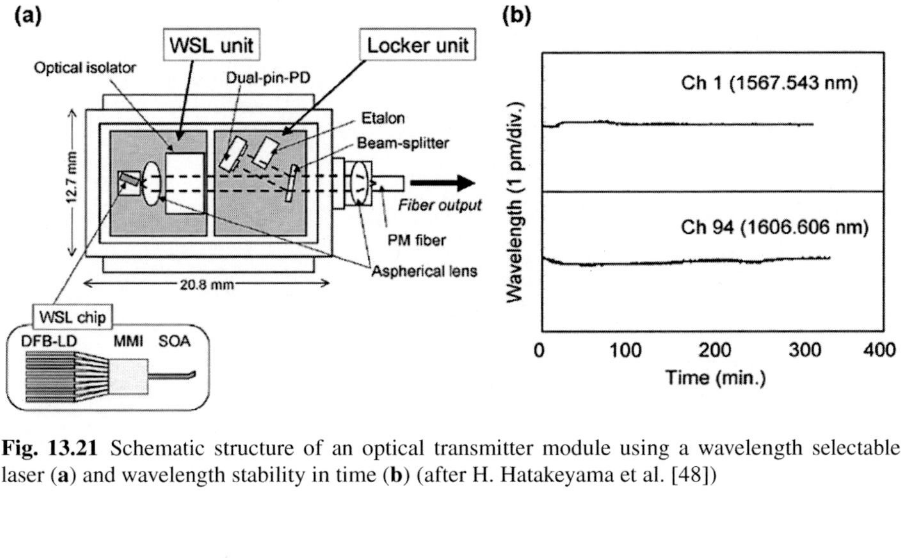

Fig. 13.21 Schematic structure of an optical transmitter module using a wavelength selectable laser (**a**) and wavelength stability in time (**b**) (after H. Hatakeyama et al. [48])

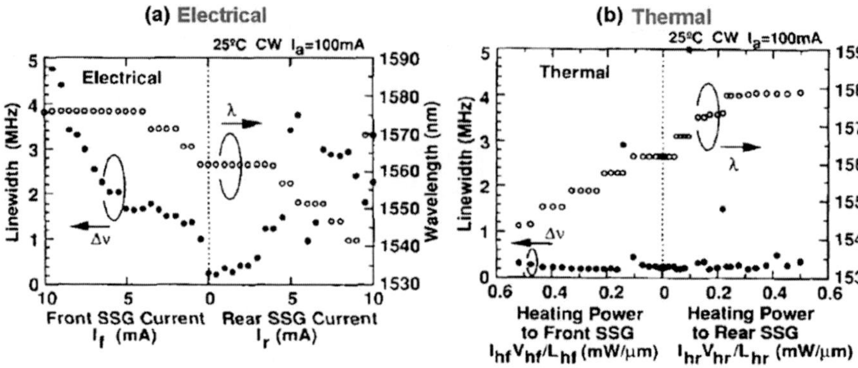

Fig. 13.22 Spectral linewidth of an SSG-DBR-LD under (**a**) electrical and (**b**) thermal wavelength tuning (after H. Ishii et al. [33])

Fig. 13.22a and b. In case of electrical tuning, linewidth increases with increasing current injected into the wavelength tuning regions, and the linewidth exceeds 1 MHz. In case of thermal tuning, on the other hand, linewidth was kept almost constant, and it was around 400 kHz. Then, we should care about the tuning scheme.

13.5.6 Wavelength Stability

Wavelength stability is also an important feature. Figure 13.23 shows the root Allan variance of frequency fluctuation of an SSG-DBR laser as a function of time [33]. It is about 10^{-9} to 10^{-8} in case of 1 s and 10^{-10} to 10^{-9} in case of 10^{-3} s. To suppress the wavelength fluctuation, absolute frequency criterion such as gas absorption lines is used.

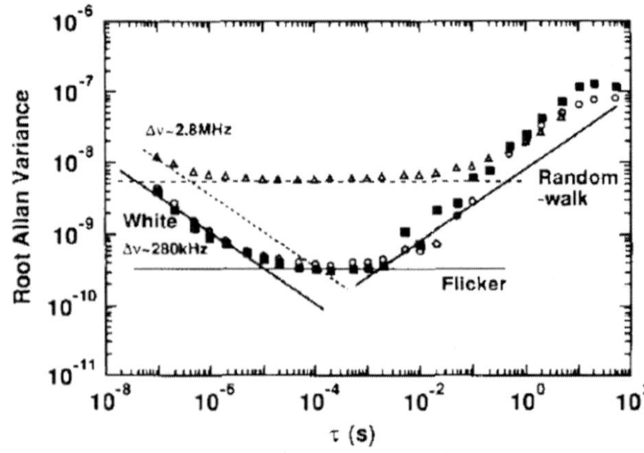

Fig. 13.23 Root Allan variance of frequency fluctuation of an SSG-DBR-LD as a function of time (after H. Ishii et al. [33])

13.6 Summary

Semiconductor lasers for high-density optical communication systems are reviewed. Narrow spectral linewidth is essential for coherent detection, and the efforts for spectral narrowing and several schemes have been described. In addition, wavelength tunability is also desired in the large-capacity optical transmission, and several types of tunable lasers have been explained. To realize narrow spectral linewidth and wide tenability at the same time, selection of the laser structure and tuning scheme should be selected carefully.

References

1. M. Seimetz, Laser linewidth limitations for optical systems with high-order modulation employing feed forward digital carrier phase estimation. Paper presented at optical fiber communication conference 2008 (OFC2008), OTuM2, San Diego, CA, 24-28 Feb 2008
2. M. Okai, M. Suzuki, T. Taniwatari, Strained multiquantum-well corrugation pitch-modulated distributed feedback laser with ultranarrow (3.6 kHz) spectral linewidth. Electron. Lett. 29(19), 1696 (1993)
3. G.M. Smith, J.S. Hughes, R.M. Lammert, M.L. Osowski, G.C. Papen, J.T. Verdeyen, J.J. Coleman, Very narrow linewidth asymmetric cladding InGaAs-GaAs ridge waveguide distributed Bragg reflector lasers. IEEE Photon. Technol. Lett. 8(4) 476 (1996)
4. T. Fujita, J. Phya, K. Matsuda, M. Ishino, H. Sato, H. Serizawa, "Narrow spectral linewidth characteristics of monolithic integrated-passive-cavity InGaAsP/lnP semiconductor lasers". Electron. Lett. 21(9), 374 (1985)
5. M. Kourogi, C.-H. Shin, M. Ohtus, A 250 Hz spectral linewidth 1.5 μm MQW-DFB laser diode with negative-electrical-feedback. IEEE Photon. Technol. Lett. 3(6), 496 (1991)
6. B. Dahmani, L. Hollberg, R. Drullinger, Frequency stabilization of semiconductor lasers by resonant optical feedback. Opt. Lett. 12(11), 876 (1987)
7. C.-H. Shin, M. Ohtsu, Stable semiconductor laser with a 7-Hz linewidth by an optical-electrical double-feedback technique. Opt. Lett. 15(24), 1455 (1990)
8. S. Ogita, Y. Kotaki, M. Matsuda, Y. Kuwahara, H. Ishikawa, Long-cavity, multiple-phase-shift, distributed feedback laser for linewidth narrowing. Electron. Lett. 25(10), 629 (1989)
9. M. Takahashi, Y. Michitsuji, M. Yoshimura, Y. Yamazoe, H. Nishizawa, T. Sugimoto, Narrow spectral linewidth 1.5 μm GaInAsP/InP distributed Bragg reflector (DBR) lasers. IEEE J. Quantum Electron. 25(6), 1280 (1989)
10. T. Kunii, Y. Ogawa, H. Wada, T. Nonaka, Y. Kawai, Narrow-spectral-linewidth, high-output-power operation, and FM response characteristics in 1.5 μm butt-jointed DBR lasers. IEEE J. Quantum Electron. 27(6), 1773 (1991)
11. M. Kitamura, H. Yamazaki, T. Sasaki, N. Kida, H. Hasumi, I. Mito, 250 kHz spectral linewidth operation of 1.5 μm multiple quantum well DFB-LD's. IEEE Photon. Technol. Lett. 2(5), 310 (1990)
12. M. Okai, T. Tsuchiya, K. Uomi, N. Chinone, T. Harada, Corrugation-pitch modulated MQW-DFB lasers with narrow spectral linewidth. IEEE J. Quantum Electron. 27(6), 1767 (1991)
13. T. Kunii, Y. Matsui, H. Horikawa, T. Kamijoh, T. Nonaka, Narrow linewidth (85 kHz) operation in long cavity 1.5 μm-MQW DBR laser. Electron. Lett. 27(9), 691 (1991)
14. F. Kano, T. Yamanaka, N. Yamamoto, H. Mawatari, Y. Tohmori, Y. Yoshikuni, Linewidth enhancement factor in InGaAsP/InP modulation-doped strained multiple-quantum-well lasers. IEEE J. Quantum Electron. 30(2), 533 (1994)
15. C.H. Henry, Theory of the linewidth of semiconductor lasers. IEEE J. Quantum Electron. QE-18, 259 (1982)

16. N. Hatori, M. Ishida, H. Ebe, M. Sugawara, Y. Arakawa, Measurement and evaluation of chirp and linewidth enhancement factor of a 1.3 μm quantum dot laser. Presented on conference on lasers and electrooptics 2004 (CLEO2004), vol.1, San Francisco, CA, 2004.
17. A.V. Uskov, E.P. O'Reilly, D. McPeake, N.N. Ledentsov, D. Bimberg, G. Huyet, Carrier-induced refractive index in quantum dot structures due to transitions from discrete quantum dot levels to continuum states. Appl. Phys. Lett. 84(2), 272 (2004)
18. K. Kotaki, S. Ogita, M. Matsuda, Y. Kuwahara, H. Ishikawa, Tunable, narrow-linewidth and high-power /4-shifted DFB laser. Electron. Lett. 25(15), 990 (1989)
19. H. Yamazaki, M. Yamaguchi, M. Kitamura, Spectral linewidth rebroadening in MQW-DFB LDs caused by spontaneous emission noise in SCH barrier layers. IEEE Photo. Technol. Lett. 6(3), 341 (1994)
20. K. Kikuchi, Effect of 1/f-type FM noise on semiconductor-laser linewidth residual in high-power limit. IEEE J. Quantum Electron. 25(4), 684 (1989)
21. Redfern Integrated Optics Inc., http://www.rio-inc.com/
22. L.A. Coldren, G.A. Fish, Y. Akulova, J.S. Barton, L. Johansson, C.W. Coldren, Tunable semiconductor lasers: a tutorial. IEEE J. Lightwave Technol. 22(1), 193 (2004)
23. J. Buus, E.J. Murphy, Tunable lasers in optical networks. IEEE J. Lightwave Technol. 24(1), 5 (2006)
24. P.I. Kuindersma , W. Scheepers, J.H.M. Cnoops, P.J.A. Thijs, G.L.A.v.d. Hofstad, T.v. Dongen, J.J.M. Binsma, Tunable three-section, strained MQW, PA-DFB's with large single mode tuning range (72Å) and narrow linewidth (around 1 MHz). Presented on 12th IEEE Intrenational Semiconductor Laser Conference(ISLC1990), Davos, M-4 (1990)
25. N. Nunoya, H. Ishii, Y. Kawaguchi, Y. Kondo, H. Oohashi, Wideband tuning of tunable distributed amplification distributed feedback laser array. Electron. Lett. 44(3), 205 (2008)
26. A. Hayakawa , K. Takabayashi, S. Tanaka, S. Tomabechi, M. Ekawa, K. Morito, Tunable twin-guide distributed feedback laser with 8-nm mode-hop-free tuning range. Presented on CLEO/PacificRim 2005, CWJ2-3 (2005)
27. S. Murata, I. Mito, K. Kobayashi, Over 720GHz (5-8 nm) frequency tuning by a 1.5 μm DBR laser with phase and Bragg wavelength control regions. Electron. Lett. 23(8), 403 (1987)
28. C.J. Chang-Hasnain, Tunable VCSEL. IEEE J. Select. Topics Quantum Electron. 6, 978 (2000)
29. W. Janto, K. Hasebe, N. Nishiyama, C. Caneau, T. Sakaguchi, A. Matsutani, P.B. Dayal, F. Koyama, C.-E. Zah, Athermal operation of 1.55 μm InP-based VCSEL with thermally-actuated cantilever structure. Presented on 20th IEEE International Semiconductor Laser Conference(ISLC2006), PD1.1, Hawaii, USA, Sept. 2006
30. D. Anthon, J.D. Berger, A. Tselikov, C+L band MEMS tunable external cavity semiconductor laser. Presented on optical fiber communication conference 2004(OFC2004), WL2, San Jose, CA, 2004
31. H. Hatakeyama, K. Naniwae, K. Kudo, N. Suzuki, S. Sudo, S. Ae, Y. Muroya, K. Yashiki, K. Satoh, T. Morimoto, K. Mori, T. Sasaki, Wavelength-selectable microarray light sources for S-, C-, and L-band WDM systems. IEEE Photon. Technol. Lett. 15(7), 903 (2003)
32. T. Mukaihara, Y. Nakagawa, H. Nasu, H. Kambayashi, M. Oike, S. Yoshimi, T. Kurobe, T. Kimoto, K. Muranushi, T. Nomura, A. Kasukawa, High power, low noise, low power consumption, 25 GHz × 180 ch thermally tunable DFB laser module integrated with stable wavelength monitor. Presented in 29th European conference on optical communication(ECOC2003), Rimini, We.P81, 2003
33. H. Ishii, F. Kano, Y. Tohmori, Y. Kondo, T. Tamamura, Y. Yoshikuni, Narrow spectral linewidth under wavelength tuning in thermally tunable super-structure-grating (SSG) DBR lasers. IEEE J. Sel. Top. Quantum Electron. 1(2), 401 (1995)
34. Y.A. Akulova, G.A. Fish, P.-C. Koh, C.L. Schow, P. Kozodoy, A.P. Dahl, S. Nakagawa, M.C. Larson, M.P. Mack, T.A. Strand, C.W. Coldren, E. Hegblom, S.K. Penniman, T. Wipiejewski, L.A. Colren, Widely tunable electroabsorption-modulated sampled-grating DBR laser transmitter. IEEE J. Sel. Top. Quantum Electron. 8(6), 1349 (2002)

35. A.J. Ward, D.J. Robbins, G. Busico, E. Barton, L. Ponnampalam, J.P. Duck, N.D. Whitbread, P.J. Williams, D.C.J. Reid, A.C. Carter, M.J. Wale, Widely tunable DS-DBR laser with monolithically integrated SOA: design and performance. IEEE J. Sel. Top. Quantum Electron. 11(1), 149 (2005)
36. S. Matsuo, S.-H. Jeong, T. Segawa, H. Okamoto, Y. Kawaguchi, Y. Kondo, H. Suzuki, Y. Yoshikuni, Digitally tunable ring laser using ladder filter and ring resonator. IEEE J. Sel. Top. Quantum Electron. 11(5), 924 (2005)
37. T. Segawa, S. Matsuo, T. Kakitsuka, T. Sato, Y. Kondo, H. Suzuki, Full C-band tuning operation of semiconductor double-ring resonator-coupled laser with low tuning current. IEEE Photon. Technol. Lett. 19(17), 1322 (2007)
38. K. Takabayashi, K. Takada, N. Hashimoto, M. Doi, S. Tomabechi, T. Nakazawa, K. Morito, Widely (132 nm) wavelength tunable laser using a semiconductor optical amplifier and an acousto-optic tunable filter. Electron. Lett. 40(19), 1187 (2004)
39. R. Todt, S. Watanabe, Y. Deki, M. Takahashi, T. Takeuchi, S. Takaesu, T. Miyazaki, M. Horie, H. Yamazaki, Widely tunable resonated-ring-reflector lasers covering C- and L-bands. Presented on 33rd European conference on optical communication (ECOC2007), Berlin, PD.2.5, 2007
40. B. Pezeshki, E. Vail, J. Kubicky, G. Yoffe, S. Zou, J. Heanue, P. Epp, S. Rishiton, D. Ton, B. Faraji, M. Emanuel, X. Hong, M. Sherback, V. Agrawal, C. Chipman, T. Razazan, 20-mW widely tunable laser module using DFB array and MEMS selection. IEEE Photon. Technol. Lett. 14(10), 1457 (2002)
41. J. De Merlier, K. Mizutani, S. Sudo, K. Naniwae, Y. Furushima, S. Sato, K. Sato, K. Kudo, Full C-band external cavity wavelength tunable laser using a liquid-Crystal-based tunable mirror. IEEE Photon. Technol. Lett. 17(3), 681 (2005)
42. Y. Deki, T. Hatanaka, M. Takahashi, T. Takeuchi, S. Watanabe, S. Takaesu, T. Miyazaki, M. Horie, H. Yamazaki, Wide-wavelength tunable lasers with 100 ghz fsr ring resonators. Electron. Lett. 43(4), 225 (2007)
43. H. Tanobe, F. Koyama, K. Iga, Spectral linewidth of AlGaAs/GaAs surface-emitting laser. Electron. Lett. 25(21), 1444 (1989)
44. F. Monti di Sopra, H.P. Zappe, M. Moser, R. Hövel, H.-P. Gauggel, K. Gulden, Near-infrared vertical-cavity surface-emitting lasers with 3-MHz linewidth. IEEE Photon. Technol. Lett. 11(12), 1533 (1999)
45. R. Shau, H. Halbritter, F. Riemenschineider, M. Ortsiefer, J. Rosskopf, G. Böhm, M. maute, P. Meissner, and M.-C. Amann, Linewidth of InP-based 1.55 ?m VCSELs with buried tunnel junction. Electron. Lett. 39(24), 1728 (2003)
46. N. Fujiwara, R. Yoshimura, K. Kato, H. Ishii, F. Kano, Y. Kawaguchi, Y. Kondo, K. Ohbayashi, H. Oohashi, 140-nm quasi-continuous fast sweep using SSG-DBR lasers. IEEE Photon. Technol. Lett. 20(12), 1015 (2008)
47. N. Fujiwara, H. Ishii , H. Okamoto, Y. Kawaguchi, Y. Kondo, H. Oohashi, Suppression of thermal wavelength drift in super-structure grating distributed Bragg reflector (SSG-DBR) laser with thermal drift compensator. IEEE J. Sel. Top. Quantum Electron. 13(5), 1164 (2007)
48. H. Hatakeyama, K. Kudo, Y. Yokoyama, K. Naniwae, T. Sasaki, Wavelength-selectable microarray light sources for wide-band DWDM applications. IEEE J. Sel. Top. Quantum Electron. 8(6), 1341 (2002)

Chapter 14
Monolithic InP Photonic Integrated Circuits for Transmitting or Receiving Information with Augmented Fidelity or Spectral Efficiency

C.R. Doerr

Abstract This chapter discusses monolithic photonic integrated circuits in InP for generating and receiving fiber-optic signals. These transceivers have advanced features, such as high spectral efficiency and optical equalization. InP monolithic integration can aid in reducing their footprint, cost, and power consumption.

14.1 Introduction

As the demand for bandwidth increases on each optical fiber, there is a significant pressure for optical transceivers to be cheaper, smaller, and more power efficient, yet generate and receive an increasing number of bits per second. As the bit rate increases, factors such as optical dispersion tolerance, optical filtering tolerance, and spectral occupation become more important, requiring more complex transmitters and receivers. A potential way to make transceivers that have all these characteristics is monolithic integration of the various functions onto a single photonic integrated circuit (PIC) [1]. The only material platform known today that can provide all the necessary optical functions in the fiber's low-loss wavelength window without needing external optical sources is indium phosphide (InP). This chapter discusses InP PIC transmitters and receivers with advanced features. We focus on monolithically integrated devices, meaning that all elements of interest are on one chip, which lowers packaging cost and minimizes footprint.

The main advanced feature we discuss involves advanced digital modulation formats. An advanced digital modulation format is any signal format other than two-state on–off keying (OOK). Figure 14.1 shows constellations of some popular formats. Each dark circle represents a symbol, and the symbol is a function of magnitude, phase, and polarization. For example, the left-most constellation, labeled OOK, shows conventional two-state OOK, in which a "zero" is the absence of light,

C.R. Doerr (✉)
Bell Laboratories, Alcatel-Lucent, Holmdel, NJ 07748, USA
e-mail: crdoerr@alcatel-lucent.com

M. Nakazawa et al. (eds.), *High Spectral Density Optical Communication Technologies*, Optical and Fiber Communications Reports 6, DOI 10.1007/978-3-642-10419-0_14,

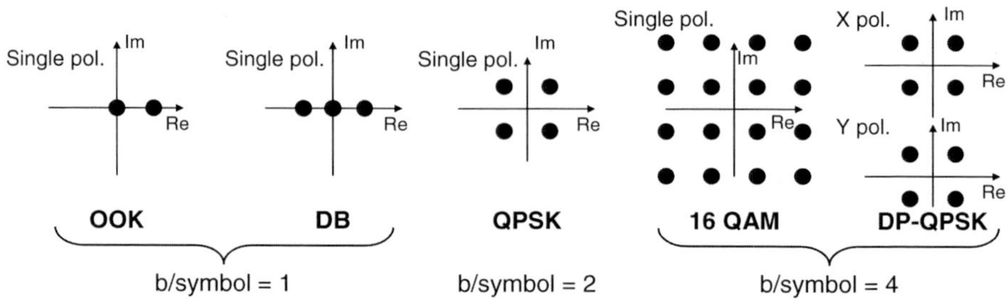

Fig. 14.1 Constellations of some common modulation formats. *OOK* on–off keying, *DB* duobinary, *QPSK* quadrature phase-shift keying, *QAM* quadrature amplitude modulation, and *DP* dual polarization

and a "one" is the presence of light with an approximately constant carrier phase between neighboring "ones."

The constellations of Fig. 14.1 do not show the transitions between symbols. If the signal always goes to the origin before going to the next symbol, the format is prefixed with "return-to-zero" (RZ). Otherwise it is prefixed with "non return-to-zero" (NRZ). If the signal follows a straight line between symbols, it is "non-chirped," otherwise it is "chirped." A strict definition of non-chirped is that the phase remains constant over time whenever the power is non-zero, implying that all transitions follow lines that pass through the origin. However, by convention, signals can be considered non-chirped even if some transitions follow non-radial lines, such as in NRZ-QPSK, as long as the transitions are straight lines [2].

With advanced digital modulation formats, one often uses the terminology bit rate and symbol rate. Bit rate is the total delivered information, given in bits per unit time. Symbol rate is the delivered symbols per unit time, given in baud. One symbol may contain more than one bit of information. For example, in single-polarization quadrature phase-shift keying (QPSK), each symbol can be one of four possible phases, and thus each symbol contains two bits of information.

We will use the concept of spectral efficiency, which is unit-less but is often expressed as bit/s/Hz. For example, a signal with a spectral efficiency of 1 means that if you want to transmit one bit per second of digital information, then you could do it by occupying only 1 Hz of optical bandwidth. If you are transmitting only a single wavelength channel and do not use optical filters, it will not be clear how to measure the spectral efficiency. However, if you have many wavelength channels, then one can define an average spectral efficiency for the channels by taking the bit rate per channel and dividing by the channel spacing in frequency. The most common signal format, two-state OOK, can potentially achieve a spectral efficiency of 1, unless one filters it asymmetrically [3]. The spectral efficiency can be increased by three known ways: using phase, using more power levels, or using polarization. In principle, the achievable spectral efficiency is unlimited. However, as the spectral efficiency increases, the data "constellation" becomes denser, and thus the required signal-to-noise ratio increases. This in turn means that more optical power is required. Ultimately nonlinearity in the transmission fiber or other components will limit the optical power. One study predicts a limit of 5.6 bit/s/Hz for one polarization using a constellation with over a thousand points over 2,000 km of standard single-mode fiber (SSMF) [4]. Using both polarizations, transmitting with a spectral efficiency of ~10 bit/s/Hz over 2,000 km of SSMF may be possible.

14.2 InP basics

InP is a III–V material. It crystallizes in a zinc blende structure, which contains two intersecting face-centered cubic lattices, one for In and one for P. InP is valuable because it has a direct bandgap with a wavelength near the low-loss wavelength of silica optical fiber, and because it has a very high electron mobility. The first capability allows for efficient telecommunication semiconductor lasers, including sources and amplifiers, and the second capability allows for very fast transistors, optical modulators, and photoreceivers. Indeed, some of the fastest electronics known today are made in InP. The drawbacks of InP are that it is expensive, has more growth defects than other popular semiconductors, such as Si, is not as mechanically strong as Si and has a larger bandgap energy than Si (~1.35 eV as compared to ~1.2 eV), thus requiring more voltage for transistors and consequently more power (unless a high voltage is required for a driver, in which case InP may require less power than Si). Also, although P is a very common element on Earth, found in many organic materials such as fertilizers, In is rare, estimated to exist in only 0.1 parts per million on Earth, which is about the same as silver. Today's liquid-crystal displays consume large quantities of In in the form of indium tin oxide (ITO), which is optically transparent yet electrically conducting. Some projections estimate that all the readily accessible In on Earth will be used by circa 2020 [5]. Hence, the recycling of displays could become important to the future of InP PICs, although semiconductor industries' In consumption is estimated to be only ~15% of the total In consumption.

InP PIC optical waveguides consist of InP-based compounds on top of a 2"- or 3"-diameter InP wafer [6]. InP can be combined with Ga and As to form ternary, such as InGaAs, and quaternary, such as InGaAsP, materials. Also, P can be replaced by Al, useful for modulators and amplifiers. These materials form the same crystal structure, and with the proper ratios have the same lattice constant, as InP. Using compositions slightly off from these results in strain in the lattice, either compressive or tensile, which affects the optical polarization dependence. Ternary and quaternary materials have a narrower bandgap for electrons and holes than InP and thus a higher permittivity for photons than InP, allowing the construction of waveguides with an InP cladding, optical amplifiers, optical modulators, etc.

Dopants can be added to change the semiconductor from a weak insulator to a conductor or, alternatively, make it strongly insulating. Adding S or Si makes the material conducting via electron transport, i.e., n-type; and adding C, Be, or Zn makes the material conducting via hole transport, i.e., p-type. p-type doping increases the optical loss significantly. For instance, InP with a p-type doping of $10^{18}/\text{cm}^3$ has an approximate optical loss of 10 dB/mm at 1550 nm [7]. This loss

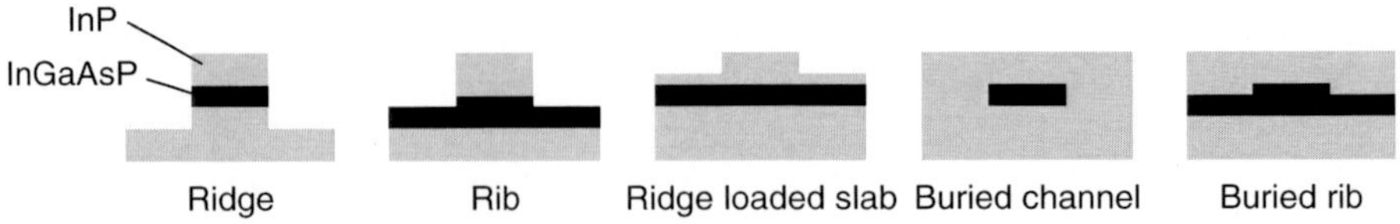

Fig. 14.2 Typical waveguide cross-sections in InP

is due to electron transitions between the split-off band and the top of the valence band. n-Type doping has much less effect on the optical loss. For instance, InP with an n-type doping of $10^{18}/\text{cm}^3$ has an approximate loss of 0.4 dB/mm at 1550 nm [8]. Adding Fe makes the semiconductor highly insulating, because the Fe atoms trap free carriers. Adding Fe increases the optical loss somewhat.

Cross-sections of popular types of "heterostructure" optical waveguides are shown in Fig. 14.2. Heterostructures were one of the keys to achieving efficient semiconductor lasers, for which a Nobel Prize was awarded in 2000. From right to left in Fig. 14.2, the lateral effective index contrast between the core and the cladding becomes higher, allowing for tighter confinement. The index contrast is defined as

$$\frac{n_{\text{core}}^2 - n_{\text{cladding}}^2}{2n_{\text{core}}^2}. \tag{14.1}$$

Tighter confinement allows for bends with smaller radii of curvature and narrower, lower capacitance, and modulators. However, the higher the index contrast, the higher the waveguide loss due to sidewall roughness, and the larger the phase errors in optical filters due to increased sensitivity to waveguide width. Undoped InGaAsP/InP ridge waveguides typically have 0.5 dB/mm loss, whereas buried rib waveguides typically have 0.02 dB/mm loss.

InP is a direct bandgap material. A direct bandgap means that the minimum energy in the conduction band has the same momentum as the maximum energy in the valence band. This means that electrons can drop across the bandgap without needing coupling from phonons. The drop across the bandgap emits a photon, and this photon has a minimum chance of being reabsorbed because the drop was as small as possible. The fact that InP is a direct bandgap material and that it can form heterostructures using Ga and As means it can provide efficient optical amplification. These semiconductor optical amplifiers (SOAs) have many uses, from acting as lasers to providing optical gain.

Three of many possible SOA structures are shown in Fig. 14.3. The top contact is p-doped, and the bottom contact is n-doped. The Fermi level (the top level of electrons at zero Kelvin—electrons are fermions, which means two electrons with the same spin cannot exist in the same energy level in a given crystal) is high in the p-doped material, which means that the carriers injected by the current will end up in the conduction band. To get through to the other contact of the current sources, the carriers must drop across the bandgap, ending up in the valence band in the n-doped material. As each electron and hole drop across the bandgap, they can

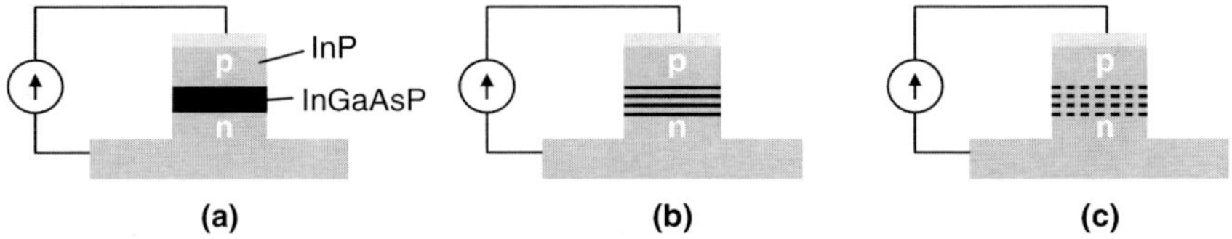

Fig. 14.3 SOA structures. **a** Bulk SOA, **b** multiple quantum well (MQW) SOA, and **c** quantum dot SOA

combine and emit a photon. Figure 14.3a shows a bulk SOA. Figure 14.3b shows a multiple quantum well (MQW) SOA, which can have a higher gain than a bulk SOA, but with a narrower spectrum. Figure 14.3c shows a quantum dot SOA, which can have a very wide spectrum due to variations in the dot sizes and low temperature dependence due to its atom-like nature.

InP can also absorb photons and generate electron–hole pairs, acting as a photodetector. The photodetector is usually configured as a photodiode, with an absorber region in a p–n junction. The absorber is a material with a bandgap wavelength longer than the wavelength of the photons that one needs to detect. The absorber is usually InGaAs, which has a bandgap wavelength of $\sim$1.7 μm.

When a photon is absorbed in the absorber, a free electron and a free hole are generated. The hole must drift to the p contact, and the electron must drift to the n contact. High-speed detectors, greater than about 20 GHz, are often limited in speed by the transport time of the electrons and holes. A voltage is usually applied across the p–n junction to create an electric field to speed up the electron and hole drifting. The electron mobility is significantly higher than that of the hole. Thus one often designs the photodetector such that the p-doping extends into the absorber, minimizing the distance the hole must travel before reaching the conducting p-doped material, which is called a partially depleted absorber waveguide photodiode [9]. In fact, sometimes the entire absorber is p-doped. Such a case is called a uni-traveling carrier (UTC) photodiode [10], because the hole does not have to drift anywhere, and the electron must drift to a "collector", which is a lightly or undoped region between the absorber and the n-doped region. The collector doping is light so that a strong electric field can exist there. Figure 14.4a shows a conventional p–i–n photodiode; Fig. 14.4b shows a partially depleted absorber photodiode; and Fig. 14.4c shows a UTC photodiode.

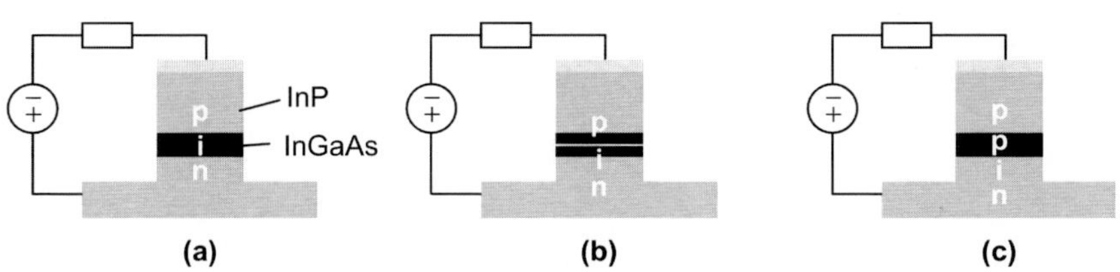

Fig. 14.4 Photodiode structures. **a** p–i–n photodiode, **b** partially depleted absorber photodiode, and **c** a UTC photodiode

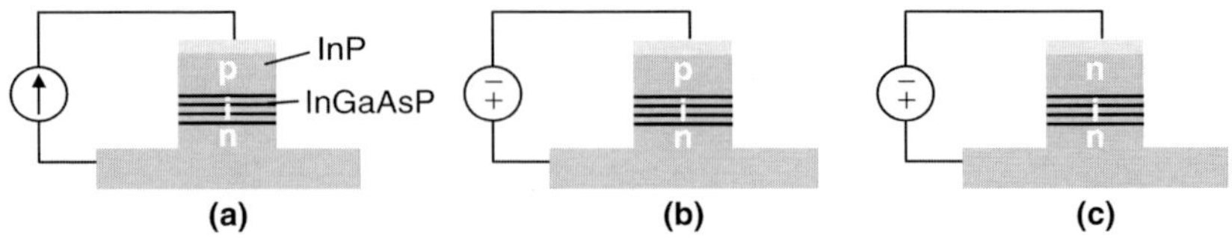

Fig. 14.5 Electro-refraction and absorption structures. **a** current injection phase-shifter/absorber, **b** reverse voltage phase-shifter/absorber with a p–i–n junction, and **c** phase-shifter/absorber with an n–i–n junction

The InP material system can also provide an electro-absorption and/or refraction modulator, i.e., the optical insertion loss and/or phase changes with an applied electrical signal. There are two main types of operation. The first is current injection, shown in Fig. 14.5a, which involves passing a current through a p–i–n junction with a material having a bandgap wavelength that is shorter than the signal wavelength. This causes the refractive index to decrease with increasing current, i.e., a blue shift. This effect provides a very strong index change, but it is relatively slow, ~1-ns time constant, because it involves the movement of carriers between bands. The lower the doping one can start with, the larger the change for a given current. Such a slow phase modulator is typically called a phase shifter and is used for direct-current (DC) adjustments. The absorption change is very small; although at very high injection currents, free carriers may be generated, increasing the insertion loss. However, such an increase in loss is typically negligible compared to that experienced in indirect bandgap materials such as Si. The second type of operation is reverse voltage [11], which involves applying a reverse voltage across a p–i–n or n–i–n junction, shown in Fig. 14.5b and c. There are two main effects. One is the Pockels effect, which is an ultra-fast electro-optic effect, similar to that found in $LiNbO_3$. The sign of the index change depends on the orientation of the electric field with respect to the crystal axis. This effect is weak and typically requires cm-long devices to achieve a π–phase shift. The absorption change is negligible. The other effect is a shifting of the band edge, which is called the Franz–Keldysh effect in bulk material and the quantum-confined Stark effect in MQWs. The electric field moves the band edge toward longer wavelengths. When the band edge reaches the signal wavelength, it causes an increase in absorption. This is called the electro-absorption (EA) effect and is used to make EA modulators (EAMs). The Franz–Keldysh and quantum-confined Stark effects have time constants <10 ps. As expected from the Kramers–Kronig relations, the change in the absorption spectrum causes a phase change [12], schematically shown in Fig. 14.6. The phase increases with increasing voltage when the signal wavelength is longer than the band-edge wavelength. If the band-edge wavelength moves past the signal wavelength, then the phase change decreases with increasing voltage.

Depending on the structure and composition, strong excitonic effects may be present. Excitons are electron–hole pairs that are attracted to each other, one in the conduction band and one in the valence band, but are prevented by the bandgap from combining. They usually cause sharp peaks in the absorption spectrum. They are often too weak to notice in InGaAsP, but can be dominant in InGaAsAl.

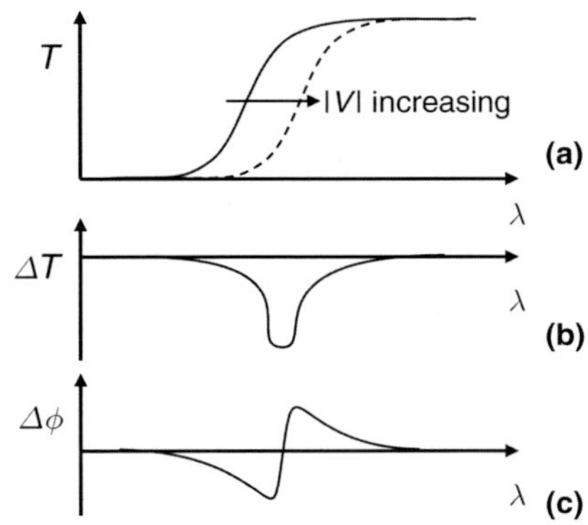

Fig. 14.6 Very approximate chematic of electro-refraction and absorption change in InGaAsP due an applied voltage V. **a** The transmissivity (T) of the modulator vs. wavelength (λ) near the band edge and shows the movement of the band edge with voltage. **b** The change in T from **a**. **c** The corresponding phase change

To summarize InP electro-optic modulators, one can make a strong DC phase shifter by injecting current into undoped InGaAsP, can make a fast intensity modulator by applying a voltage across InGaAsP with a band-edge wavelength very near the signal wavelength, and can make a fast phase modulator by applying a voltage across InGaAsP with a band-edge wavelength relatively far from the signal wavelength. We show the InGaAsP as MQWs in Fig. 14.5, but it could also be bulk material, however, generally having a less sharp band edge.

The refractive index of InP is temperature dependent such that any filter will tune $\sim +0.1$ nm/C. This is about 10 times that of silica. This large sensitivity often means that the InP device must be temperature controlled. This temperature dependence can be used to make a thermooptic phase shifter by placing a current-driven heater over a portion of an InP waveguide. Such thermooptic phase shifters have a response time on the order of 1 ms and are often used as DC phase shifters because they avoid the potential for increased optical loss due to p-doped material required for a current injection phase shifter.

14.3 Transmitters

The most successful InP PIC today is arguably the EAM laser (EML). It consists of a distributed feedback (DFB) laser monolithically integrated with an EAM. The first EML was reported in 1987 and is shown in Fig. 14.7 [13]. It is used as a low-cost on–off keying (OOK) transmitter for intermediate (40 km) and long reach (80 km or more) links. As can be surmised from Fig. 14.6, the transitions between the two points in the OOK constellation generated by an EML are curved lines. When the EAM is driven so that the band edge passes the laser wavelength (see Fig. 14.6), this chirp becomes negative (blue) [14, 15]. The blue chirp increases the positive dispersion tolerance in fiber transmission by about a factor of two. SSMF has positive dispersion. For example, whereas a chirp-free 10-Gb/s OOK transmitter can transmit only ~40 km in SSMF without dispersion compensation, an appropriately driven 10-Gb/s EML can transmit ~80 km in SSMF without dispersion compensa-

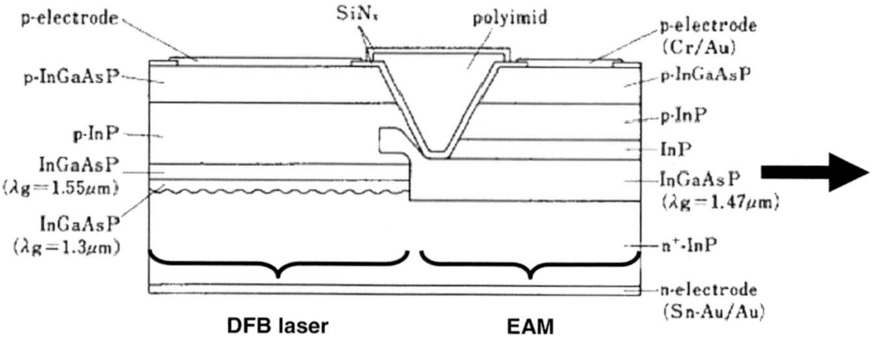

Fig. 14.7 First reported EML, from [1]

tion. A significant issue with EAMs is that their extinction ratio degrades as the optical power is increased limiting the output power of EMLs. This is because EAMs absorb the light in the off-state, creating electron–hole pairs which can shield out the applied electric field if they are not swept out fast enough. EMLs of 10 Gb/s are commonly sold today in tiny transmitter optical sub-assemblies (TOSAs), which go inside small pluggable transceivers called XFPs costing in the order of $1,500, consume ~3.5W, and have an output power of ~ -1 to $+3$ dBm.

EAM speeds up to 80 Gb/s [16] and EML speeds up to 100 Gb/s [17] have been reported. The EAM in such a case was only 50-μm long in order to have a low capacitance.

14.3.1 Increasing the Spectral Efficiency via Polarization

The spectral efficiency of OOK using an EML is far from the ultimate limit. A straightforward way to double the spectral efficiency using an EML-based transmitter is to employ both polarizations of the light, which is equivalent to changing the polarization of the signal at the symbol rate. Figure 14.8 shows a dual-polarization (DP) OOK modulator [18]. It produces a signal format with four symbols. It uses a polarization splitter, two polarization-independent EAMs, and a polarization combiner. It does not have an integrated laser and instead is coupled to an off-chip laser whose polarization is adjusted such that half the light is transverse-electric (TE) polarized, and the other half is transverse-magnetic (TM) polarized on the PIC. The first polarization beam splitter (PBS) sends the TE light to one EAM and the TM light to another. The two EAMs are driven with separate digital data streams. The second PBS recombines the light. The output carries the information of two data streams using the optical bandwidth of only a single-data stream.

A photograph and waveguide layout of the actual dual-polarization OOK modulator are shown in Fig. 14.9. The PBSs consist of Mach–Zehnder interferometers with one arm longer than the other by ΔL. The waveguides are

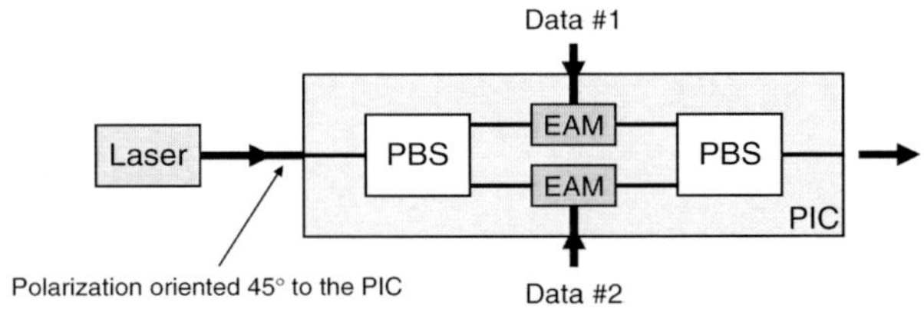

Fig. 14.8 Schematic of a dual-polarization OOK modulator

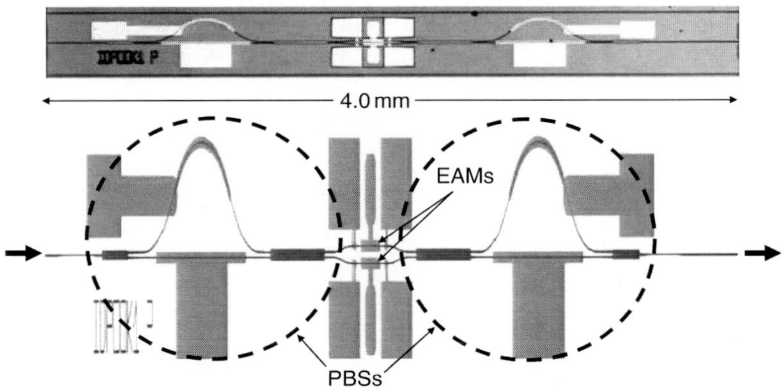

Fig. 14.9 Photograph (*upper*) and waveguide layout (*lower*, stretched vertically for clarity) of the dual-polarization OOK modulator

benzocyclobutene- (BCB-) clad ridge waveguides (left-most cross-section in Fig. 14.2). Such waveguides have a large birefringence. When

$$\Delta L = \frac{\lambda_0 n}{2(n_{\mathrm{TE}} - n_{\mathrm{TM}})n_{\mathrm{g}}}, \tag{14.2}$$

where n is the average phase index, n_{g} is the average group index, and n_{TE} and n_{TM} are the effective refractive indices for TE and TM, then the interferometer acts as a PBS for λ_0, where λ_0 is the free-space wavelength. This PBS design is facile to fabricate but is wavelength sensitive. To tune the PBS to the signal wavelength, there are current injection phase shifters on both arms of the interferometer, as shown in Fig. 14.9.

Experimental results are shown in Fig. 14.10. The two EAMs were driven with independent 40-Gb/s data streams. The results when looking at one polarization in the output are nearly unchanged whether only one EAM is driven at a time or if both EAMs are driven simultaneously, showing that the cross talk is low.

To receive a dual-polarization signal, the receiver must be able to track the polarization, because the polarization will change in an uncontrolled fashion in the fiber link. Also, polarization mode dispersion (PMD) and polarization-dependent loss (PDL) will cause cross talk between the two polarization signals. Polarization tracking can be done either optically, using a series of polarization converters, or electronically, using a polarization diversity coherent receiver. Neither is simple,

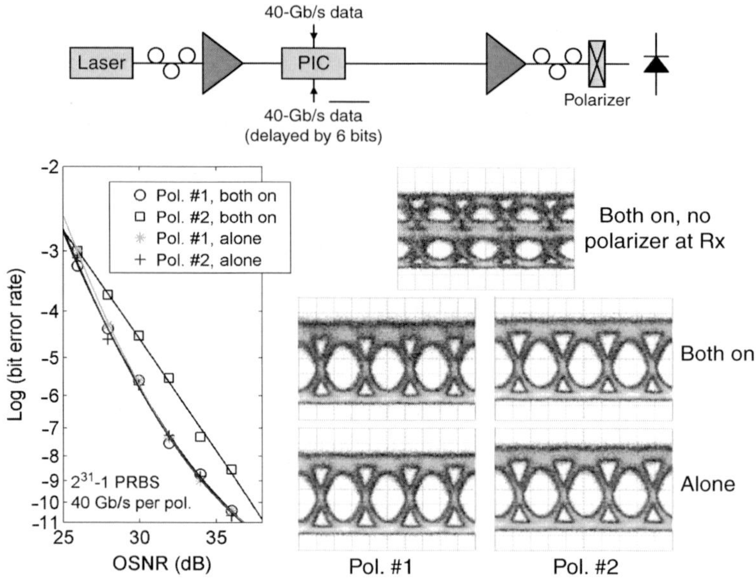

Fig. 14.10 Experimental results from the dual-polarization OOK modulator

and thus polarization multiplexing is often a last resort for improving spectral efficiency. An easier method is usually utilizing the optical phase, the subject of the next section.

14.3.2 Increasing the Spectral Efficiency via Phase

Many advanced modulation formats use the optical phase to carry information. The phase is almost always used differentially, meaning that the information is encoded in the phase difference between neighboring symbols rather than the absolute phase. When the information is encoded differentially, the format name is often prefixed by "D."

A format that allows direct binary detection yet increases the spectral efficiency by about a factor of two is optical duobinary (DB), sometimes called phase-shaped binary transmission (PSBT), whose constellation is shown in Fig. 14.1. Conventionally, duobinary is generated by phase modulators driven push–pull in an MZI with a π-phase bias, as shown in Fig. 14.11a. If there is no filtering, the apparatus of Fig. 14.11a produces binary phase-shift keying (BPSK). Likewise, BPSK can be produced by intensity modulators driven push–pull in an MZI with a π-phase bias, as shown in Fig. 14.11b. If one filters the BPSK electrical driving signals with a bandwidth of approximately one quarter the data rate or the BSPK optical signal with a bandwidth of approximately one half the data rate, BPSK is changed into DB. Looking at the optical power in the time domain, DB is nearly indistinguishable

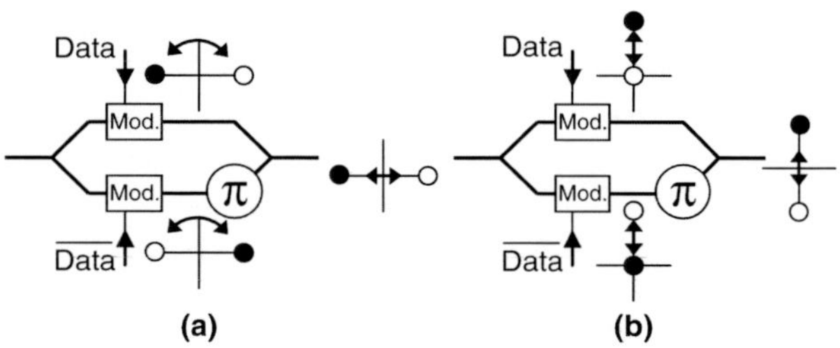

Fig. 14.11 Schematics of MZI binary phase modulators generating BPSK using (**a**) phase modulators and (**b**) intensity modulators

from two-state OOK; except that the data are now differentially encoded. However, in the optical spectral domain, DB's spectral width is about half that of two-state OOK.

DB of 10 Gb/s has been generated by an InP MZI PIC with phase modulators, and a transmission distance of ∼180 km in SSMF, approximately four times that of a non-chirped OOK modulator, was achieved [19]. The filtering to generate DB was done electronically. DB of 40 Gb/s was demonstrated using an InP MZI PIC in which the electrical filtering was accomplished by the limited modulator bandwidth [20]. More recently a tunable laser was monolithically integrated with a 10-Gb/s MZI DB modulator in InP using phase modulators [21]. Again, the filtering was done electronically.

Figure 14.12 shows an 80-Gb/s MZI DB modulator PIC that uses EAMs operated part way between intensity modulation and phase modulation [22]. If the EAMs were operated as just phase modulators, they would have to be longer than ∼500 μm, making it difficult to achieve operation at speeds much above 10 Gb/s without using a traveling wave structure. A traveling wave structure is one in which the electrical drive and optical signal co-propagate along the modulator. Such a structure is more difficult to design and fabricate than a lumped element structure

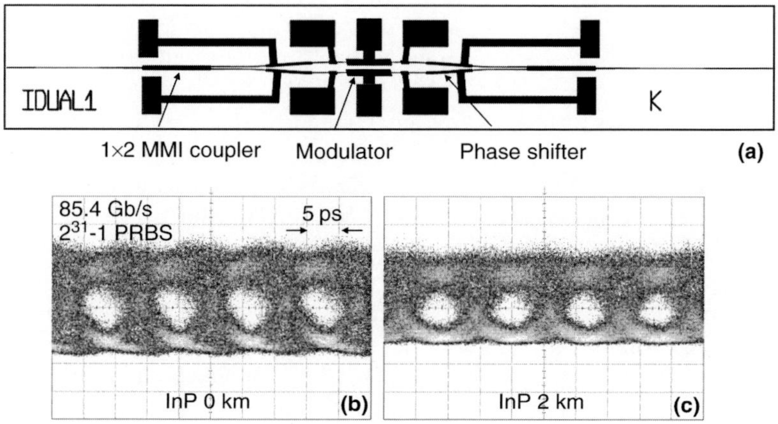

Fig. 14.12 Waveguide layout (**a**) of the MZI modulator PIC using electro-absorption/refraction modulators. Measured 85.4-Gb/s duobinary eye diagrams from the MZI modulator after 0 km (**a**) and 2 km (**b**) of SSMF

because difficult steps must be taken to slow down the radio frequency (RF) electrical drive signal and to avoid excessive RF loss of the electrical drive signal. Achieving pure intensity modulation with an EAM is nearly impossible because of the Kramers–Kronig relations, as shown in Fig. 14.6. In the device in Fig. 14.12, the modulators were driven as intensity modulators with large red chirp. This allowed the modulators to be only 115-μm long, making 80-Gb/s operation possible with a lumped-element design. The filtering in this case was done electronically, and actually no additional electronic filter was employed—the electronic driver and the modulator and its packaging provided the appropriate electronic filtering.

DB is a signal format with three symbols. A four-symbol format that uses only phase is quadrature phase-shift keying (QPSK), whose constellation is shown in Fig. 14.1. Unlike OOK or DB, QPSK cannot be received directly by direct detection. It must be either interfered with a separate continuous-wave (CW) signal, which is called a local oscillator (LO), or with a delayed copy of itself. QPSK is usually differentially encoded, called DQPSK. The advantages of QPSK are that the optical bandwidth is reduced by a factor of two compared to OOK at the same bit rate, the electronics for the transmitter and receiver run at only half the bit rate, and when balanced detection is used, the receiver sensitivity is increased by nearly 3 dB as compared to OOK at the same bit rate.

QPSK is usually generated by two MZIs driven push–pull inside a large MZI with a 90° phase difference between them, called a "nested MZI modulator" [23]. The MZIs "digitize" the optical phase, removing most of the intersymbol interference (ISI) from the electronic drivers. One driving data stream is called the "in-phase" (I) data stream and the other is the "quadrature" (Q) data stream. Figure 14.13 shows an InP-nested MZI modulator PIC and results from using it to generate an 80-Gb/s QPSK signal [24]. It uses phase modulators with traveling wave electrodes to achieve the high speed.

Figure 14.14 shows an InP PIC that monolithically integrates 10 DFB lasers at 10 different wavelengths, 10 nested MZI modulators, and an arrayed waveguide grating (AWG) multiplexer [25]. Performance is shown for one channel of 43-Gb/s QPSK. When all 10 channels are operated simultaneously, this PIC will be able to transmit 430 Gb/s of data.

Figure 14.15 shows an InP QPSK modulator PIC that uses EAMs rather than phase modulators, and its generation of 80-Gb/s QPSK is shown in Fig. 14.16 [2]. It works by using the EAMs as digital shutters. The two arms with EAMs have a 90° phase difference between them. The center arm has no EAM and is used to center the constellation on the origin. Using EAMs to generate an optical phase-shift keying format was first demonstrated in [26]. The advantage of using EAMs is that the modulators can be operated at high speed without needing a traveling wave structure, because the operation takes place at the band edge. In fact, the entire chip in Fig. 14.15 is shorter than the phase modulator inside the MZIs in Fig. 14.13. However, the drawbacks to using EAMs are that the intrinsic insertion loss is high, ∼12 dB in the ideal case; the signal is chirped; and the signal is not as insensitive to electronic driver ISI as the nested MZI structure.

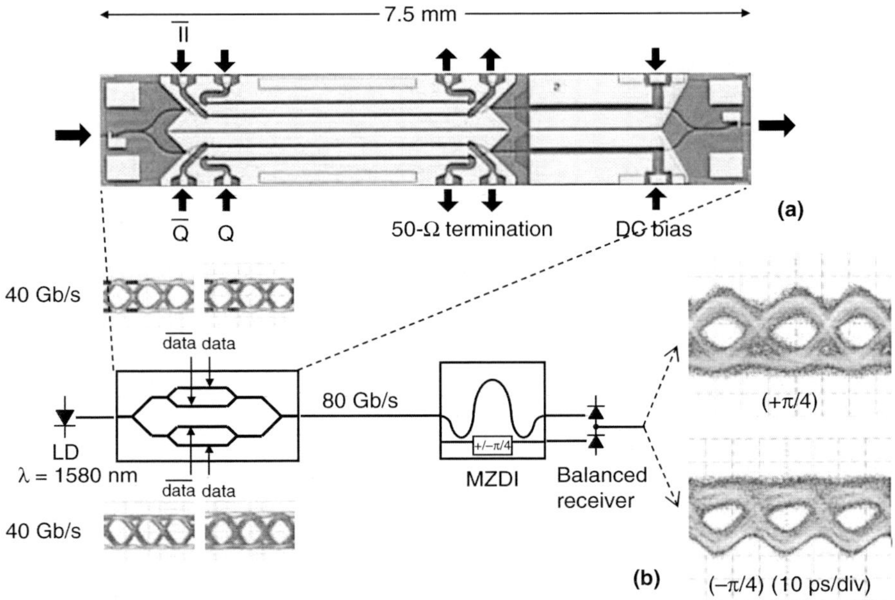

Fig. 14.13 (**a**) Nested MZI modulator PIC and (**b**) demonstration of 80-Gb/s DQPSK generation. Courtesy of N. Kikuchi

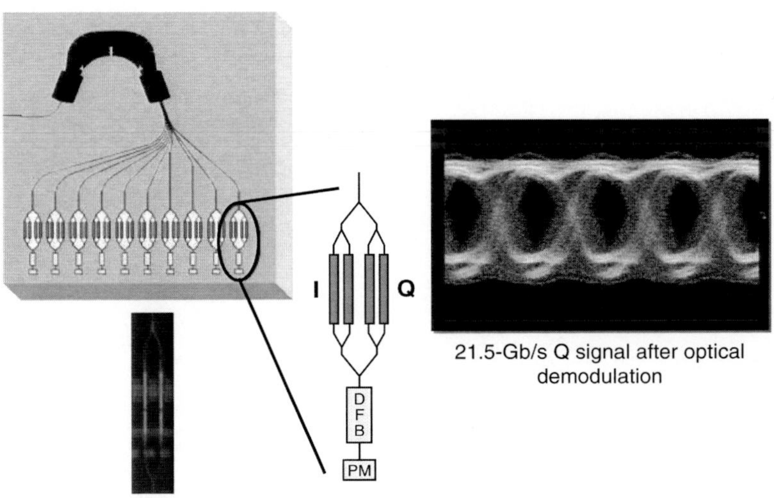

Fig. 14.14 Ten-wavelength transmitter PIC with a nested MZI modulator for each channel and demonstration of 43-Gb/s QPSK generation for one channel (Courtesy of C.H. Joyner)

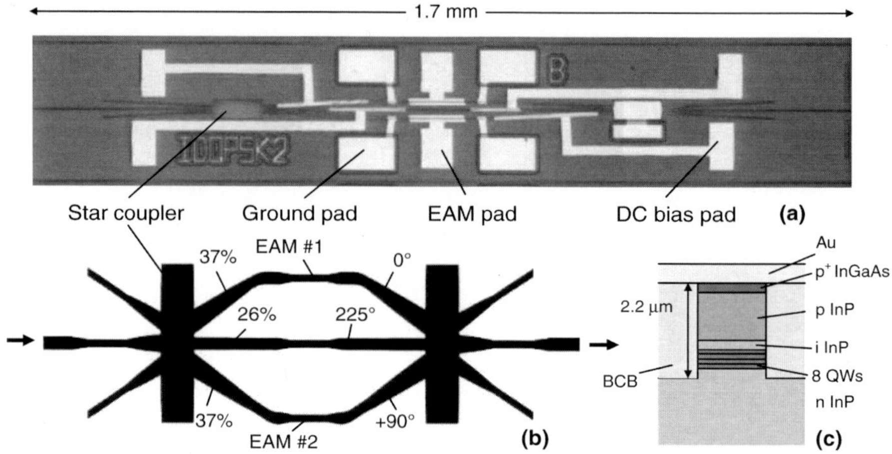

Fig. 14.15 QPSK modulator PIC using a three-arm interferometer with two EAMs. (**a**) photograph of chip, (**b**) waveguide layout, stretched vertically for clarity, and (**c**) cross-section of the waveguide structure

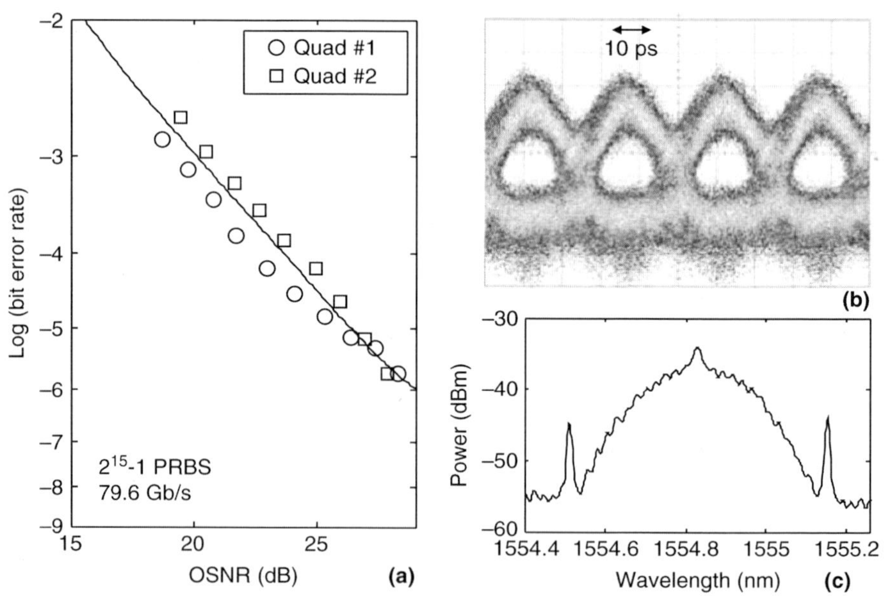

Fig. 14.16 QPSK performance of 80 Gb/s using the modulator of the previous figure. **a** Measured bit-error rate (BER) vs. optical signal-to-noise ratio (OSNR), **b** measured optically demodulated eye diagram, and **c** measured optical spectrum

14.3.3 Increasing the Spectral Efficiency via Multiple Levels

So far we have shown increasing the spectral efficiency via polarization and optical phase. Here we show the final method of increasing spectral efficiency: using multiple levels of optical magnitude. One such format is 16 quadrature amplitude modulation (QAM), whose constellation is shown in Fig. 14.1. It employs both magnitude and phase changes to achieve 16 possible symbols, reducing the spectral bandwidth by a factor of four compared to OOK.

One way to generate 16 QAM is to use four EAMs in a four-arm interferometer [27], as shown in Fig. 14.17. As with the EAM-based QPSK modulator of Fig. 14.16, the EAMs are used as shutters. The magnitudes and phases in each arm are shown in Fig. 14.17. The principle of operation is detailed in Fig. 14.18. Unlike the QPSK modulator, an arm without an EAM is not included, and thus the constellation is not centered on the origin, resulting in a residual carrier. The carrier is

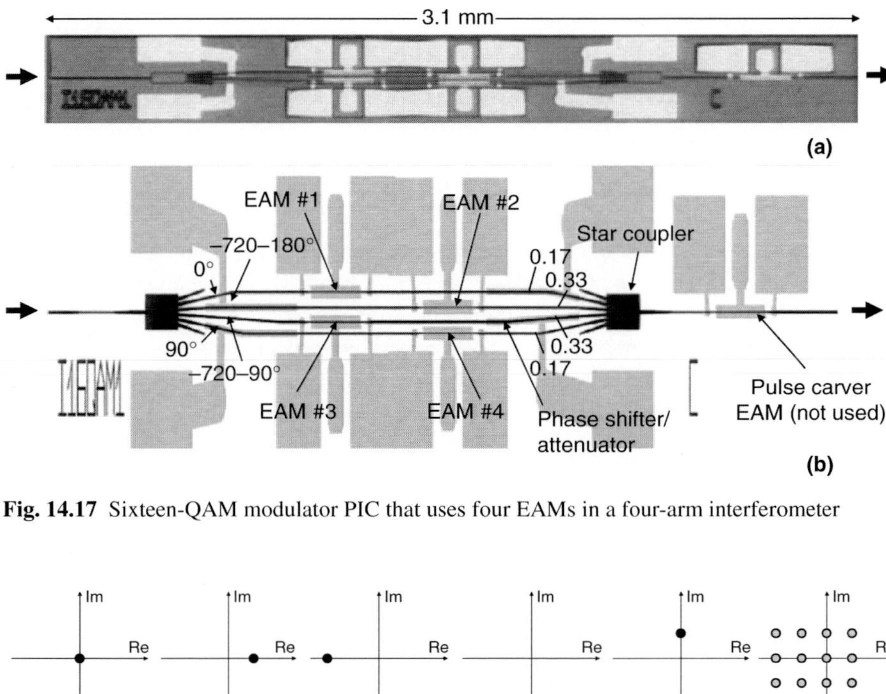

Fig. 14.17 Sixteen-QAM modulator PIC that uses four EAMs in a four-arm interferometer

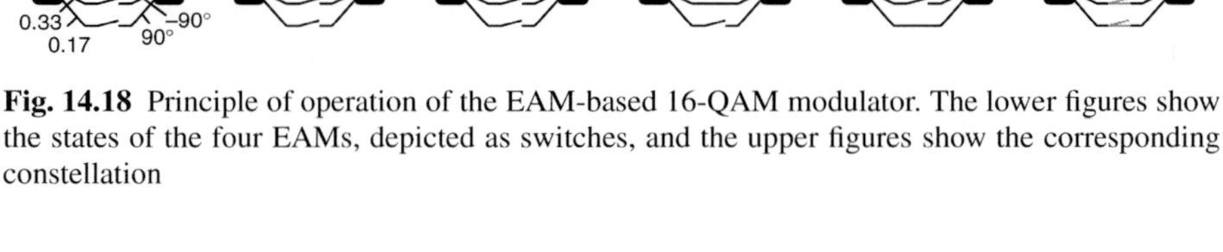

Fig. 14.18 Principle of operation of the EAM-based 16-QAM modulator. The lower figures show the states of the four EAMs, depicted as switches, and the upper figures show the corresponding constellation

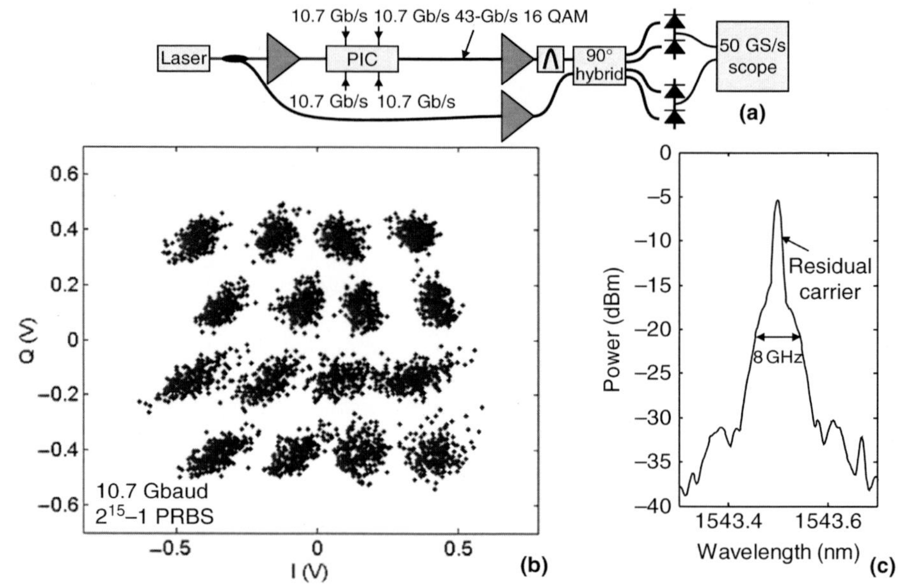

Fig. 14.19 Demonstration of 42.8-Gb/s 16-QAM generation using the modulator PIC shown in the previous figure. **a** experimental setup, **b** measured constellation, and **c** measured optical spectrum

relatively small in this case, and thus does not cause a significant sensitivity penalty. The measured performance at 40-Gb/s is shown in Fig. 14.19. The advantage of the EAM-based 16-QAM modulator is that it is very compact and simple to operate. The disadvantage is that the signal is chirped, and any electronic driver ISI is only partially eliminated.

14.3.4 Compensating for Fiber Chromatic Dispersion

We have so far given examples of transmitter PICs that produce signals with advanced modulation formats. A different feature that can be added to a PIC transmitter is an optical equalizer. An optical equalizer is a passive device that pre- or post-compensates ISI in an optical signal [28, 29]. A common cause of ISI is fiber chromatic dispersion. As we mentioned earlier, a chirped 10-Gb/s NRZ-OOK signal will become too distorted to receive after ~80 km of SSMF. For a given format, dispersion tolerance decreases as the square of the symbol rate, so a chirped 40-Gb/s NRZ-OOK signal can travel only ~5 km. By integrating an adjustable optical equalizer with a modulator, one can pre-compensate for chromatic dispersion in the transmission fiber and extend the distance.

Figure 14.20 shows such a PIC [30]. It contains an EAM followed by a three-arm interferometer. The three arms have different path lengths, and the attenuation and phase of each arm are controllable via electrical drive signals. The impulse response

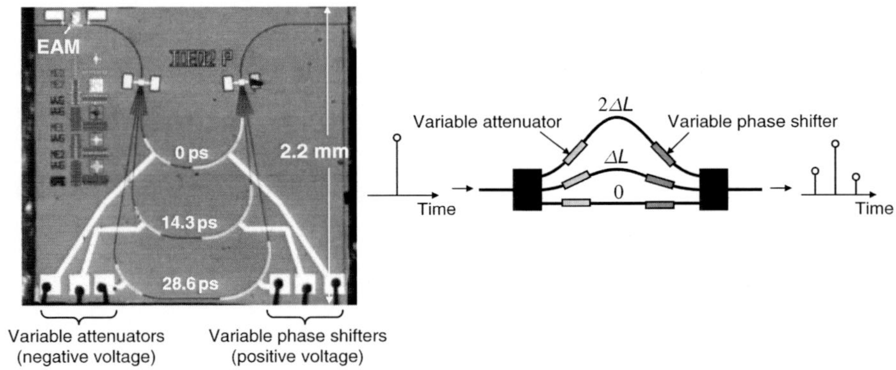

Fig. 14.20 Concept of a three-tap optical equalizer, and waveguide layout of a PIC with an EAM integrated with a three-arm optical equalizer

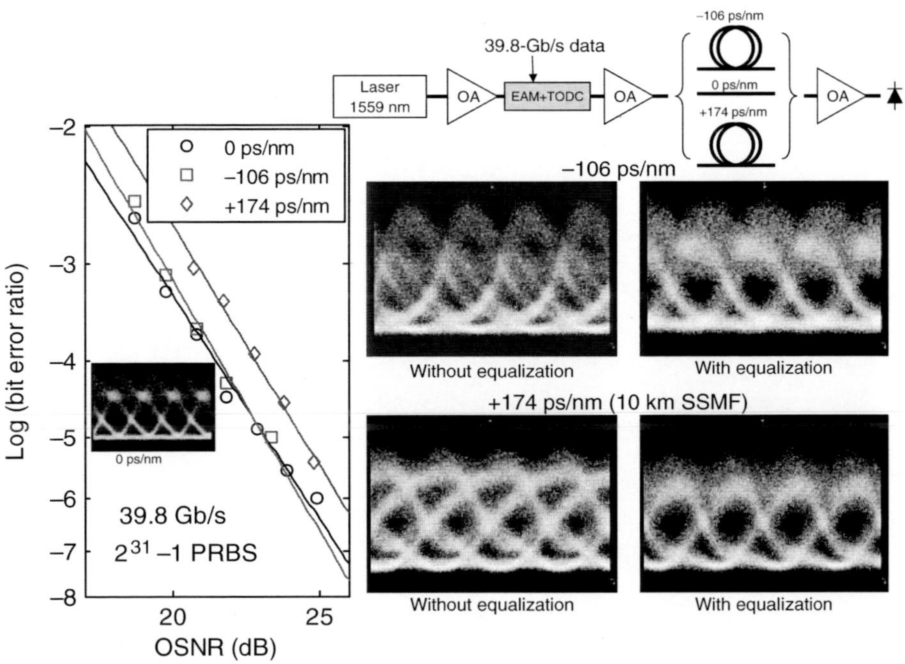

Fig. 14.21 Results of optical dispersion compensation from the integrated optical equalizer for a 39.8-Gb/s NRZ signal

of the equalizer consists of three impulses separated by 14.3 ps, the magnitude and phase of each impulse being independently controllable.

This optical equalizer was used as a tunable optical dispersion compensator (TODC) in the experiment shown in Fig. 14.21. Driving the EAM with a 40-Gb/s signal, the OEQ could extend the transmission distance from ~5 to 10 km of SSMF. Operation of the OEQ required is less than 500 μW of total electrical power. This is thousands of times less power than what would be required for the same compensation with digital electronics.

14.4 Receivers

The first reported PIC receiver that integrated multiple functions is shown in Fig. 14.22 [31], and a similar PIC was reported nearly simultaneously in [32]. It is a heterodyne coherent receiver for detecting frequency-shift-keyed signals. It consists of a distributed Bragg reflector laser, a 3-dB optical coupler, and two photodiodes. The laser serves as the local oscillator (LO) for the coherent detection.

Figure 14.23 shows a DQPSK receiver PIC [33]. It consists of a Mach–Zehnder delay interferometer, a 2 × 4 star coupler serving as a 90° hybrid [34], and four photodiodes. The principle of operation of this star coupler is shown in Fig. 14.24. Such a PIC is thousands of times smaller in footprint area than the corresponding discrete optical elements to perform the same function. The delay interferometer

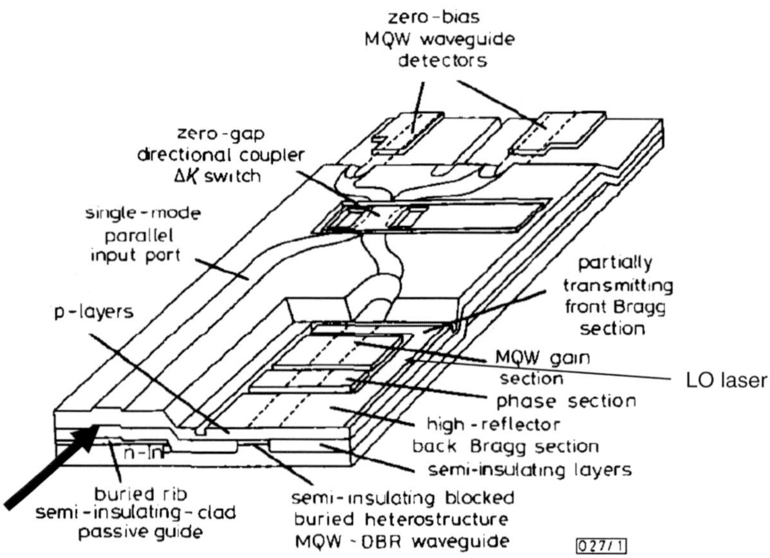

Fig. 14.22 First reported coherent receiver, for heterodyne detection (From [31])

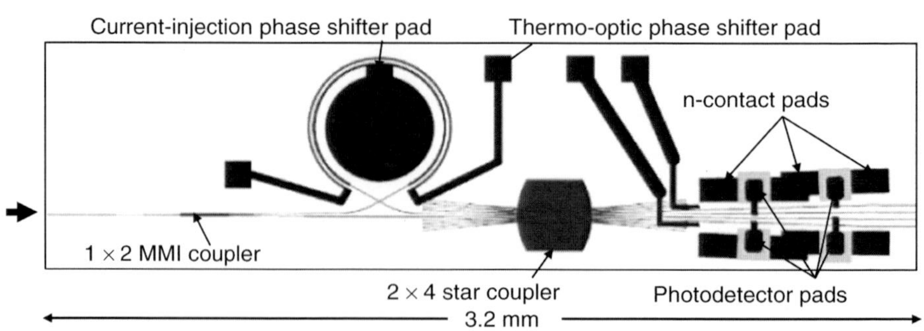

Fig. 14.23 DQPSK receiver PIC waveguide layout

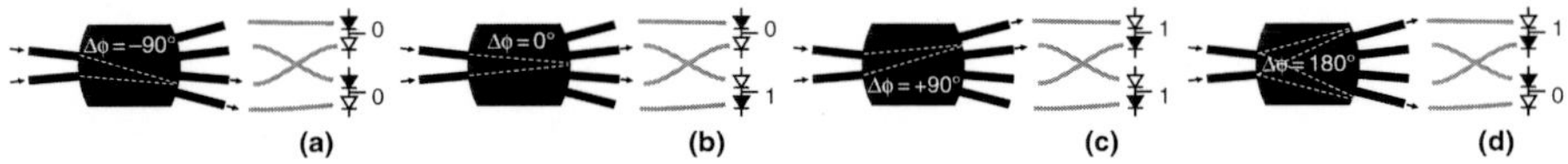

Fig. 14.24 Operation principle of the 90° hybrid in the DQPSK demodulator. $\Delta\phi$ is the phase difference between successive symbols, and DQPSK has four possible values for $\Delta\phi$: $-90°$, $0°$, $90°$, and $180°$

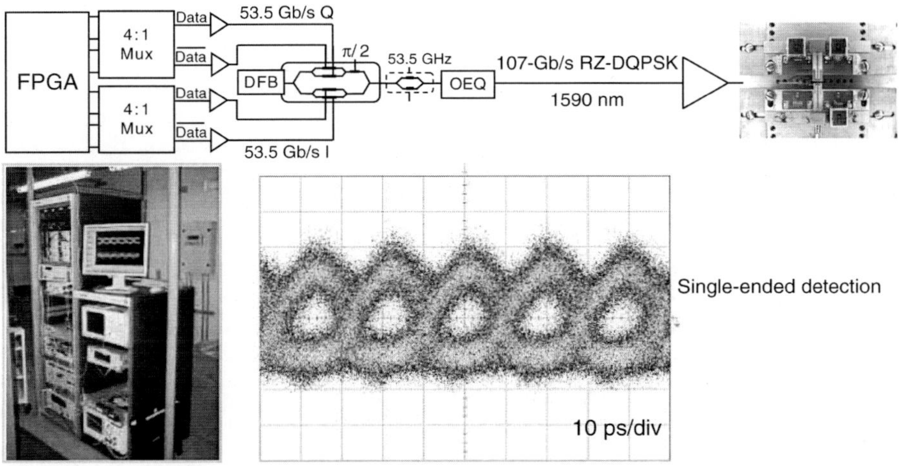

Fig. 14.25 Measured results of DQPSK receiver PIC receiving a 107-Gb/s RZ-QPSK signal

has a delay of 18.7 ps and demodulates symbols at a rate of 53.5 Gbaud. The delay interferometer contains a current injection phase shifter for controlling the polarization dependence of the interferometer and a thermooptic phase shifter for adjusting the overall phase of the interferometer. Figure 14.25 shows successful reception of a 107-Gb/s RZ-DQPSK signal using the PIC of Fig. 14.23.

14.5 Conclusions

We discussed InP monolithically integrated transmitter and receiver PICs that have advanced features. We gave several examples in which spectral efficiency was increased by polarization, phase, and/or multiple levels of magnitude and phase. We believe that with more research and development and as the demand for more information transfer yet lower cost, smaller footprint, and lower power consumption continues, the case for advanced feature PICs will become more and more compelling.

Acknowledgements I am especially indebted to L. Zhang. I also thank L. L. Buhl, P. J. Winzer, N. J. Sauer, G. Raybon, A. H. Gnauck, P. Bernasconi, D. T. Neilson, C. Kazmierski, Y. K. Chen, and M. Zirngibl.

References

1. R. Nagarajan, et al., Large-scale photonic integrated circuits. J. Sel. Top. Quant. Electron. **11**, 50–65, Jan–Feb (2005).
2. C.R. Doerr, L. Zhang, A. Adamiecki, N. Sauer, J. Sinsky, P. Winzer, Compact EAM-based InP DQPSK modulator and demonstration at 80 Gb/s. In optical fiber communication conference and exposition and the national fiber optic engineers conference, OSA Technical Digest Series (CD) (Optical Society of America, 2007), paper PDP33.
3. C.X. Yu, S. Chandrasekhar, T. Zhou, D.T. Neilson, 0.8 bit/s/Hz spectral efficiency at 10 Gbit/s via vestigial-sideband filtering, Electron. Lett. 39(2) 225–227, 23 Jan (2003).
4. R.-J. Essiambre, G.J. Foschini, G. Kramer, P.J. Winzer, Capacity limits of information transport in fiber-optic networks. Phys. Rev. Lett. **101** 163901–1–163901–4 (2008).
5. A. Chipman, A commodity no more. *Nature* **449**, 131, 13 September (2007).
6. A. Sneh, C.R. Doerr, Indium phosphide-based photonic circuits and components. In *Integrated Optical Circuits and Components, Design and Applications*. (Marcel Dekker, New York, 1999).
7. C.H. Henry, R.A. Logan, F.R. Merritt, J.P. Luongo, The effect of intervalence band absorption on the thermal behavior of InGaAsP lasers. IEEE J. Quantum Electron. **QE-19**, 947–952 (1983).
8. W.P. Dumke, M.R. Lorenz, G.D. Pettit, Intra- and interband free-carrier absorption and the fundamental absorption edge in n-type InP. Phys. Rev. B **1**, 4668–4673 (1970).
9. S. Demiguel, X. Li, N. Li, H. Chen, J.C. Campbell, J. Wei, A. Anselm, Analysis of partially depleted absorber waveguide photodiodes. J. Lightwave Technol. **23**, 2505–2512, Aug. (2005).
10. N. Shimizu, Y. Miyamoto, A. Hirano, T. Ishabashi, RF saturation mechanism of InP/InGaAs unitravelling-carrier photodiode. Electron. Lett. **36**, 750–751, Apr. (2000).
11. J.-F. Vinchant, J.A. Cavailles, M. Erman, P. Jarry, M. Renaud, InP/GaInAsP guided-wave phase modulator based on carrier-induced effects: theory and experiment. J. Lightwave Technol. **18**, 63–70, Jan. (1992).
12. N. Susa, Electric-field-induced refractive index changes in three-step asymmetric coupled quantum wells. J. Appl. Phys. **73**, 8463–8470 (1993).
13. M. Suzuki, Y. Noda, H. Tanaka, S. Akiba, Y. Kushiro, H. Isshiki, Monolithic integration of InGaAsP/InP distributed feedback laser and electroabsorption modulator by vapor phase epitaxy. Lightwave Technol. J. **5**(9), 1277–1285, Sep. (1987).
14. F. Dorgeuille, F. Devaux, On the transmission performances and the chirp parameter of a multiple-quantum-well electroabsorption modulator. IEEE J. Quantum Electron. **30**, 2565–2572 (1994).
15. J.C. Cartledge, B. Christensen, Optimum operation points for electroabsorption modulators in 10 Gb/s transmission systems using nondispersion shifted fiber. J. Lightwave Tech. **16**, 349–357, Mar. (1998).
16. Y. Yu, R. Lewen, S. Irmscher, U. Westergren, L. Thylen, U. Eriksson, W.S. Lee, 80Gb/s ETDM transmitter with a traveling-wave electroabsorption modulator. In optical fiber communication conference, paper OWE1, 2005.
17. C. Kazmierski, et al., 100 Gb/s operation of an AlGa In As Semi-insulating buried heterojunction EML Optical Fiber Comm. Cont., OThT7, 2009.
18. C.R. Doerr, L. Zhang, Monolithic 80-Gb/s dual-polarization on-off-keying modulator in InP. In optical fiber communication conference, PDP19, 2008.
19. R.A. Griffin, A. Tipper, I. Betty, Performance of MQW InP Mach–Zehnder modulators for advanced modulation formats. In optical fiber communication conference, OTuL5, 2005.
20. P.J. Winzer, S. Chandrasekhar, C.R. Doerr, D.T. Neilson, A. Adamiecki, R.A. Griffin, 42.7-Gb/s modulation with a compact InP Mach–Zehnder transmitter. In European conference on optical communications, We1.6.1, 2006.
21. L.A. Johansson, L.A. Coldren, P.C. Koh, Y.A. Akulova, G.A. Fish, Transmission of 10 Gbps duobinary signals using an integrated laser Mach–Zehnder modulator. In optical fiber

communication / national fiber optic engineers conference, OFC/NFOEC, 1–3, 24–28 February 2008.

22. C.R. Doerr, G. Raybon, L. Zhang, L.L. Buhl, A.L. Adamiecki, J.H. Sinsky, N.J. Sauer, Compact and low-chirp 80-Gb/s duobinary modulator in InP using electroabsorption/refraction modulators. IEEE Photon. Technol. Lett. 21, 1199–1201, 2009
23. R.A. Griffin, R.I. Johnstone, R.G. Walker, J. Hall, S.D. Wadsworth, K. Berry, A.C. Carter, M.J. Wale, J. Hughes, P.A. Jerram, N.J. Parsons, 10 Gb/s otpical differential quadrature phase shift key (DQPSK) transmission using GaAs/AlGaAs integration. In optical fiber communication conference, paper FD6, 2002.
24. N. Kikuchi, H. Sanjoh, Y. Shibata, K. Tsuzuki, T. Sato, E. Yamada, T. Ishibashi, H. Yasaka, 80-Gb/s InP DQPSK modulator with an n-p-i-n structure. In European conference on optical communication, 2008.
25. S. Corzine, P. Evans, M. Kato, G. He, M. Fisher, M. Raburn, A. Dentai, I. Lyubomirsky, R. Nagarajan, M. Missey, V. Lal, A. Chen, J. Thomson, W. Williams, P. Chavarkar, S. Nguyen, D. Lambert, T. Butrie, M. Reffle, R. Schneider, M. Ziari, C. Joyner, S. Grubb, F. Kish, D. Welch, 10-channel x 40Gb/s per channel DQPSK monolithically integrated InP-based transmitter PIC. In optical fiber communication conference and exposition and the national fiber optic engineers conference, OSA Technical Digest (CD) (Optical Society of America, 2008), paper PDP18.
26. Kang, Phase-shift-keying and on-off-keying with improved performances using electroabsorption modulators with interferometric effects. Opt. Express **15**, 1467–1473 (2007).
27. C.R. Doerr, P. Winzer, L. Zhang, L. Buhl, N. Sauer, Monolithic InP 16-QAM modulator. In national fiber optic engineers conference, OSA Technical Digest (CD) (Optical Society of America, 2008), paper PDP20.
28. M. Bohn, G. Mohs, C. Scheerer, C. Glingener, C. Wree, W. Rosenkranz, An adaptive optical equalizer concept for single channel distortion compensation. In European conference on optic communication, paper Mo.F.2.3, 2001.
29. C.R. Doerr, S. Chandrasekhar, P.J. Winzer, A.R. Chraplyvy, A.H. Gnauck, L.W. Stulz, R. Pafchek, E. Burrows, Simple multichannel optical equalizer mitigating intersymbol interference for 40-Gb/s nonreturn-to-zero signals. J. Lightwave Technol. **22**, 249–256, Jan. (2004).
30. C.R. Doerr, L. Zhang, L. Buhl, N. Sauer, A. Adamiecki, 40-Gb/s modulator with monolithically integrated tunable optical dispersion compensator. In optical fiber communication conference and exposition and the national fiber optic engineers conference, OSA Technical Digest Series (CD) (Optical Society of America, 2007), paper PDP45.
31. T.L. Koch et al., GaInAs/GaInAsP multiple-quantum-well integrated heterodyne receiver. Electron. Lett. **25**, 1621–1623, Nov. (1989).
32. H. Takeuchi, K. Kasaya, Y. Kondo, H. Yasaka, K. Oe, Y. Imamura, Monolithic integrated coherent receiver on InP substrate. IEEE Photon. Technol. Lett. **1**, 398–400, Nov. 1989.
33. C.R. Doerr, L. Zhang, L. Buhl, J. Sinsky, A. Gnauck, P. Winzer, A. Adamiecki, N. Sauer, High-Speed InP DQPSK Receiver. In optical fiber communication conference and exposition and the national fiber optic engineers conference, OSA Technical Digest (CD) (Optical Society of America, 2008), paper PDP23..
34. C.R. Doerr, D.M. Gill, A.H. Gnauck, L.L. Buhl, P.J. Winzer, M.A. Cappuzzo, A. Wong-Foy, E.Y. Chen, L.T. Gomez, Monolithic demodulator for 40-Gb/s DQPSK using a star coupler. J. Lightwave Tech. **24**, 171–174, Jan. (2006).

Chapter 15
Integrated Mach–Zehnder Interferometer-Based Modulators for Advanced Modulation Formats

Tetsuya Kawanishi

Abstract This chapter describes high-speed lightwave modulation using electro-optic effect. A Mach–Zehnder interferometer can control amplitude of lightwaves precisely. By using a pair of Mach–Zehnder interferometers, the real and imaginary components of the lightwave can be manipulated independently. This technique is called vector modulation, which is commonly used in wireless radio communication systems. As well known, lightwave state can be mathematically expressed by a point on a two-dimensional complex plane, thus we can generate arbitrary lightwave states by using the vector modulation. Multi-level quadrature amplitude modulation signal can be synthesized by using parallel Mach–Zehnder modulators. We can generate complicated multi-level optical signals from binary data streams, where binary modulated signals are vectorically summed in optical circuits. Frequency response of each Mach–Zehnder interferometer is also very important to achieve high-speed signals. We can enhance the bandwidth of the response by using thin substrate. Modulation of 87 Gbaud was demonstrated with a dual-parallel Mach–Zehnder modulator.

15.1 Background

Modulation technologies play important roles in lightwave transmission systems, where digitized information should be described by some parameters of lightwave, such as amplitude, phase, frequency [1]. Intensity modulation (IM) and on off keying (OOK) are commonly used in commercial systems [2]; however, recently, various types of modulation techniques, for example, differential phase-shift keying (DPSK) [3], differential quadrature phase-shift keying (DQPSK) [4–8], amplitude- and phase-shift keying (APSK) [9, 10], frequency-shift keying (FSK) [11–15], single-sideband (SSB) modulation techniques [16–18], n-level quadrature amplitude modulation (QAM) [19–24] were investigated to obtain enhanced spectral

T. Kawanishi (✉)
National Institute of Information and Communications Technology (NICT),
4-2-1 Nukui-Kitamachi, Koganei, Tokyo 184-8795, Japan
e-mail: kawanish@nict.go.jp

M. Nakazawa et al. (eds.), *High Spectral Density Optical Communication Technologies*, Optical and Fiber Communications Reports 6, DOI 10.1007/978-3-642-10419-0_15,

efficiency or receiver sensitivity in optical transmission systems. As well known, lightwave state can be mathematically expressed by a point on a two-dimensional complex plane which has a real and imaginary axes. If the real and imaginary components of the lightwave can be manipulated independently, we can generate arbitrary lightwave states. This technique is called vector modulation, which is commonly used in wireless radio communication systems. We note that the carrier frequency in radio systems should be typically lower than 100 GHz, while that should be higher than 100 THz in lightwave systems. The significant difference in carrier frequencies causes difficulties in precise lightwave control on the complex plane. The linewidths of light sources are much larger than radio wave sources, so that it is very difficult to fix a lightwave state on the complex plane. Thus, precise and high-speed lightwave control is very important in transmitter side, while high-performance lightwave estimation is needed in receivers, to achieve vector modulation based high-speed transmission systems [25].

Direct modulation of injected current to laser diodes provides a simple transmitter configuration, however, the high-speed operation is limited by the response of the lasers, and the output would have undesired frequency or phase shift (chirp, henceforth) [26]. Optical external modulators with electro-optic (EO) or electro-absorption (EA) effect can provide high-speed modulation, so that they are commonly used in high-performance transmission systems. The size of EA modulators can be much smaller than that of EO modulators, however, the output of EA modulators still would have some chirp [2]. Although EA modulators can basically control only optical intensity, complicated modulation is possible by using integrated EA modulators [27]. On the other hand, in a material having the EO effect, the refractive index (RI) depends on the voltage applied on the material, so that an optical phase modulator can be constructed with an optical waveguide and an electrode on EO material. Various types of modulators for advanced modulation formats can be built with optical phase modulators. Mach–Zehnder interferometer-based modulators (MZMs) consisting of two high-speed optical phase modulators can control lightwave amplitude, precisely [28]. Dual-parallel MZMs (DPMZMs), which can control the real and imaginary components independently, are commonly used for high-speed DQPSK signal generation [6, 8]. A quad-parallel MZM (QPMZM) can synthesize four-level signals optically both in the real and imaginary components, so that a 16-QAM signal can be generated from four electric binary data streams [20].

In this section, we review EO effect based-optical modulator technologies. EO modulators, which can control the phase and frequency of lightwave, as well as the intensity. First, basic concept of vector modulation scheme is described, to discuss requirement of optical components in transmission systems with advanced modulation formats. The principle of optical phase modulators using EO effect is provided, to discuss mathematical expression of MZMs which can control intensity or amplitude of lightwaves. The concept of parallel MZMs is described briefly. The total expected bitrate generated by an integrated modulator depends on the number of MZI elements and the frequency response. Some experimental results of DQPSK and 16-QAM are also shown, where multi-level optical signals can be synthesized in integrated parallel MZMs. The modulation speed can be higher than 40 Gbaud with a DPMZM, so that the bitrate can be higher than 80 Gb/s. By using a QPMZM, we can achieve 50 Gb/s transmission where the modulation speed is 12.5 Gbaud. We also investigate frequency response of MZMs. MZMs with travailing-wave co-planer waveguide (CPW) electrodes on thin substrate can provide enhanced frequency response over 40 GHz, where the response in a range from 5 to 63 GHz was measured by using two different methods: electric frequency domain network analyzer technique and optical spectral measurement of high-speed modulated lightwaves. The latter method would be useful in very high-frequency region.

15.2 Optical Components for Vector Modulation Schemes

Phase control of lightwaves is more difficult than that of radio waves as described in the previous section, however, recently, high-speed transmission techniques using vector modulation schemes have attracted significant attention in research and development during the last several years, because the phase rotation due to phase noise of the laser source can be compensated by digital signal processing [24, 29–31]. This scheme is called digital coherent, whose setup consists of a diode laser as a light source, an optical modulator, a 90° optical hybrid, an analog-to-digital converter (ADC), digital-to-analog converter (DAC), a digital signal processor, a pair of balanced photodetectors, and another diode laser as a local oscillator, as shown in Fig. 15.1. In conventional OOK systems, wavelength-domain multiplexing (WDM) is commonly used to increase total transmission capacity. Optical multiplexing techniques including WDM can be applied to digital coherent systems in which particular components are narrow linewidth laser light sources, optical modulators for vector modulation, 90° optical hybrids for vector demodulation, balanced photodetectors, ADCs for multi-level modulation, and DACs for multi-level demodulation.

Narrow linewidth laser light sources are indispensable for advanced modulation formats. The fluctuation of symbols on the complex plane depends on the linewidth of the light sources. The typical linewidth of laser diodes is a few megahertz, but, recently, lasers with narrow linewidths less than a few kilohertz are commercially

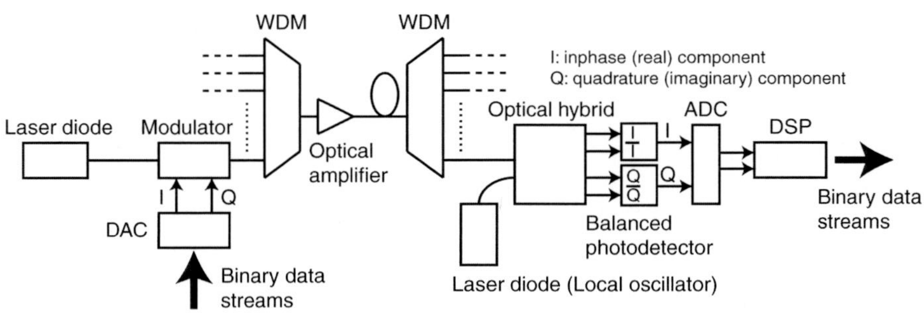

Fig. 15.1 Optical transmission system with vector modulation

available [32–34]. The impact on the signal waveform depends on the ratio between baud rate and linewidth, so that the requirement for the linewidth would be relaxed if the baud rate is higher than 10 Gb/s. A laser diode is also used in a digital coherent receiver, as a local oscillator. The phase or frequency of lightwave generated by the local oscillator would have some difference from that of the light source, due to the finite linewidths of the laser diodes. A 90° optical hybrid can mix a received signal and a local oscillator, to discriminate the real and imaginary components [35–37]. The optical hybrid would deliver four lightwave signals. Two of them form a differential signal pair of the real component, while the other two are for the imaginary component. In addition, an optical hybrid with eight output ports provides polarization diversity operation [38], where polarization mode dispersion (PMD) can be compensated by digital signal processing [29]. Balanced photodetectors can convert the optical differential signal into an electric output [39]. For high baud rate systems, high-speed monolithically integrated balanced photodetectors which shows high symmetry within 0.5 dB were reported, where the 3 dB cutoff frequency was 40 GHz [40].

As shown in Fig. 15.2, the state of the received signal is mapped onto a complex plane. The local oscillator acts as a reference to determine the real and imaginary axes, where the local oscillator signal corresponds to the unit vector on the real axis. The phase noise of the light source and local oscillator would have large impact on phase fluctuation, while the amplitude noise of the lasers would affect the amplitude fluctuation on the complex plane. The phase fluctuation is dominant on the complex plane and is compensated by using digital signal processing. However, the bandwidth of the phase fluctuation should be narrow, in order to reduce required computational power for the signal processing. Thus, narrow linewidth lasers would be useful for advanced modulation formats, even if we can use digital coherent techniques.

As described in the previous section, components for lightwave modulation play very important roles in the systems with vector modulation. Direct modulation

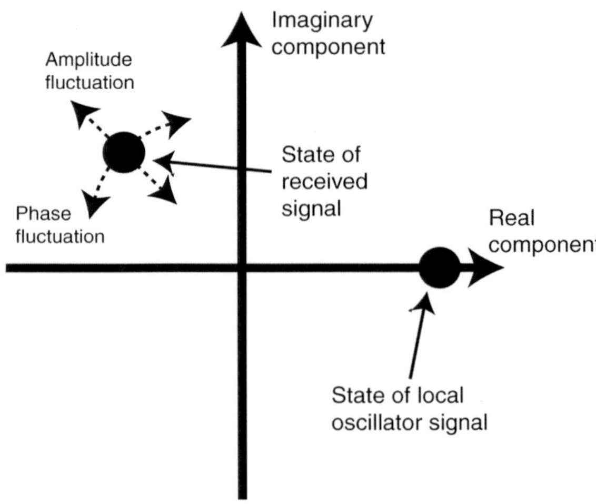

Fig. 15.2 State of received signal mapped onto a two-dimensional complex plane, where a local oscillator acts as a reference to determine the real and imaginary axes

provides a simple modulation scheme. By changing injection electric current to the laser, we can control the amplitude, phase, and frequency of the output lightwave, however, it is very difficult to control the real and imaginary components on the complex plane, independently. On the other hand, external modulation can provide precise and high-speed lightwave control in the two-dimensional complex plane. Mach–Zehnder interferometer based modulators (MZMs) consisting of two high-speed optical phase modulators can control lightwave amplitude, precisely. Recently, various integrated MZMs have been developed for advanced modulation formats. Dual-parallel MZMs, which can control the real and imaginary components independently, are commonly used for high-speed DQPSK signal generation. Typical 3 dB bandwidth of electro-optic response of MZMs is 30 GHz. By using high-speed dual-parallel MZMs, 100 Gb/s DQPSK transmission was demonstrated, where the modulation speed was 50 Gbaud, and a pair of binary data streams would be fed to the modulator [41, 42]. For more complicated modulation formats, such as 16-QAM, multi-level modulation is needed both in the real and imaginary components. Multi-level electric signals can be generated by using a high-speed DAC. Recently, over 20 Gsa/s DACs and ADCs have been developed [31], so that digital coherent scheme can be applied to high baud rate systems [22]. However, the modulation speed would be limited by the performance of electric amplifiers as well as the sampling rate of the DAC, because linear, large-amplitude and high-speed amplification is required to drive high-speed external modulators. One of the solution for this issue is photonic signal synthesis. A quad-parallel MZM can synthesize four-level signals optically both in the real and imaginary components, so that a 16-QAM signal can be generated from four electric binary data streams [20].

One of the simplest modulation schemes using vector modulation is quadrature phase-shift keying (QPSK), which is identical to 4-QAM, where the real and imaginary components are modulated by binary data streams. When the signal is modulated by QPSK with differential coding (DQPSK), optical delay interferometers can convert the optical signal into a pair of intensity-modulated binary signals [43, 44]. In addition to vector modulation devices and optical delay interferometers, we need some precorders or decorders for the differential coding, However, conventional optical components designed for binary modulation formats, such as high-speed photodetectors, electric amplifiers, optical amplifiers, digital interfaces can be used for generation, transmission, and detection of DQPSK signals. Thus, DQPSK signals can be demodulated without using the digital coherent, and the baud rate can be easily increased to achieve huge capacity transmission [45–47].

15.3 Phase Modulator

Mathematical expressions for optical phase modulators are given to describe the basic principle of modulators using the EO effect. Electrical control of optical phase can be achieved by using EO materials, such as lithium niobate (LN) [8, 48–50], lithium tantalate (LT) [51], gallium arsenide (GaAs) [6], indium phosphite (InP)

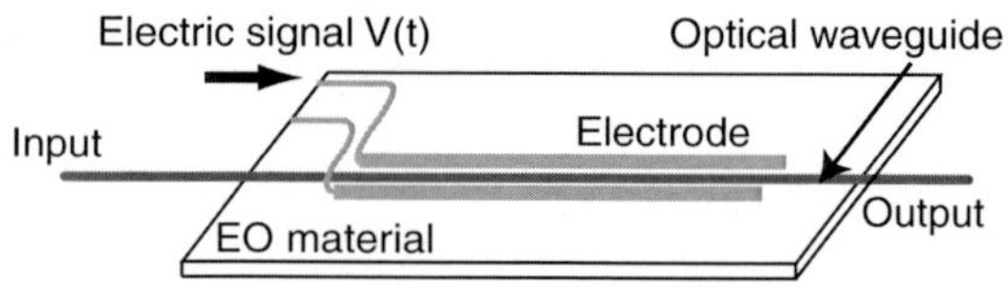

Fig. 15.3 Optical phase modulator using electro-optic effect, where an electrode is placed along an optical waveguide

[52, 53]. Figure 15.3 shows a schematic of an optical modulator using EO effect. An electrode is placed along with an optical waveguide. By applying electric voltage on the electrode, we can change the RI in the waveguide, so that the phase of the lightwave can be controlled. For simplicity, the input lightwave is assumed to be monochromatic and be described by $E_\mathrm{i}\mathrm{e}^{\mathrm{i}\omega_0 t}$. The output lightwave R can be expressed by

$$R = K E_\mathrm{i}\mathrm{e}^{\mathrm{i}\omega_0 t + \mathrm{i} f(t)}, \tag{15.1}$$

$$f(t) \equiv \Xi V(t), \tag{15.2}$$

where $V(t)$ and Ξ, respectively, denote the electric voltage on the electrode and the coupling coefficient between the electric signal and the lightwave signal. K is the optical transmittance in the waveguide. When $V(t)$ has high-frequency components, the frequency response of Ξ should be taken into account. The coefficient Ξ can be assumed to be a constant if the wavelengths of the high-frequency components on the electrode are much longer than the length of the electrode. The response of EO materials is much faster than the frequency response of Ξ, which is dominated by that of the electrode and by the phase mismatch between the electric and lightwave signals [48–50]. The halfwave voltage defined by

$$V_\pi \equiv \frac{\pi}{\Xi} \tag{15.3}$$

is the voltage for π (180°) optical phase shift.

15.4 Mach–Zehnder Intensity Modulator

Mach–Zehnder (MZ) structures are commonly used to achieve IM by EO effects. MZ structures for IM are called MZ modulators (MZMs). As shown in Fig. 15.4 the MZM consists of two optical phase modulators. We can control the output optical intensity by applying a pair of electric signals to the two phase modulators. Balanced push–pull scheme is ideal for IM, where the induced phase of the modulator A is

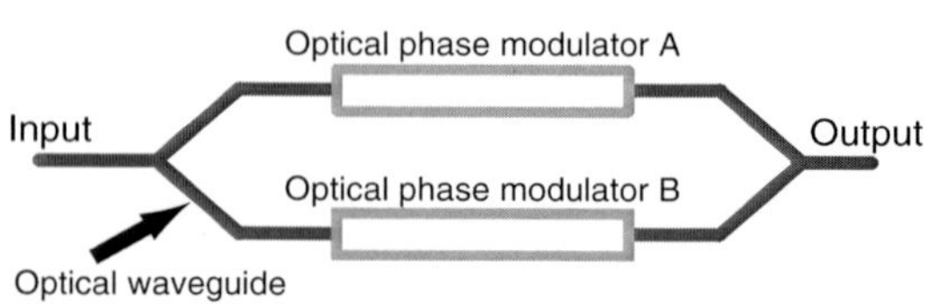

Fig. 15.4 A schematic of a MZ structure consisting of two optical phase modulators (A and B). Each optical phase modulator has an electrode for optical phase control

$g(t)/2$ and that of the modulator B is $-g(t)/2$. $g(t)$ describes the optical phase difference between the two arms of the MZ structure. When the lightwaves in the two arms are in phase $g(t) = 2n\pi$, $n = \ldots, -1, 0, +1, \ldots$, the modulator is in "on" state. At the junction of the output, the lightwaves interfere constructively. On the other hand, when the lightwaves in the two arms are in opposite phase $g(t) = (2n + 1)\pi$, the lightwaves are converted into high-order radiative mode waves which cannot propagate through the waveguide for output. Thus, we turn the modulator "on" and "off" by changing the voltage on the electrode. The output lightwave intensity from the MZM can be given by

$$|R| = \frac{|KE_{\mathrm{i}}|}{2} \left| \left(\mathrm{e}^{\mathrm{i}g(t)/2} + \mathrm{e}^{-\mathrm{i}g(t)/2} \right) \mathrm{e}^{\mathrm{i}\omega_0 t} \right| \tag{15.4}$$

$$= \left| KE_{\mathrm{i}} \cos\left[g(t)/2 \right] \right|. \tag{15.5}$$

V_π of the MZM is defined by the voltage required to switch the modulator from the "on" state to the "off" state and is equal to a half of V_π of each phase modulator in the MZM. When $g(t) = 2n\pi, n = \ldots, -1, 0, +1, \ldots$, $|R|$ becomes the maximum, where $\left|\cos\left[g(t)/2\right]\right| = 1$. However, the polarity of R, i.e., $\cos\left[g(t)/2\right]$, depends on n. When n is an odd number, the polarity is plus. When n is an even number, the polarity is minus. Thus, amplitude of the optical output can be continuously controlled by changing $g(t)$ from $2n\pi$ to $2(n + 1)\pi$. For example, when $g(t) = 0$, $\cos\left[g(t)/2\right] = +1$, while it becomes -1 with $g(t) = 2\pi$. Thus, we can flip the polarity of the waveform and achieve binary PSK (BPSK) with 180° phase deviation. This technique itself is very useful for DPSK. In addition, more complicated optical signals can be synthesize from binary PSK signals by using integrated parallel MZMs.

15.5 Integrated Modulators for High Data Rate Signal Generation

A DPMZM consists of two MZI elements embedded in a photonic circuit, as shown in Fig. 15.5. For DQPSK modulation, two binary data streams are applied to the two MZI elements ($\mathrm{MZ_A}$ and $\mathrm{MZ_B}$), to achieve control of in-phase (I) and quadrature (Q) components. Figure 15.6 shows a schematic of 16-QAM using a QPMZM, where a pair of 2-bit codes for in-phase (I_1, I_2) and for quadrature (Q_1, Q_2) components are fed to four MZI elements. While a DPMZM can generate a 2-bit/symbol DQPSK signal from two binary data streams, a QPMZM can achieve 4 bit/symbol from four binary data streams. In general, a parallel MZM with N MZI elements can generate an N bit/symbol lightwave signal from N binary data streams. Bit rate of optical modulated output generated from N electric binary data streams can be expressed by $B = N \times R$ (R : symbol rate). For example, by using an octet-parallel MZM (OPMZM) we can generate 256-QAM from eight binary data streams (I_1, I_2, I_3, I_4, Q_1, Q_2, Q_3, and Q_4,) where $N = 8$, as shown in Fig. 15.7. To achieve

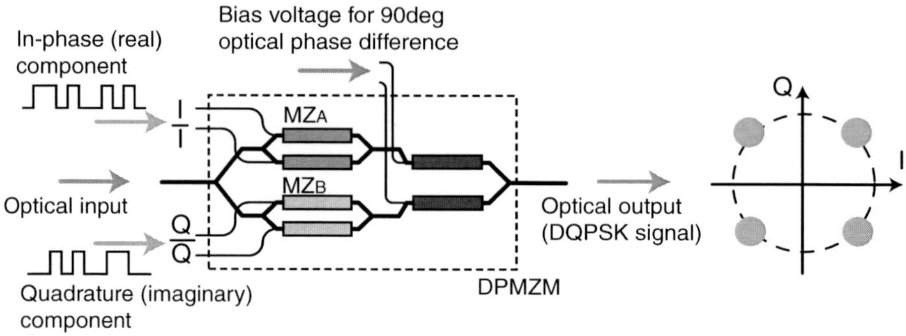

Fig. 15.5 QPSK using DPMZM

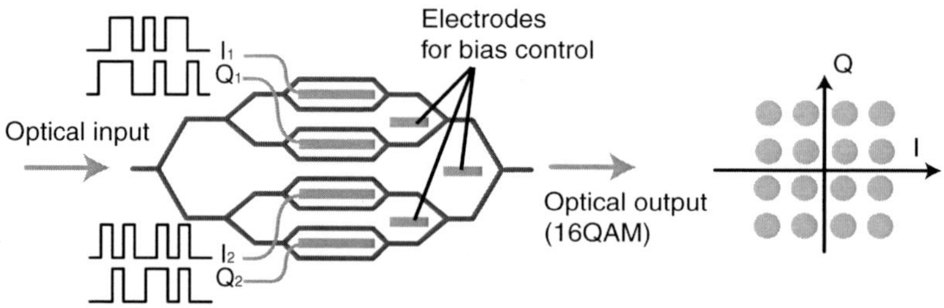

Fig. 15.6 16-QAM using QPMZM

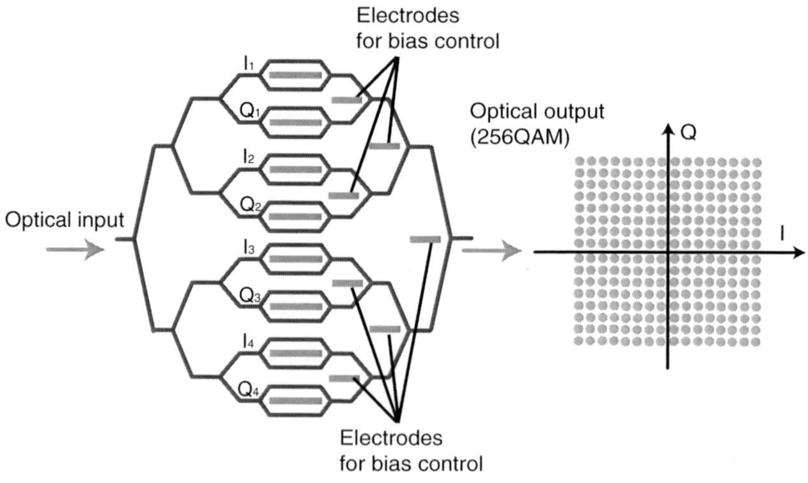

Fig. 15.7 256QAM using OPMZM

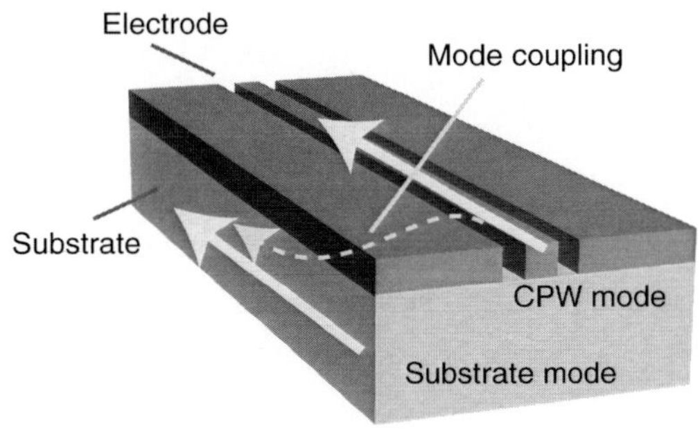

Fig. 15.8 Undesired mode coupling between CPW and substrate modes

high bit rate signal generation, we need integrated modulators with an array of MZI elements ($N \geq 2$) designed for large symbol rate R. To increase the symbol rate R higher than 40 Gbaud, we need flat modulator frequency response up to millimeter wave region where mode coupling between CPW and substrate modes may occur as shown in Fig. 15.8. The response would have some dips or large loss due to the mode coupling. When the substrate thickness is small, the cutoff frequency of the substrate mode goes up to high, so that the dips and loss can be suppressed in a millimeter region over 40 GHz.

15.6 High-Speed Optical Multi-level Modulation Using DPMZM and QPMZM

For QPSK modulation, a pair of data signals are applied to the two MZI elements in a DPMZM, to achieve control of in-phase and quadrature components [8, 20, 21]. DQPSK (80 Gb/s) was denibstrated by using a fabricated DPMZM on a z-cut LN substrate [1, 8], where V_π of each MZI elements was 2.5 V in push–pull operation at low frequency. Optical 3 dB bandwidth of each electrode was larger than 27 GHz. The total optical insertion loss of the packaged modulator was 5.1 dB. Each of MZI elements was in a minimum transmission bias point, where optical phase difference between the two MZI elements was adjusted to $\pi/2$. The amplitude of I and Q signals at the input ports of the modulator was 6.5 V (peak-to-peak), corresponding to $2V_\pi$ at 40 Gb/s, to generate an 80 Gb/s optical DQPSK signal at the output port of the modulator. As shown in Fig. 15.9, we measured an optical spectrum of a DQPSK signal at the output port of the modulator, without using any optical filters, where full spectral width measured 20 dB down from the maximum of the central wavelength peak was 60 GHz. As shown in, clear eye openings were observed for the two tributaries whose symbol rate was 40 Gbaud, where the receiver sensitivity at the BER of 10^{-9} was -20 dBm (Fig. 15.10).

By using a QPMZM consisting of two DPMZMs, we demonstrated high-speed 16-QAM [20, 21]. Figure 15.11 shows principle of optical 16-QAM modulation in our scheme, where the 16-QAM signal can be synthesized by superposing two QPSK signals with different amplitudes. The intensity difference between the QPSK

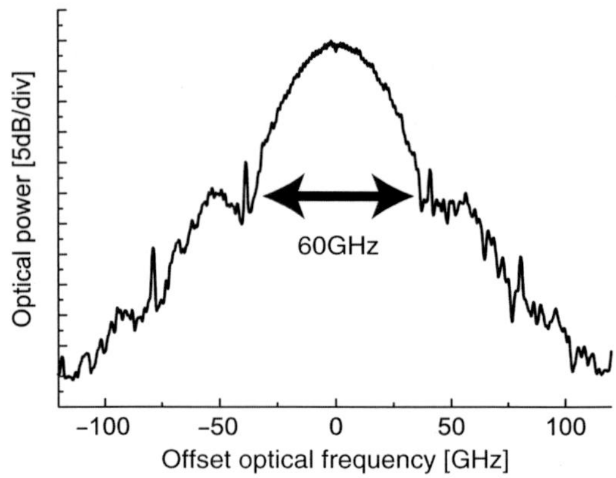

Fig. 15.9 80 Gb/s optical DQPSK signal [1]

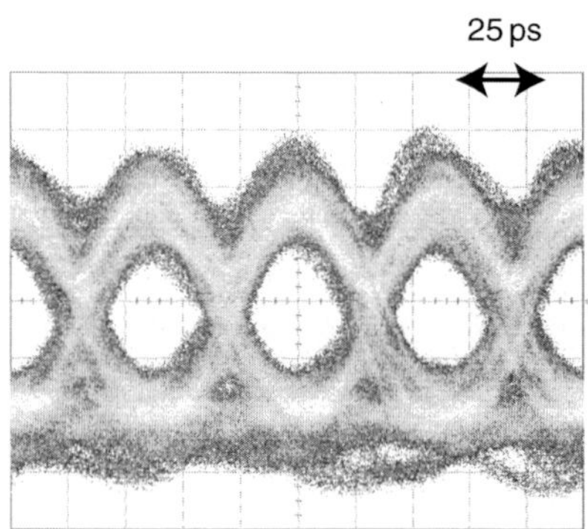

Fig. 15.10 Eye diagrams of demodulated optical DQPSK signal [1]

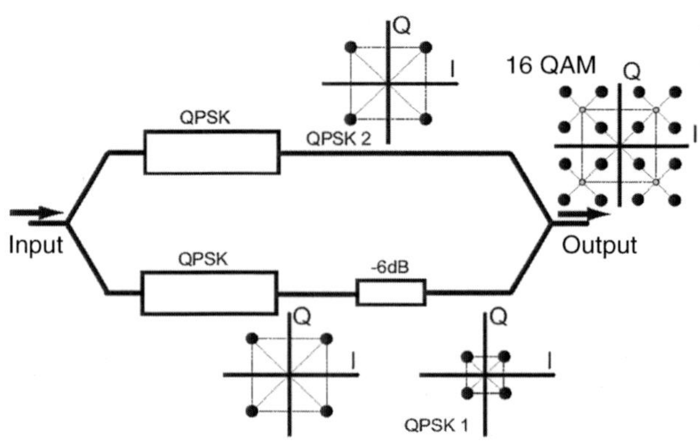

Fig. 15.11 16-QAM synthesized from two QPSK signals [20]

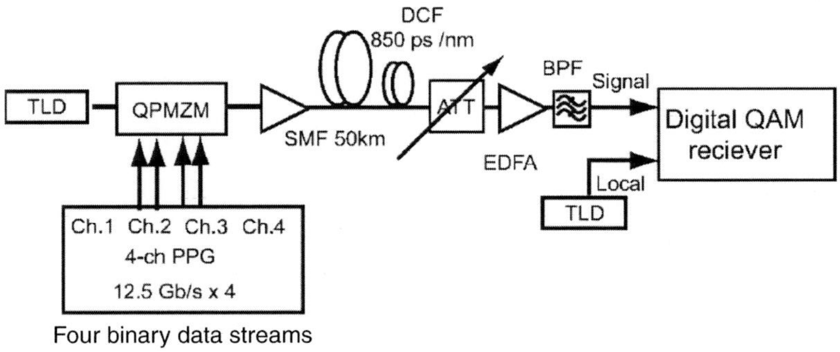

Fig. 15.12 Experimental setup for for 16-QAM using a QPMZM; TLD: tunable laser diode, PPG: pulse-pattern generator, BPF: bandpass filter [21]

signals is set at 6 dB. The large amplitude QPSK (QPSK 2 in Fig. 15.11) determines the quadrant where the symbol is mapped, while the small amplitude QPSK (QPSK 1) fixes the position of the mapping in each quadrant. By the combination of QPSK signals, totally 16 symbols are mapped with equal spacing in a phaser diagram. Note that such a 16-QAM mapping can be achieved from binary data sequences without handling multi-level electric signals.

The QPMZM was fabricated by the hybrid integration of LN waveguide modulators and silica-based planner lightwave circuit (PLC)-based couplers [20]. Optical 3 dB bandwidth of each electrode was about 10 GHz. V_π of each MZI was 2.9 V. The total optical insertion loss of the fabricated device was 10 dB. Figure 15.12 shows experimental setup. In the transmitter side, a CW light generated from an external cavity laser diode was 16-QAM modulated with the QPMZM. Each arm of the modulator was push–pull driven with 12.5-Gb/s binary NRZ data which were generated from a typically used 4-ch pulse pattern generator, where one DPMZM was driven in the range between $-V_\pi$ and V_π to generate large-amplitude QPSK; the other DPMZM was driven in the range between $-V_\pi/2$ and $V_\pi/2$ to generate small ampli-

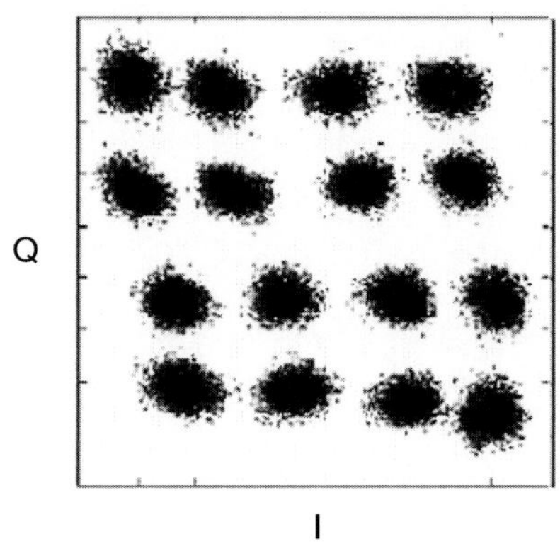

Fig. 15.13 Constellation maps of 12.5 Gbaud 16 QAM [21]

tude QPSK. The 16-QAM signal was transmitted over 50-km SMF and demodulated with a QAM receiver. In the QAM receiver, another external cavity laser diode was used as the local oscillator (LO) and frequency detuning between LO and signal was tuned within less than 500 MHz. Carrier phase and data recovery functions were implemented by a so-called off-line processing, i.e., emulated by a numerical computer. Figure 15.13 shows a typical constellation map of the received QAM signal after 50-km transmission. With the received power, larger than -25 dBm, the bit error ratio (BER) was always less than the FEC limit (BER $= 2 \times 10^{-3}$).

Acknowledgments We would like to express our appreciation to Dr. T. Sakamoto and Dr. A. Chiba of National Institute of Information and Communications Technology (NICT), Dr. P. J. Winzer of Bell laboratories, Alcatel-Lucent, and Dr.T. Lee of SHF Communication Technologies for fruitful discussion. We also wish to acknowledge the support of Mr. J. Ichikawa and Mr. K. Higuma of Sumitomo Osaka Cement.

References

1. T. Kawanishi, S. Sakamoto, M. Izutsu, High-speed control of lightwave amplitude, phase, and frequency by use of electrooptic effect. IEEE J. Select. Top. Quantum Electron. **13**, 79–91 (2007)
2. G.L. Li, P.K.L. Yu, Optical intensity modulators for digital and analog applications, IEEE/OSA J. Lightwave Technol. **21**, 2010–2030 (2003)
3. A.H. Gnauck, G. Raybon, S. Chandrasekhar, J. Leuthold, C. Doerr, L. Stulz, E. Burrows, 25 × 40-Gb/s copolarized DPSK transmission over 12 × 100-km NZDF With 50-GHz channel spacing IEEE Photonics Technol. Lett. **15**, 467–469 (2003)
4. N. Yoshikane, I. Morita, 1.14 b/s/Hz spectrally efficient 50 × 85.4 Gb/s transmission over 300 km using copolarized RZ-DQPSK Signals, J. Lightwave Technol. **23**, 108–114 (2005)
5. A.H. Gnauck, P.J. Winzer, S. Chandrasekher, C. Dorrer, Spectrally efficient (0.8 b/s/Hz) 1-Tb/s (25 × 42.7 Gb/s) RZ-DQPSK transmission over 28 100-km SSMF spans with 7 optical add/drops, *ECOC 2004* PD, Th4.4.1
6. R.A. Griffin, Integrated DQPSK Transmitters, *OFC 2005*, OTuM1
7. K. Ishida, K. Shimizu, T. Mizuochi, K. Motoshima, D.S. Ly-Gagnon, K. Kikuchi, Transmission of 20 × 20 Gb/s RZ-DQPSK signals over 5090 km with 0.53 b/s/Hz spectral efficiency, *OFC 2004* FM2
8. T. Kawanishi, T. Sakamoto M. Izutsu, K. Higuma, T. Fujita, S. Mori , S. Oikawa, J. Ichikawa 40 Gbit/s versatile $LiNbO_3$ lightwave modulator, *ECOC 2005,* Th2.2.6
9. N. Kikuchi, S. Sasaki, K. Sekine, T. Sugawara, Investigation of cross phase modulation (XPM) effect on amplitude- and phase- modulated multi-level signals in dense-WDM transmission, *OFC 2005*, OWA4
10. T. Miyazaki, Y. Awaji, Y. Kamio, F. Kubota, Field demonstration of 160-Gb/s OTDM signals using eight 20-Gb/s 2-bit/symbol channels over 200 km, *OFC 2005* OFF1
11. W. Idler, A. Klekamp, R. Dischler, B. Wedding, Advantages of frequency shift keying in 10-Gb/s systems, *2004 IEEE/LEOS Workshop on Advanced Modulation Formats* FD3 (2004)
12. T. Kawanishi, T. Sakamoto, S. Shinada, M. Izutsu, K. Higuma, T. Fujita, J. Ichikawa, $LiNbO_3$ high-speed optical FSK modulator, Electron. Lett. **40**, 691–692 (2004)
13. T. Kawanishi, T. Sakamoto, S. Shinada, M. Izutsu, K. Higuma, T. Fujita, J. Ichikawa, High-speed optical FSK modulator for optical packet labeling, J. Lightwave Technol. **23**, 87–94 (2005)
14. T. Sakamoto, T. Kawanishi, T. Miyazaki, M. Izutsu, Novel modulation scheme for optical continuous-phase frequency-shift keying, *OFC 2005* OFG2

15. T. Sakamoto, T. Kawanishi, M. Izutsu, Optical minimum-shift-keying with external modulation scheme, Opt. Express. **13**, 7741–7747 (2005)
16. M. Izutsu, S. Shikamura, T. Sueta, Integrated optical SSB modulator/frequency shifter, J. Quantum. Electron. **17**, 2225–2227 (1981)
17. T. Kawanishi, M. Izutsu, Linear single-sideband modulation for high-SNR wavelength conversion, Photon. Tech. Lett. **16**, 1534–1536 (2004)
18. D.D. Fonseca, P. Monteiro, A.V.T. Cartaxo, M. Fujita, Single sideband demonstration using a four phase-modulators structure, *2004 IEEE/LEOS Workshop on Advanced Modulation Formats* FC2 (2004)
19. M. Yoshida, H. Goto, K. Kasai, M. Nakazawa, 64 and 128 coherent QAM optical transmission over 150 km using frequency-stabilized laser and heterodyne PLL detection, Opt. Express **16**, 829–840 (2008)
20. T. Sakamoto, A. Chiba, T. Kawanishi, 50-Gb/s 16 QAM by a quad-parallel Mach-Zehnder modulator, *ECOC 2007*, Tu.1.E.3, postdeadline paper
21. T. Sakamoto, A. Chiba, T. Kawanishi, 50-km SMF transmission of 50-Gb/s 16 QAM generated by quad-parallel MZM, *ECOC 2008,* Tu.1.E.3
22. P.J. Winzer, A.H. Gnauck, 112-Gb/s Polarization-Multiplexed 16-QAM on a 25-GHz Grid, *ECOC 2008*, Th.3.E.5, postdeadline paper
23. M. Nakamura, Y. Kamio, T. Miyazaki, Linewidth-tolerant 10-Gbit/s 16-QAM transmission using a pilot-carrier based phase-noise cancelling technique, Opt. Express **16**, 10611–10616 (2008)
24. Y. Mori, C. Zhang, K. Igarashi, K. Katoh, K. Kikuchi, Unrepeated 200-km transmission of 40-Gbit/s 16-QAM signals using digital coherent receiver, Opt. Express **17**, 1435–1441 (2009)
25. E. Ip, A P.T. Lau, D.J.F. Barros, J.M. Kahn, Coherent detection in optical fiber systems, Opt. Express **16**, 753–791 (2008)
26. F. Koyama, K. Iga, Frequency chirping in external modulators, IEEE/OSA J. Lightwave Technol. **6**, 87–93 (1988)
27. C.R. Doerr, P.J. Winzer, L. Zhang, L. Buhl, N.J. Sauer, Monolithic InP 16-QAM modulator, *OFC 2008*, PDP20
28. T. Kawanishi, K. Kogo, S. Oikawa, M. Izutsu, Direct measurement of chirp parameters of high-speed Mach–Zehnder-type optical modulators, Opt. Commun. **195**, 399–404 (2001)
29. T. Pfau, S. Hoffmann, O. Adamczyk, R. Peveling, V. Herath, M. Porrmann, R. Noe, Coherent optical communication: Towards realtime systems at 40 Gbit/s and beyond, Opt. Express **16**, 866–872 (2008)
30. C.R.S. Fludger, T. Duthel, D. van den Borne, C. Schulien, E.D. Schmidt, T. Wuth, J. Geyer, E.D. Man, G.-D. Khoe, H. de Waardt, Coherent equalization and POLMUX-RZ-DQPSK for robust 100-GE transmission, IEEE/OSA J. Lightwave Technol. **26**, 64–72 (2008)
31. S.J. Savory, Digital signal processing options in long haul transmission, *OFC/NFOEC 2006*, OTuO3
32. K. Sato, Y. Kondo, M. Nakao, M. Fukuda, 1.55-μm Narrow-Linewidth and high-power distributed feedback lasers for coherent transmission systems, IEEE/OSA J. Lightwave Technol. **7**, 1515–1519 (1989)
33. T. Kunii, Y. Matsui, H. Horikawa, T. Kamijoh, T. Nonaka, Narrow line width (85 kHz) operation in long cavity 1.5 μm-MQW DBR laser, Electron. Lett., **27**, 691–692 (1991)
34. A.J. Ward, D.J. Robbins, G. Busico, E. Barton, L. Ponnampalam, J.P. Duck, N.D. Whitbread, P.J. Williams, D.C.J. Reid, A.C. Carter, M.J. Wale, Widely tunable DS-DBR laser with monolithically integrated SOA: Design and performance, IEEE J. Select. Top. Quantum Electron. **11**, 149–156 (2005)
35. L.G. Kazovsky, L. Curtis, W.C. Young, N.K. Cheung, All-fiber 90° optical hybrid for coherent communications, Appl. Opt. **26**, 437–439 (1987)
36. Y. Wang, W.R. Leeb, A 90° optical fiber hybrid for optimal signal power utilization, Appl. Opt. **26**, 4181–4184 (1987)

37. H.-G. Bachl, A. Matiss, C.C. Leonhardt, R. Kunkel, D. Schmidt, M. Schell, A. Umbach, Monolithic 90° hybrid with balanced PIN photodiodes for 100 Gbit/s PM-QPSK receiver applications, *OFC/NFEOC 2009*, OMK5
38. R. Langenhorst, W. Pieper, M. Eiselt, D. Rohde, H. G. Weber, Balanced phase and polarization diversity coherent optical receiver, IEEE Photonics Technol. Lett. **3**, 80–82 (1991)
39. C. Schramm, H.-G. Bach, A. Beling, G. Jacumeit, S. Ferber, R. Ludwig, R. Ziegler, G.G. Mekonnen, R. Kunkel, D. Schmidt, W. Schlaak, G. Unterborsch, High-bandwidth balanced photoreceiver suitable for 40-Gb/s RZ-DPSK modulation formats, IEEE J. Select. Top. Quantum Electron. **11**, 127–134 (2005)
40. A. Beling, H.-G. Bach, G.G. Ludwig, S. Ferber, C. Schubert, C. Boerner, J. Berger, C. Schmidt, U. Troppenz, and H.G. Weber, Monolithically integrated balanced photodetector and its application in OTDM 160 Gbitls DPSK transmission, *OFC 2003*, PDP4-1
41. M. Daikoku, I. Morita, H. Taga, T. Tanaka, T. Kawanishi, T. Sakamoto, T. Miyazaki, T. Fujita 100-Gb/s DQPSK transmission experiment without OTDM 100G ethernet transport, IEEE/OSA J. Lightwave Technol. **25**, 139–145 (2007)
42. P.J. Winzer, G. Raybon, H. Song, A. Adamiecki, S. Corteselli, A.H. Gnauck, D.A. Fishman, C.R. Doerr, S. Chandrasekhar, L.L. Buhl, T.J. Xia, G. Wellbrock, W. Lee, B. Basch, T. Kawanishi, K. Higuma, Y. Painchaud, 100-Gb/s DQPSK transmission: From laboratory experiments to field trials IEEE/OSA J. Lightwave Technol. **26**, 3388–3402 (2008)
43. Y.K. Lize, M. Faucher, E. Jarry, P. Ouellette, E. Villenueve, A Wetter, F. Sequin, Phase-tunable lowloss, S-, C-, and L-band DPSK and DQPSK demodulator, IEEE Photonics Technol. Lett. **19**, 1886–1888 (2007)
44. A.G. Steffan, M.L. Nielsen, G. Tsianos, A. Umbach, A. Boutin, L. Fulop, F. Verluise, 40 Gb/s DPSK receiver module with integrated free-space delay-line-interferometer, *OFC/NFEOC 2009*, OWX4
45. A. Sano, H. Masuda, Y. Kisaka, S. Aisawa, E. Yoshida, Y. Miyamoto, M. Koga, K. Hagimoto, T. Yamada, T. Furuta, H. Fukuyama, 14 Tb/s (140 × 111-Gb/s PDM/WDM) CSRZDQPSK transmission over 160 km using 7-THz bandwidth extended L-band EDFAs, *ECOC2006*, Th4.1.1
46. H. Masuda, A. Sano, T. Kobayashi, E. Yoshida, Y. Miyamoto, Y. Hibino, K. Hagimoto, Y. Yamada, T. Furuta, H. Fukuyama, 20.4 Tb/s (204 × 111 Gb/s) transmission over 240 km using bandwidth-maximized hybrid raman/EDFA, *OFC/NFOEC2007*, PDP20
47. A.H. Gnauck, G. Charlet, P. Tran, P.J. Winzer, C.R. Doerr, J.C. Centanni, E.C. Burrows, T. Kawanishi, T. Sakamoto, K. Higuma 25.6 Tb/s WDM transmission of polarization-multiplexed RZ-DQPSK signals, IEEE/OSA J. Lightwave Technol. **26**, 79–84 (2009)
48. K. Noguchi, H. Miyazawa, O. Mitomi, 75GHz broadband Ti: LiNbO3 optical modulator with ridge structure, Electron. Lett. **30**, 949–950 (1989)
49. R.C. Alferness, Waveguide electro-optic modulators, IEEE Trans. Micro. Theory Technol. **30**, 1121–1137 (1982)
50. M. Izutsu, Y. Yamane, T. Sueta, Broad-band traveling-wave modulator using a $LiNbO_3$ optical waveguide, IEEE J. Quantum Electron. **QE-13**, 287–290 (1977)
51. H. Murata, K. Kinoshita, G. Miyaji, A. Morimoto, T. Kobayashi, Quasi-velocity-matched $LiTaO_3$ guided-wave optical phase modulator for integrated ultrashort optical pulse generators, Electron. Lett. **36**, 1459–1460 (2000)
52. S. Corzine, P. Evans, M. Kato, G. He, M. Fisher, M. Raburn, A. Dentai, I. Lyubomirsky, R. Nagarajan, M. Missey, V. Lal, A. Chen, J. Thomson, W. Williams, P. Chavarkar, S. Nguyen, D. Lambert, T. Butrie, M. Reffle, R. Schneider, M. Ziari, C. Joyner, S. Grubb, F. Kish, D. Welch, 10-channel × 40Gb/s per channel DQPSK monolithically integrated InP-based transmitter PIC, *OFC 2008*, PDP18
53. K. Tsuzuki, N. Kikuchi, Y. Shibata, W. Kobayashi, H. Yasaka, Surface mountable 10-Gb/s InP Mach–Zehnder modulator module for SFF transponder, IEEE Photonics Technol. Lett. **20**, 54–56 (2008)

Chapter 16
Key Devices for High-Speed Optical Communication and Their Application to Transceiver Module

Hiroyuki Matsuura

Abstract High-speed semiconductor device technologies and optical device technologies are shown in this section. The integration technologies of these devices and their packaging technologies are also important to utilize them in communication equipments.

Recently the optical phase modulation formats, for example, DPSK (differential phase shift keying) and DQPSK (differential quadrature phase shift keying), have been used in high-performance transceiver modules. The example of the transceiver module is also shown in this section.

16.1 High-Speed Electrical Devices and Optical Devices

There are several kinds of transistors for high-speed integrated circuits. HBT (heterojunction bipolar transistor), HEMT (high electron mobility transistor), and MESFET (metal semiconductor field effect transistor) on a compound semiconductor substrate, like GaAs (gallium arsenide) or InP (indium phosphor), are used very often, because of their speed characteristics, high output voltage, and low noise characteristics. Also CMOS (complimentary metal oxide semiconductor) with short gate length and SiGe (silicon germanium) HBT are suitable for integration of the large amount of devices.

Figure 16.1 shows the cross sectional view of the InP HBT [1–4]. The structure of the device is mesa type and the collector region under the base electrode is removed by the etching technology to reduce the stray capacitance between the base and the collector of the transistor. There is a ledge layer to protect the surface region between the emitter and the base of the transistor. This ledge layer is very important to keep in high reliability for long-term use.

Figure 16.2 shows the high-speed characteristics of the InP HBT with 1×10-μm emitter. F_T (transition frequency or cut-off frequency at grounded emitter circuit) is

H. Matsuura (✉)
Department of Marketing, Communication Equipment Headquarters, Yokogawa Electric Corporation, Tokyo, Japan
e-mail: hiroyuki.matsuura@jp.yokogawa.com

M. Nakazawa et al. (eds.), *High Spectral Density Optical Communication Technologies*, Optical and Fiber Communications Reports 6, DOI 10.1007/978-3-642-10419-0_16,

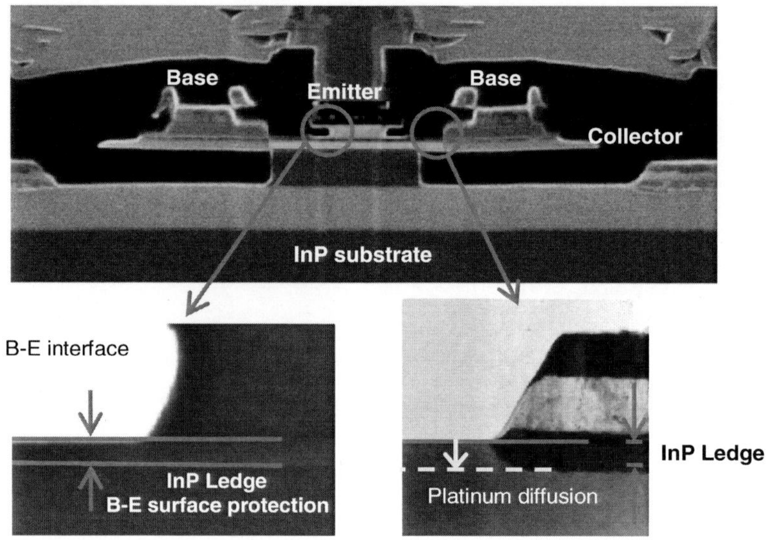

Fig. 16.1 Cross sectional view of InP HBT

150 GHz and $F\text{max}$ (maximum oscillation frequency) is 200 GHz. In the literature there are many reports with faster speed of the F_T and $F\text{max}$, but the yield and the reliability for long term are also important in the application of the long-haul optical communication, so the F_T and $F\text{max}$ value shown above are in a good balance for this application. These devices are suitable for 40-Gb/s application. For 100 Gb/s or more higher application, phase modulation technology, coherent modulation technology, dual polarization technology, or their combinations are needed.

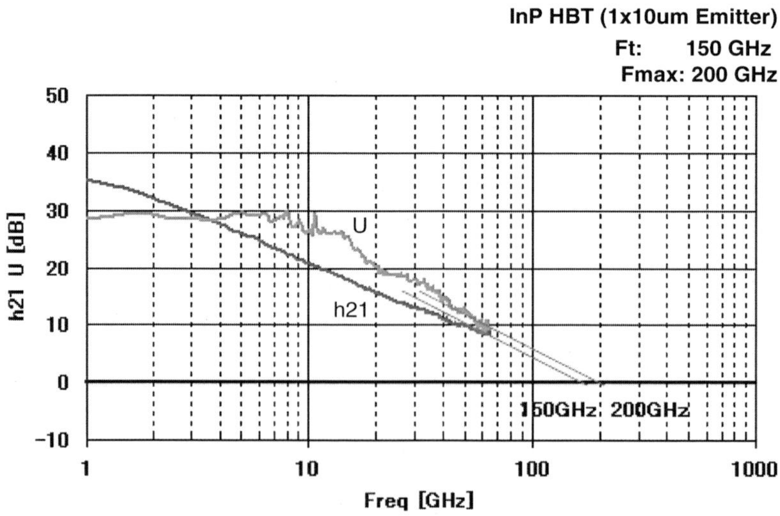

Fig. 16.2 Frequency characteristics of InP HBT

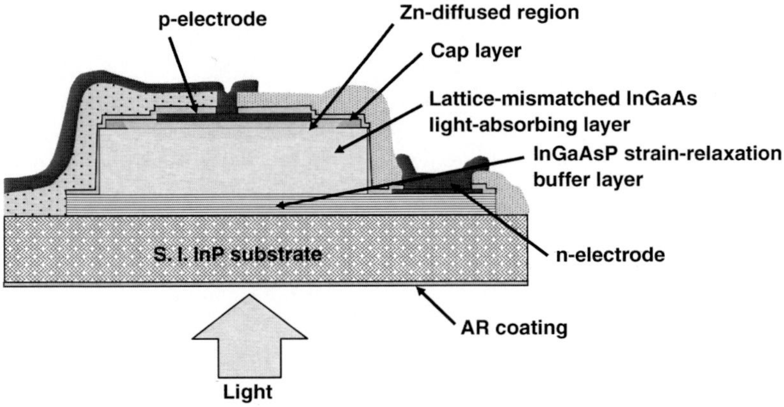

Fig. 16.3 Cross sectional view of PIN photodiode

To make high-speed integrated circuits, Schottky diodes using collector layer of transistor with proper metal, thin film resisters, MIM (metal insulator metal) capacitors, and inter connection layers are integrated with the transistor.

Photodiodes are used for the conversion device from the high-speed optical signal to the electrical signal. Figure 16.3 shows the cross sectional view of the PIN photodiode [2–4]. "PIN" means the intrinsic layer between p-type and n-type semiconductor. The photodiode is illuminated from the substrate side to keep high sensitivity, because both the light from backside and the light reflected at the surface electrode are caught by the device. The photodiode with this structure also performs better PDL (polarization-dependent loss) characteristics than the optical waveguide-type photodiode. The responsivity, means the sensitivity of the photodiode, and its temperature dependency are shown in Fig. 16.4. The square point shows the characteristics of the full-WDM(wavelength division multiplexing)-band photodiode with the optimized lattice-matched InGaAs (indium gallium arsenide) layer in Fig. 16.3. This photodiode covers up to a WDM wavelength area calls U-band; it can transmit communication data or can be assigned so-called service channels. Also this device has a very wide bandwidth of 50 GHz or more within a package. It can be used for 40 Gb/s or more higher bit-rate applications.

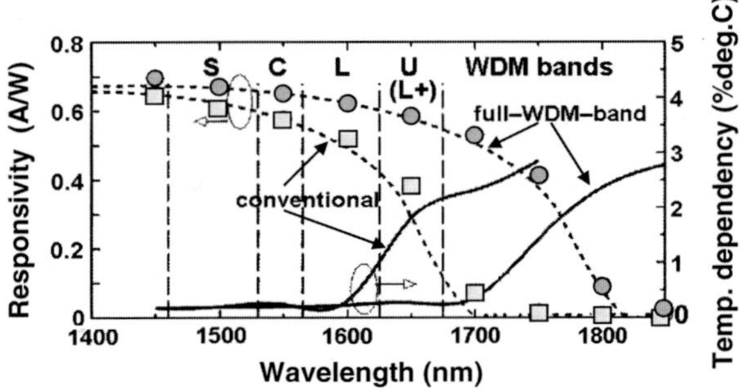

Fig. 16.4 Responsivity of PIN photodiode

Recently, optical phase modulation formats, for example, DPSK, DQPSK, are adopted. To receive these signals, a dual photodiode (sometimes called a twin photodiode or a balanced photodiode) is used [5]. This device consists of two photodiodes connected in serial and incoming light is divided in opposite optical phase by delay interferometer and they applied to each photodiodes to get good sensitivity. This is a very simple integration of the optical device but very important at the phase modulation detector.

To achieve more high-speed electrical circuit, an RTD (resonant tunneling diode) is the candidate for the next step [1]. The RTD is also compound semiconductor device which has an energy band designed to form a quantum well which causes the tunnel effect of electrons. The cross sectional view and functions are shown in Fig. 16.5. This RTD has an InGaAs/AlAs (aluminum arsenide, 2.4 nm)/InGaAs (3.8 nm)/AlAs (2.5 nm)/InGaAs quantum well structure on a InP substrate.

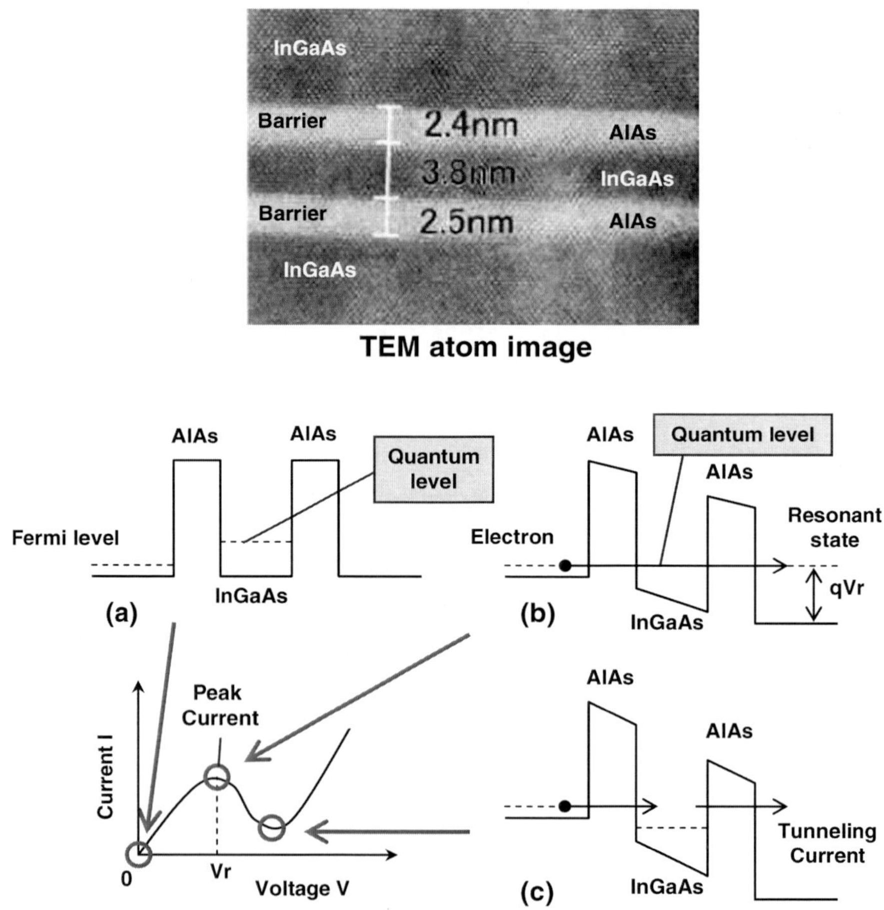

Fig. 16.5 Cross sectional view of RTD

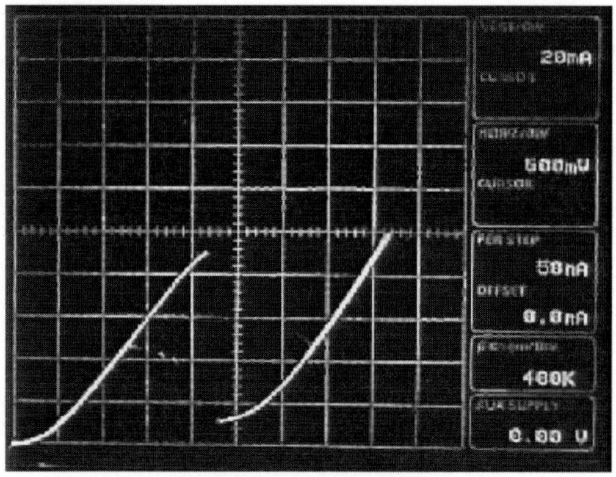

Horizontal: 0.5V/div.
Vertical: 20mA/div.

Fig. 16.6 I-V characteristic of RTD

The I–V characteristic of the RTD shows negative resistance (Fig. 16.6). The switching speed of RTD during negative resistance region with proper bias voltage is below a few pico-second, so the RTD can be used in a high bit-rate circuit of 100 Gb/s or more without a parallel circuit or a system configurations. The RTD technologies can be merged with HBT technology on a compound semiconductor substrate. The layers of the RTD can be inserted at the emitter and base region to make ultra-fast three-terminal device and this device is called RBT, resonant bipolar transistor [1].

Optical devices that can control optical paths are also important for the next-generation equipment. Especially, an optical switch is one of the key elements for optical transmission system. There are many configurations of the optical switch. For example, using MEMS (micro electro mechanical system) technology, moving mirrors are used for changing direction of optical path. Or, utilizing the thermal characteristics of the optical waveguide, the control of optical path is achieved by the heater that changes temperature of the optical waveguide. The switching speed of these devices is slow because of the moving part or the thermal response time. To get ultra-fast switching time the moving part and thermal control cannot be used. Figure 16.7 shows the top view of the optical switch that has optical waveguides for the transmission of the optical signal and an electrode for the control of the optical path [2]. Two optical waveguides are crossed and the electrode is placed at the cross section of the waveguides. The material of the waveguides is InP semiconductor with proper impurities and layers. When no electric current is applied to the

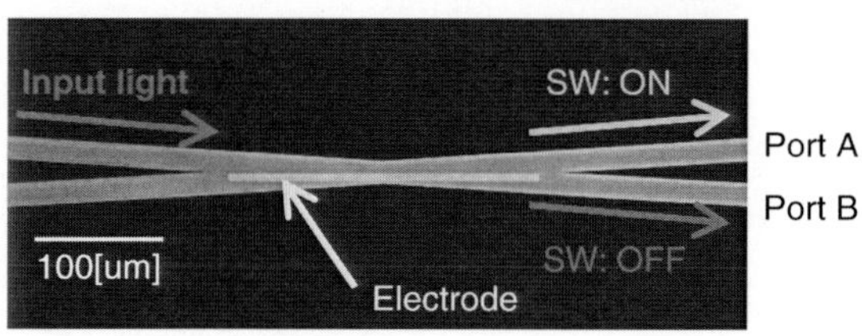

Fig. 16.7 Top view of optical switch (SEM image)

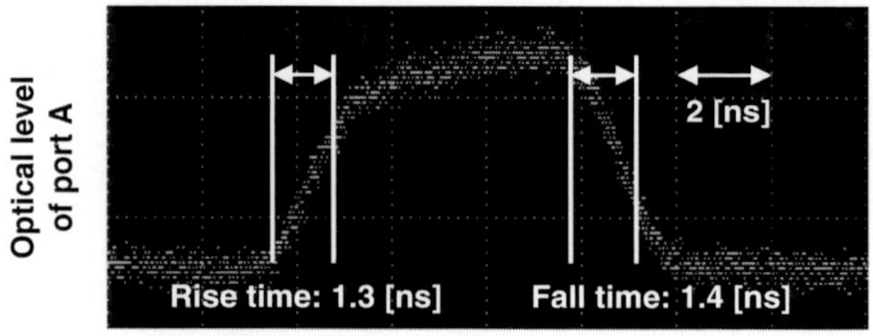

Fig. 16.8 Response of optical switch

electrode, the light applied from the upper left waveguide in Fig. 16.7 goes straight and is propagated to the lower right waveguide. When the electric current is applied to the electrode, the permittivity of the waveguide under the electrode is changed and the light applied from the upper left waveguide is reflected at the underneath of the electrode. Then the light is propagated to the upper right waveguide. Thus, the direction of the light is controlled by the electric current to the electrode. The switching speed is very fast with the proper drive circuit for the electrode. The measured switching speed of 1.3 ns rise time and 1.4 ns fall time are shown in Fig. 16.8. This device can be applicable for the optical packet network equipments to switch each packet during the packet gap, because of its fast switching characteristics.

There are many devices for high bit-rate optical communication. Not only high-speed characteristics but also high reliability and stable design of the device are very important for the high-grade real applications. For example, one can achieve ultra-high-speed characteristics with very-high current to the device, but this may cause the degradation of the reliability. At this view point of the view, not only device technologies but also system technologies are important to realize good equipments.

16.2 Integration of High-Speed Devices

In this section, several InP HBT-integrated circuits are introduced. A multiplexer, a demultiplexer, and clock recovery-integrated circuits are the combination of logic gates. A driver amplifier for an optical modulator and a phase shifter are analog-integrated circuits. Also the IC that integrates a photodiode, an RTD, and HBTs are also shown in this section.

Figure 16.9 shows 4:1 multiplexer (sometimes called MUX or serializer) IC and 1:4 demultiplexer (sometimes called DEMUX or deserializer) IC for the 40-Gb/s applications [3, 4, 6]. The clock frequency is the half of bit-rate number, so they are called "half rate" circuits. If a DFF is added at 40-Gb/s side of the circuit, then they are called "full rate" circuits.

To receive an optical signal through optical fiber transmission line clock recovery function is needed. With recovered clock, the data signal is retimed to get good waveform using a DFF (D-type flip flop). This function is called data recovery. Figure 16.10 shows a block diagram of the clock recovery and data recovery circuit, the photograph of the clock recovery chip, and waveforms [3, 4, 6]. This configuration is a PLL (phase-locked loop). The phase of the incoming data signal is

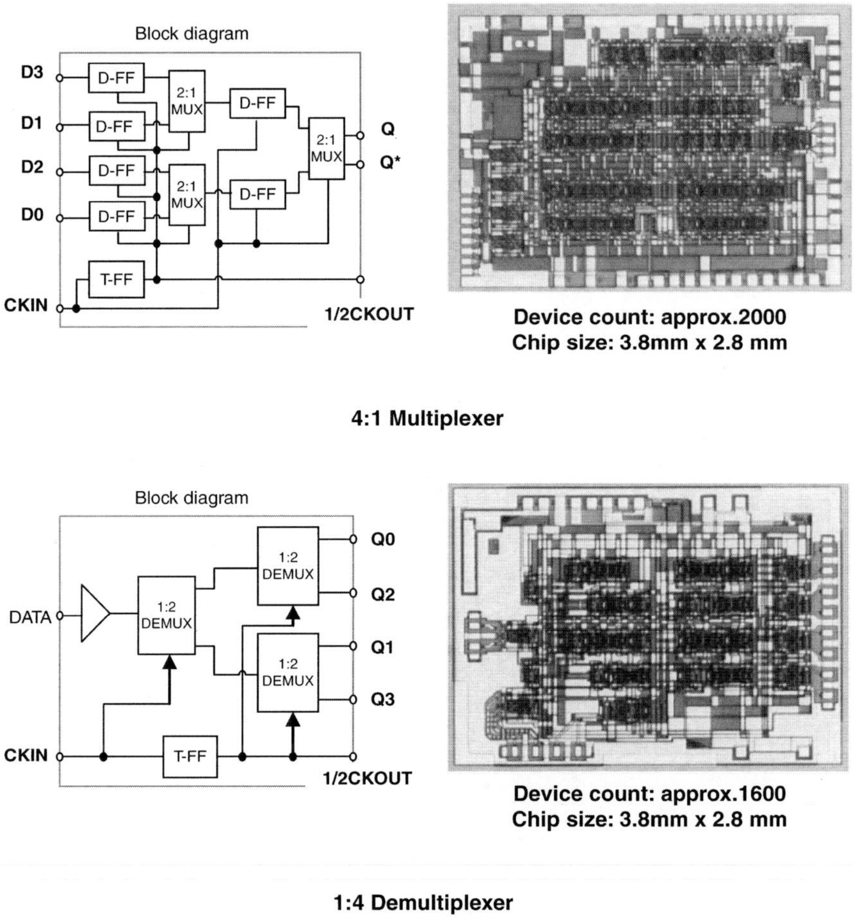

Fig. 16.9 Multiplexer and demultiplexer

compared with the internal VCO (voltage-controlled oscillator) by the phase detector. The output of the phase detector is applied to the VCO to decrease the phase error. The data recovery and the phase detector are integrated in a same chip shown in Fig. 16.10. The VCO is a separate chip, because its frequency tuning range is limited. The waveform of the input with dispersion and the data output and clock output at 43 Gb/s are shown in Fig. 16.10.

In all optical packet networks without clock synchronization function between the source terminals, the incoming packet data rate is slightly different because the source clock is different, so the burst-mode clock recovery circuit is needed at the receiving terminal. It has the feature of the fast locking time to the data signal. The 10-Gb/s burst-mode clock recovery IC and the clock output waveform are shown in Fig. 16.11 [2, 7, 8]. The circuit has two sets of gated oscillators to lock immediately

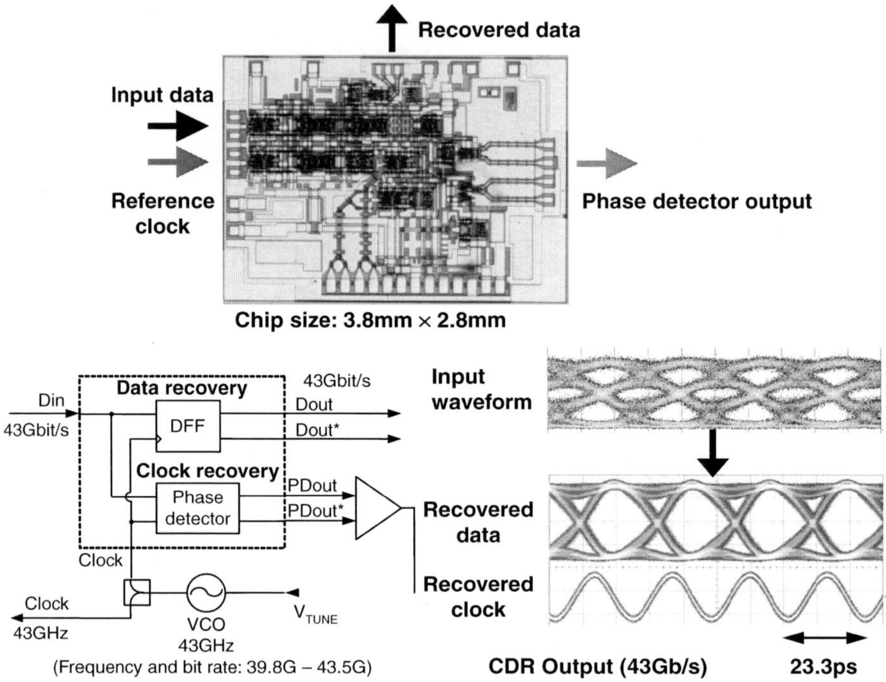

Fig. 16.10 Clock and data recovery

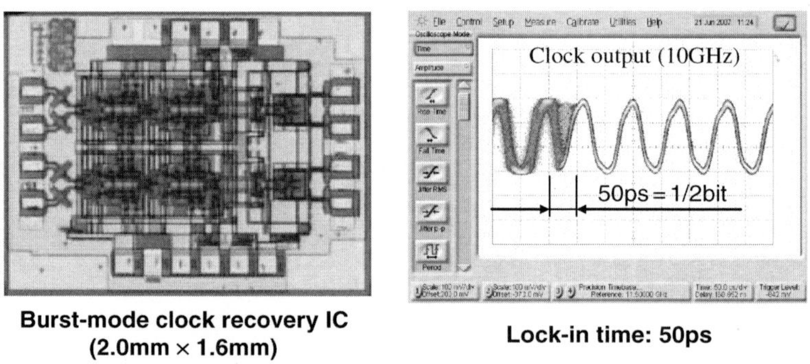

Fig. 16.11 Burst-mode clock recovery

to the incoming data. The waveform in Fig. 16.11 shows the fast locking time of 50 ps at the leading edge of the incoming packet.

Recently, the speed and the signal quality of the CMOS device are improved, so CMOS can realize by these logic circuits. To use large amount of devices, CMOS process is suitable for integration. But compound semiconductor circuits are still better in the jitter characteristics of the logic circuits and the voltage swing needed for modulator driver amplifier and another circuit that need output power.

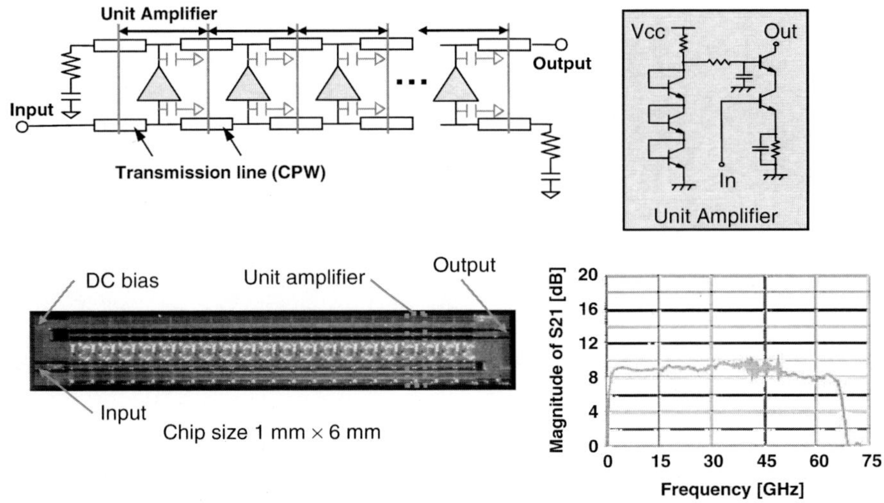

Fig. 16.12 Driver amplifier

Figure 16.12 shows the driver amplifier chip for an optical modulator [2–4, 6]. The circuit configuration is distributed amplifier that consists of cascode unit amplifiers and CPW (co-planar waveguide) transmission lines. The stray capacitance is compensated by the inductance of the transmission line. The frequency characteristic of the amplifier is shown in Fig. 16.12. The magnitude of S_{21} is very flat and the -3 dB cut-off frequency is around 67 GHz.

The control of the clock phase is important for real systems to get stable operation. The one of the key devices is the phase shifter chip that is shown in Fig. 16.13 [3, 4]. The varactor diodes on the compound semiconductor substrate, resistors, and CPW transmission lines are used in the phase shifter chip. The graph in Fig. 16.13 shows the maximum phase shift with flat frequency response.

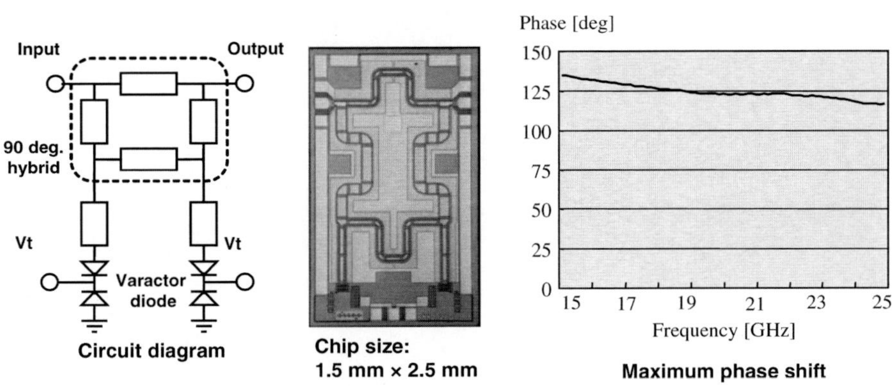

Fig. 16.13 Phase shifter

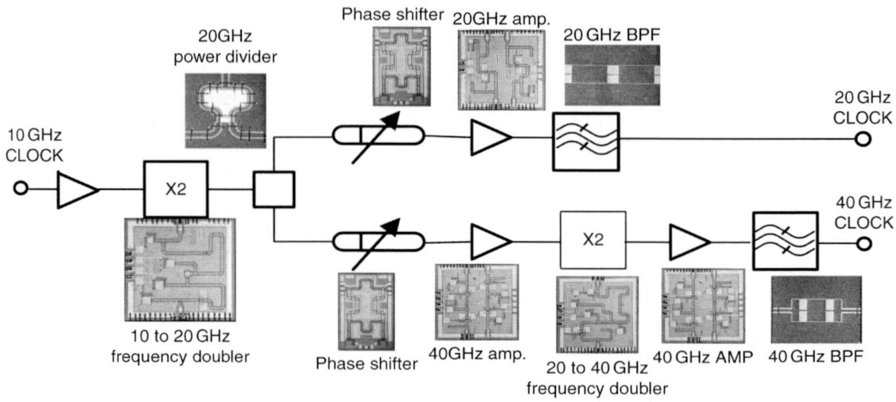

Fig. 16.14 Clock management unit

Figure 16.14 shows the block diagram of CMU (clock management unit) using multi chip configuration. The phase shifters, the frequency doublers, the amplifiers, and passive circuits are in the CMU. The passive circuits are power divider and BPFs (band pass filters). The 10-GHz input clock is multiplied to 20 GHz and 40 GHz with phase control by phase shifter chip shown in Fig. 16.13 [3, 4].

Figure 16.15 shows the chip of the integration of the photodiode, RTD, and HBTs. Its block diagram is also shown in Fig. 16.15. The interconnection length between the devices is very short with monolithic IC technology, so it works well in the high-speed application.

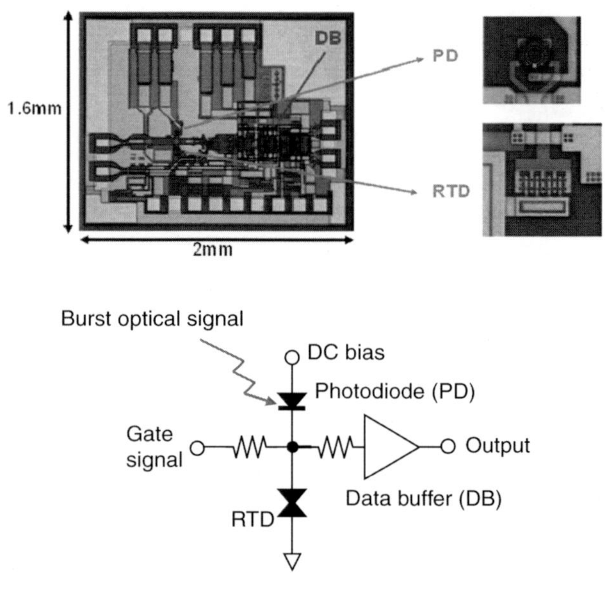

Fig. 16.15 Integration of optical and electrical elements

16.3 Packaging of Electrical Devices and Optical Devices

The packaging technology is very important for high-speed integrated circuits and optical devices shown in this section. A proper electrical interface method, for example, monolithic integration, a fillip chip, a wire bonding, connector, should be selected to achieve best result not only with speed but also with reliability and cost.

Figure 16.16 shows an example of the housing of high-speed circuit module. This is an LN modulator driver for 40-Gb/s DQPSK modulation with four-channel amplifiers and phase control circuits. The InP HBT amplifier chips, phase shifter chips, and related control chips are placed in the package. Between the semiconductor chips, the alumina circuits are used for the interconnection. The wire bonding is used for the connection between the semiconductor devices and the alumina elements. Also many passive elements for the DC bias circuits, the decoupling circuits, and the bypass capacitors are used in the package. Small-size RF connectors are used for the interface to the outside of the housing. Also, feed-through terminals are used for DC bias or slow signals. The package is hermetic sealed at the final stage of assembly.

Figure 16.17 shows the package of the photodiode module [3]. A mirror made by silicon is used for changing the direction of the light in the module. This mirror is fabricated by anisotropic etching of silicon substrates. This configuration has the features of low PDL compared to optical waveguide type. Also optical return loss is easy to reduce because the reflected light path at the device is designed properly.

Figure 16.18 shows the light path of the delay interferometer for DPSK signal demodulation using micro-optics technology. This is based on the configuration of Michelson interferometer using micro mirrors and beam splitters. So the same features mentioned at the photodiode module can be pointed out for this interferometer module. In real modules, the thermal expansion of the material should be considered. For the optical communication module, wide temperature operating range, for example, −5°C to +75°C, is required. The temperature control using a peltier device is sometimes used for this issue. On the other hand, athermal design [3, 9], which is the combination of thermal expansion of materials of the devices and the package without temperature control, is important for reducing power consumption and the cost. An optical waveguide technology is good for mass production of the

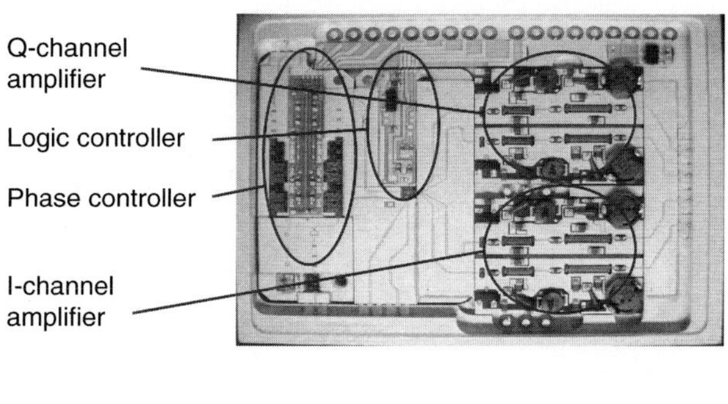

Fig. 16.16 Packaging of semiconductor circuits

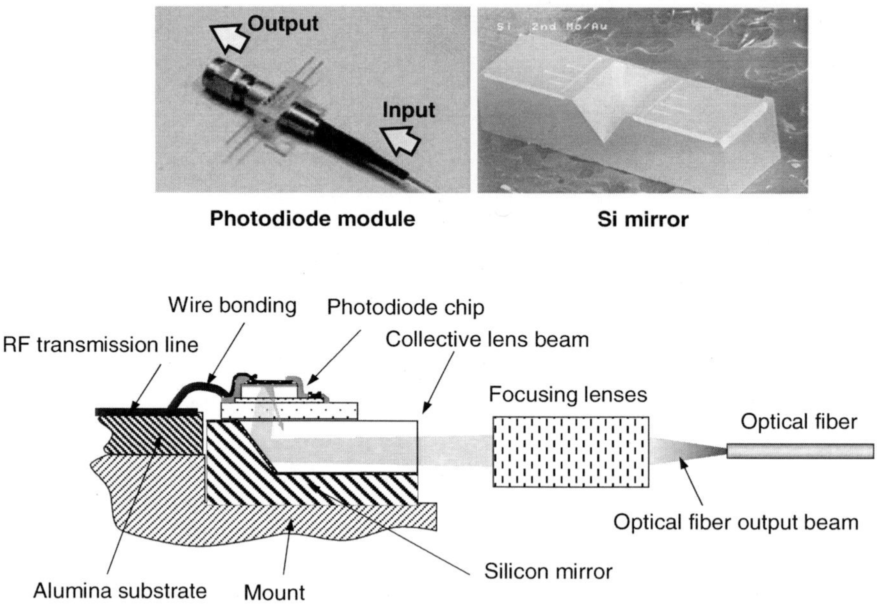

Fig. 16.17 Packaging of photodiode module

Fig. 16.18 Delay interferometer using micro-optics

devices, but micro-optics technology is also important for the high-performance optical communication modules.

16.4 High-Speed Transceivers

The demand of the network traffic has been increased and will grow continuously, so the bit rate of the transmission system should be expanded continuously, too. A phase modulation technology and other modulation technology have been used to

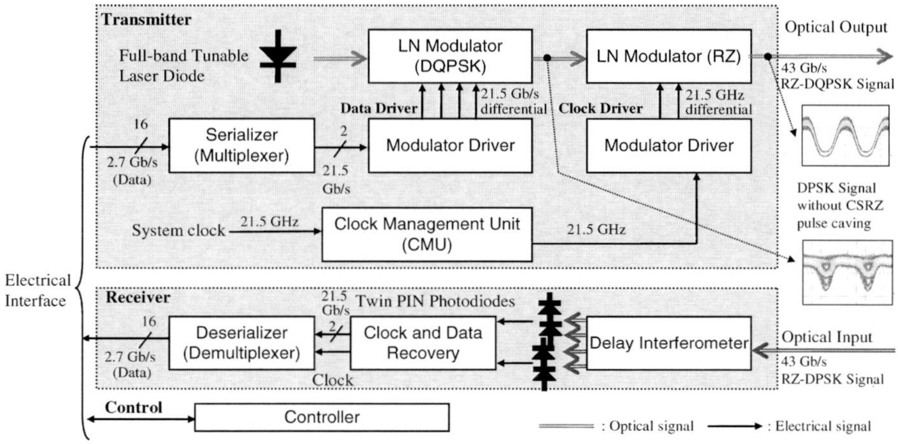

Fig. 16.19 Block diagram of 43-Gb/s RZ-DQPSK transceiver module

achieve high bit-rate transmission instead of simple NRZ (non return to zero) or RZ (return to zero) modulations.

Figure 16.19 shows the block diagram of the RZ-DQPSK transceiver for 43-Gb/s application with the FEC (forward error correction) [3, 4, 10]. The RZ-DQPSK format has the features of narrow optical bandwidth and the robustness to the distortion during the long distance optical fiber transmissions. Some of the devices and modules mentioned in this section can be used in this configuration. The control technologies for a long-term stable operation without the maintenance are also

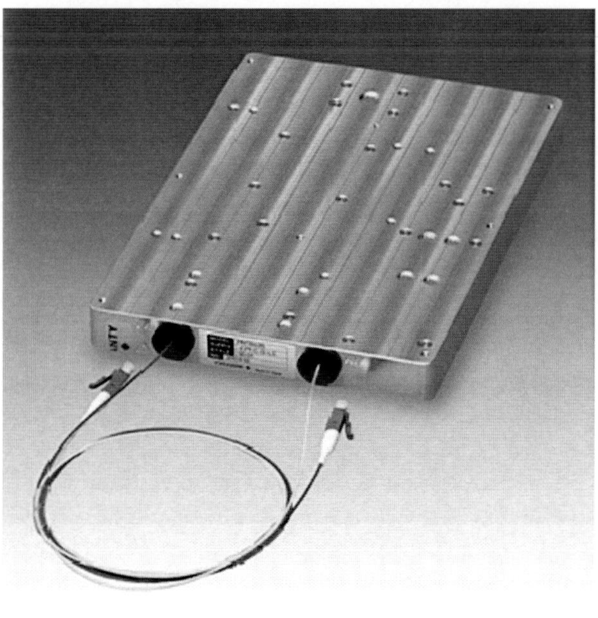

Fig. 16.20 RZ-DQPSK transceiver module of 43 Gb/s with $5 \times 7 \times 0.7$ inch size

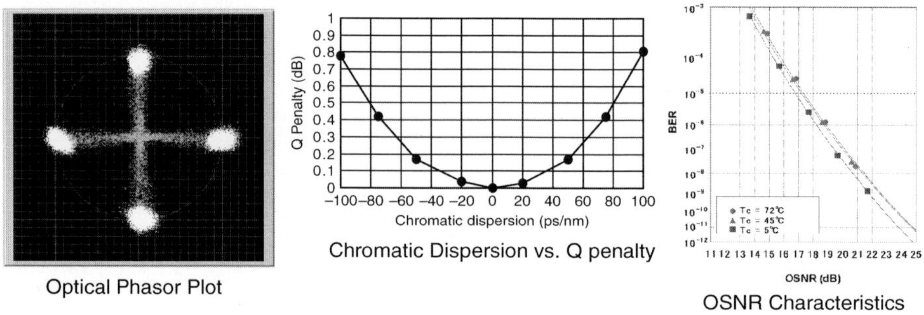

Fig. 16.21 Characteristics of 43-Gb/s RZ-DQPSK transceiver module

important. Furthermore, feedback is useful for the wide temperature range operation and the supply voltage variation. There are many feedback loops for these stable operations.

Figure 16.20 is the photograph of the example of the transceiver module. The characteristics of the transceiver module are shown in Fig. 16.21. The optical phasor plot shows the phase and the amplitude distribution of the optical output signal of the transceiver [11]. The chromatic dispersion versus Q-penalty and the OSNR (optical signal noise ratio) versus BER (bit error rate) characteristics are also shown in Fig. 16.21. Thus, the transceivers using the phase modulation technology are enabled by the high-speed devices, the ICs, and the packaging technologies with proper control technologies.

16.5 Future Directions for Devices and Transceiver Modules

As mentioned before, the demands of high bit-rate transmission should be continued. To achieve the high bit-rate transmission, the improvement of the devices and the development of the new devices, for example, RTD and RBT, will take place in every layers of the communication filed. The combination of the compound semiconductor for ultra-fast electrical optical devices and CMOS that can integrate large amount of active devices is the solution for the future network elements.

Also the network structure will be added or replaced, for example, all optical networks may be introduced in the near future. Then the optical packet networks using the optical switches are needed. Furthermore, the devices for the burst-mode transmission are also important. The burst-mode clock recovery at receiving side is already shown in this section. Furthermore, the circuits for burst-mode transmission, DC driver for LN modulator, will be needed at the transmitter side.

The hardware systems for such network applications are supported by system control technologies for the stable operations as mentioned before. The control technologies are very important forever. Transceiver modules adopting these technologies will draw the network society to the evolving human life.

References

1. I. Aoki, S. Kobayashi, T. Yakihara, H. Matsuura, A. Miura, Development of HBT-IC module for 50 Gbps optical communication systems. Yokogawa Technical Report English Edition, No. 34, 2002, pp. 1–6
2. A. Miura, H. Matsuura, M. Wada, T. Yakihara, H. Kodaka, K. Ikezawa, Photonic measurement and control technologies. Yokogawa Technical Report English Edition, No. 41, 2006, pp. 33–44
3. H. Matsuura, Optical phase modulation technologies for high speed transponders, in W orkshop on high spectral density optical communication for new generation network (HDOC-WS 2008), 25–26 June 2008, pp. 101–105
4. H. Matsuura, T. Yakihara, H. Kodaka, D. Tanimura, K. Tezuka, A. Oota, M. Ogusu, K. Ikezawa, A. Toyama, A. Miura, InP devices and their applications for high speed optical communication, in *7th topical Workshop on Heterostructure Microelectronics (TWHM2007)*, Chiba, Japan, 21–24 August 2007, pp. 93–94 (FrA-3)
5. T. Umezawa, T. Kudou, M. Hihara, M. Wada, Dual PIN photodiode for phase shift keying optical telecommunication systems. Yokogawa Technical Report English Edition, No. 46, 2008, pp. 19–22
6. S. Oka, S. Kobayashi, T. Yakihara, H. Matsuura, A. Miura, Development of HBT-based ultra-high-speed electronic devices. Yokogawa Technical Report English Edition, No. 34, 2002, pp. 7–11
7. H. Sugawara, T. Izawa, N. Iwabuchi, N. Ito, 10-Gbps burst mode CR and optical packet transceiver. Yokogawa Technical Report English Edition, No. 46, 2008, pp. 23–26
8. K. Ikezawa, H. Sugawara, T. Izawa, T. Suzuki, Y. Akasaka, A. Toyama, S. Uneme, S. Oka, T. Yakihara, A. Miura, 10 Gbps burst-mode clock recovery with synchronization time of 50 ps, in *34th European Conference and Exhibition on Optical Communications (ECOC 2008)*, Brussels, 21–25 September 2008, vol. 5, pp. 127–128 (P 3.14)
9. Y. Sanpei, Y. Suzuki, K. Iemura, J. Asano, Microoptics design technique and its application to optical devices for DWDM. Yokogawa Technical Report English Edition, No. 46, 2008, pp. 15–18
10. A. Ohta, D. Tanimura, T. Kyakuno, S. Iio, 43-Gbps RZ-DQPSK transponder for long-haul optical transmission system. Yokogawa Technical Report English Edition, No. 46, 2008, pp. 3–6
11. K. Tanimura, M. Maeda, H. Akahori, H. Ohta, NX7000 optical phase-modulation analyzer. Yokogawa Technical Report English Edition, No. 46, 2008, pp. 7–10

Chapter 17
Forward Error Correction

Takashi Mizuochi

Abstract In this chapter, forward error correction (FEC) techniques for optical communications are described. In Sect. 17.1, the basic concepts of FEC and its terminology are explained. The various FECs developed to date for optical communications are classified as belonging to three generations. In Sect. 17.2, the first-generation FEC represented by RS(255,239) is described. The second-generation FECs are explained in Sect. 17.3, where recently developed concatenated codes are discussed. In Sect. 17.4, the third-generation FECs, which are more powerful FECs based on soft-decision decoding, are analyzed in detail. A block turbo code and low-density parity-check (LDPC) codes showing a record NCG of around 10 dB are discussed in detail. In Sect. 17.5, we compare each generation of FEC with the Shannon limit. The ultimate NCG is discussed as a function of code rate. FEC error count and its application are discussed in Sect. 17.6.

17.1 Basic Concepts and Terminology

The story of forward error correction begins at the end of the 1940s as shown in Fig. 17.1. R.W. Hamming at Bell Laboratories invented the Hamming codes named after him in order to avoid read errors in punch cards for relay-based electromechanical computers [1]. Around the same time, C.E. Shannon published his epoch-making paper *A Mathematical Theory of Communication* [2] on the limits of reliable transmission of data over noisy channels. In the 1950s and 1960s, the basic concepts of several kinds of code were established, i.e., cyclic codes [3], Bose–Chaudhuri–Hocquenghem (BCH) codes [4, 5], Reed–Solomon (RS) codes [6], Reed–Müller codes [7], and convolutional codes [8]. In the 1960s and 1970s, effective decoding algorithms were developed, i.e., the Berlekamp–Massey algorithm [9, 10] and Euclid algorithm [11] for the BCH and RS codes, and the Viterbi algorithm [12] for convolutional codes. A great deal of effort has been expended in

T. Mizuochi (✉)
Mitsubishi Electric Corporation, 5-1-1 Ofuna, Kamakura, 247-8501 Japan
e-mail: mizuochi.takashi@df.MitsubishiElectric.co.jp

M. Nakazawa et al. (eds.), *High Spectral Density Optical Communication Technologies*, Optical and Fiber Communications Reports 6, DOI 10.1007/978-3-642-10419-0_17,

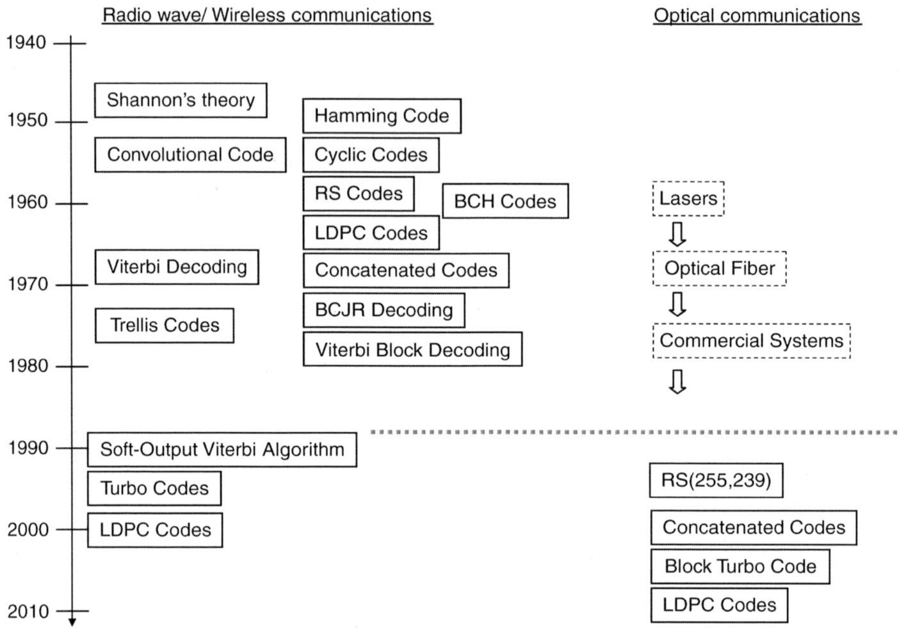

Fig. 17.1 History of forward error correction

digital radio and satellite communication systems to find efficient means of applying forward error correction. However, for a long time FEC was ignored in optical fiber communication systems, because optical communications has high data integrity with bit error ratios (BER) of the order of $10^{-9} \sim 10^{-15}$, compared with the $10^{-3} \sim 10^{-5}$ typical of radio and satellite communications.

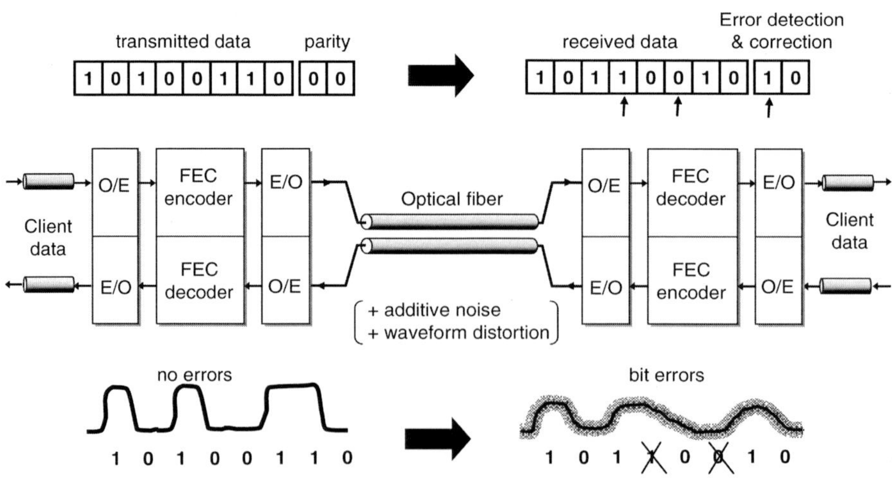

Fig. 17.2 Basic configuration of optical communication system employing forward error correction

Figure 17.2 shows a simplified configuration of an FEC-based optical transmission system. The client data is first regenerated in an optical receiver. Then it is put through an FEC encoder. The redundant bits and other overheads are added here, then the whole is converted to optical form by a line-side optical transmitter. The signal launched into the fiber suffers degradation by additive noise and/or distortion caused by dispersion. The degraded received signal is decoded by an FEC decoder, which corrects the resulting errors within the limits of its error correction capability. The extracted client signals are delivered to a far end receiver.

Figure 17.3 shows the Euclidean distance and the Hamming distance. Each signal point of quadrature phase-shift keying (QPSK) is represented in the 2-dimensional phasor diagram. The straight line distance between two points is called the Euclidean distance. With increasing the M-ary modulation, e.g., QPSK, 8-PSK, 16-QAM, 64-QAM, the Euclidean distance decreases, and it becomes more difficult to distinguish between the two. Note that the rate of decrease of the Euclidean distance is faster than the rate of noise bandwidth reduction. Therefore, higher signal-to-noise ratio is required as M increases.

From the view point of channel encoding and decoding operations, the most common channel is a binary symmetric channel (BSC). A bit error is observed by bit flipping or erasure. Let's assume that redundant bits 11 or 00 are added when 1 or 0 is sent. With no bit errors, 111 or 000 is received. When we receive 100, what is the correct information? Has 111 become 100 due to two bits errors, or has 000 become 100 due to one bit error? The difference between 100 and 000 is only 1. We call this difference the Hamming distance. When we assume the error probability of the BSC is 1%, the probability of the correct 111 changing to 100 is (1–0.01) $\times$ 0.01 $\times$ 0.01 $=$ 0.0099%. The probability of the correct 000 changing to 100 is 0.01 $\times$ (1–0.01) $\times$ (1–0.01) $=$ 0.9801%. Therefore, it is better that the received 100 is corrected to the far more probable 000. We can see that the Hamming distance between 111 and 100 is far, but that between 000 and 100 is close. The shorter distance should be corrected. In general, a code can correct any combination of $(t_{\min} - 1)/2$ errors as shown in Fig. 17.3b, where $t_{\min}$ denotes the minimum distance. In the case of 111 and 000, since the Hamming distance is 3, (3–1)/2 $=$ 1 errors can be corrected. The minimum distance of an (n, k) linear code is $t_{\min} \leq n - k + 1$.

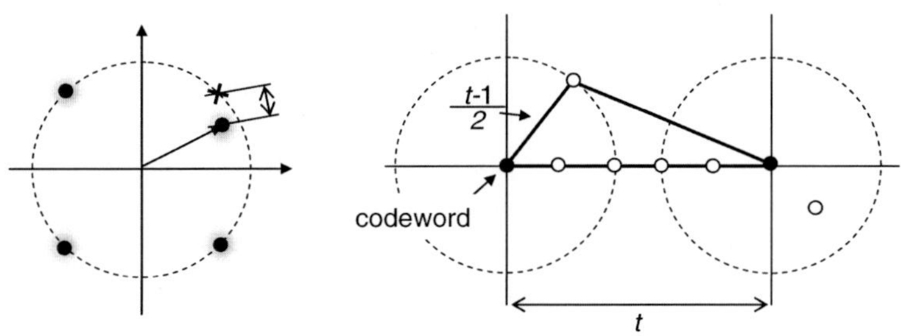

Fig. 17.3 Euclidean distance and Hamming distance. (**a**) Euclidean distance; (**b**) Hamming distance

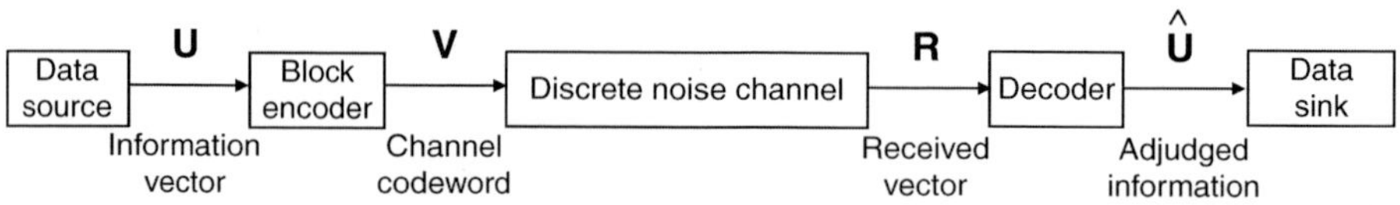

Fig. 17.4 Coded digital communication channel

Consider the coded digital communication channel illustrated in Fig. 17.4. An information vector $\boldsymbol{U} = [u_0, u_1, u_2, \ldots..u_{k-1}]$ is fed into a block encoder, then the redundancy digits are added to $\boldsymbol{U}$ and generate the channel codeword $\boldsymbol{V} = [v_0, v_1, v_2, \ldots, v_{n-1}]$. The channel codeword is transmitted through the discrete noise channel and it is subject to disturbance by noise. The receiver demodulates the received data $\boldsymbol{R} = [r_0, r_1, r_2, \ldots, r_{n-1}]$, which is not same as $\boldsymbol{V}$ due to the added noise. The decoder finds and corrects errors, then outputs the adjudged information $\hat{U} = [\hat{u}_1, \hat{u}_2, \hat{u}_3, \ldots..\hat{u}_{k-1}]$.

Here, we define the parity-check matrix $\boldsymbol{H} = [h_0, h_1, h_2, \ldots, h_{n\text{-}k-1}]$ of an (n, k) binary linear block code. Suppose that a product of codeword vectors $\boldsymbol{V}$ and $\boldsymbol{H}$ is zero, $\boldsymbol{V}\boldsymbol{H}^{\mathrm{T}} = \boldsymbol{0}$, where $\boldsymbol{H}^{\mathrm{T}}$ is the transpose of the matrix $\boldsymbol{H}$. When the error vector $\boldsymbol{E} = [e_0, e_1, e_2, \ldots, e_{n-1}]$ is added to the channel codeword $\boldsymbol{V}$, the received vector $\boldsymbol{R}$ is denoted by $\boldsymbol{R} = \boldsymbol{V} + \boldsymbol{E}$. Let us investigate $\boldsymbol{R}\boldsymbol{H}^{\mathrm{T}}$ for detecting errors. We can obtain $\boldsymbol{R}\boldsymbol{H}^{\mathrm{T}} = (\boldsymbol{V} + \boldsymbol{E})\boldsymbol{H}^{\mathrm{T}} = \boldsymbol{V}\boldsymbol{H}^{\mathrm{T}} + \boldsymbol{E}\boldsymbol{H}^{\mathrm{T}} = \boldsymbol{0} + \boldsymbol{E}\boldsymbol{H}^{\mathrm{T}}$. This means that whatever $\boldsymbol{V}$ is transmitted, we can see the $\boldsymbol{E}\boldsymbol{H}^{\mathrm{T}}$. We call $\boldsymbol{E}\boldsymbol{H}^{\mathrm{T}}$ the *error syndrome*.

Here we consider how to generate the codeword in the case of a finite number of elements, which is called a Galois field. The order of a Galois field is always a prime number or a power of a prime. GF(p) is called the prime field of order and is the field of residue classes modulo p. The finite field GF(2) consists of elements 0 and 1 which satisfy the following additions and multiplications: $0 + 0 = 0$, $0 + 1 = 1$, $1 + 0 = 1$, $1 + 1 = 0$ (mod 2), 1–1 = 0, 1–0 = 1, 0–0 = 0, 0–1 = 1.

Consider the information vector $\boldsymbol{U} = (1010)$, which is expressed as $x^3 + x$, as a generator polynomial. The parity-check matrix $\boldsymbol{H}$ is taken as $(1011) = x^3 + x + 1$. The codeword vector is $\boldsymbol{V} = \boldsymbol{U}\boldsymbol{H}^{\mathrm{T}} = (1001110) = x^6 + x^3 + x^2 + \mathrm{x}$. When the error vector $\boldsymbol{E} = (000001) = 1$ is added to $\boldsymbol{V}$, the received vector $\boldsymbol{R} = \boldsymbol{V} + \boldsymbol{E} = \boldsymbol{U}\boldsymbol{H}^{\mathrm{T}} + \boldsymbol{E} = (1001111) = x^6 + x^3 + x^2 + x + 1$. Let us calculate $\boldsymbol{R}/\boldsymbol{H}$ and $\boldsymbol{E}/\boldsymbol{H}$. The remainders of each result are both $(0000001) = 1$. An error of 1 can thus be found. For further details see *Error control block codes for communications engineers* by L.H. Charles Lee [13].

The key terminologies used in FEC are summarized in Table. 17.1. The correction capability is generally evaluated in terms of a net coding gain (NCG). This is characterized by the code rate R (the ratio of bit rate without FEC to bit rate with FEC) and the maximum allowable input BER ($\mathrm{BER_{in}}$) of the signal input to the FEC decoder which can be reduced to a reference output BER ($\mathrm{BER_{out}}$) by applying the FEC algorithm. The NCG applies to a binary input additive white Gaussian noise (AWGN) channel with:

Table 17.1 FEC terminology (ITU-T G. Supplement 39)

Term	Description
Information bit (byte)	Original digital signal to be FEC encoded before transmission
FEC parity bit (byte)	Redundant bit (byte) generated by FEC encoding
Codeword	Information bits (bytes) plus FEC parity bits (bytes)
Code rate R	Ratio of bit rate without FEC to bit rate with FEC
Redundancy ratio	$\dfrac{\text{The number of FEC redundancy bits}}{\text{The number of information bits}}$
	$=$ e.g., for RS(255,239), $\dfrac{255–239}{239} = 6.69\%$
Coding gain	Reduction of Q value for specified BER (e.g., 10^{-13}) assuming white Gaussian noise and a theoretical reference receiver
Net coding gain (NCG)	Coding gain corrected for the increased noise due to the extra bandwidth needed for the FEC bits, assuming white Gaussian noise
BER_{in}	BER of the encoded line signal (= BER of the input signal to the FEC decoder)
BER_{out}	BER of the decoded client signal (= BER of the output signal from the FEC decoder)
Q and BER	$\text{BER} = \frac{1}{2}\text{erf}c(\frac{Q}{\sqrt{2}})$
xxx (n, k) code	xxx = code class (BCH or RS) n = number of codeword bits (bytes) k = number of information bits (bytes)

$$NCG = 20\log_{10}\left[\text{erf}c^{-1}(2\text{BER}_{\text{out}})\right] - 20\log_{10}\left[\text{erf}c^{-1}(2\text{BER}_{\text{in}})\right] + 10\log_{10} R \tag{17.1}$$

with erfc^{-1} being the inverse of the complementary error function, $\text{erfc}(x) = 1–\text{erf}(x)$. The input Q factor Q_{in} is defined as a function of BER_{in} as follows:

$$Q_{\text{in}} = \sqrt{2}\text{erf}c^{-1}(2\text{BER}_{\text{in}}). \tag{17.2}$$

Note that the Q factor and BER are based on a binary input AWGN channel model which is different from the square-law photo-detected noise distribution seen in optical communication systems. The true probability density function (pdf) in optically amplified on–off keying is a chi-square distribution [14], and the pdf tail of a chi-square distribution is quite different from the Gaussian approximation. However, in the poor BER range, e.g., $10^{-1}{\sim}10^{-3}$, discussed here, the pdfs of the Gaussian approximation and the chi-square distribution show good agreement [15]. For straightforward calculation and ease of understanding using the Q parameter commonly adopted for optical communication systems, we adopt a binary input AWGN channel model in this book.

In the case of RS(255,239), a Q_{in} of 11.2 dB can be corrected to 17.3 dB. The gross coding gain is 6.1 dB (= 17.3–11.2). The NCG is calculated as 5.8 dB taking into account the code rate R at the output BER of 10^{-13}. Note that the NCG increases as the defined output BER decreases. The minimum required Q_{in} required to achieve a reference BER_{out} is called the Q limit. For a reference BER_{out} of 10^{-13}, the Q limit of RS(255,239) is 11.2 dB.

17.2 First-Generation FEC

One of the first published practical FEC experiments in optical fiber communications was reported by Grover in 1988, employing the (224, 216) shortened Hamming code [16]. The 0.964 code rate was integrated into the low speed tributaries of a 565 Mb/s signal, yielding an FEC coding gain of about 2.5 dB at an output BER of 1×10^{-13}.

After the first trial application of FEC to optical transmission, the actual motivation to introduce FEC arose in repeaterless submarine cable systems in the early 1990s. The BCH (167,151) code having 10% redundancy was introduced for the construction of a festoon system at 565 Mb/s [17]. The RS code was also selected for a 2.5 Gb/s 357 km repeaterless system [18]. Both systems aimed to increase the receiver sensitivity by using FEC. Somewhat later, erbium-doped fiber amplifiers were to be introduced to repeatered submarine systems for the first time. Researchers encountered a problem with fluctuations in the BER due to the polarization-dependent effects [19], and the RS codes were successfully adopted to mitigate the performance variation [20, 21], the FEC being used to gain system margin. In order to avoid unwanted line rate increase due to the coding, in-band FEC was studied in terrestrial systems [22]. A shortened Hamming code was embedded in the auxiliary multiplex section overhead in the Synchronous Digital Hierarchy (SDH). In the present context, these types of linear code will be classified as first-generation FEC.

The most notable first-generation FEC is the RS code designated as RS(n, k), where n is the total number of symbols per codeword and k is the number of information symbols. A symbol consists of 8 bits. The RS code can correct up to t-errored symbols in a codeword, where $t = (n\text{–}k)/2$. A decoding algorithm suited to LSI was first demonstrated by Sugiyama et al. [23].

Among the candidate codes, RS(255,239), where $n = 255$, $k = 239$, and $t = 8$, came to be used in a broad range of long-haul systems as recommended in ITU-T G.975, this being a linear cyclic block code over the Galois Field GF(2^8). The code is the systematic form of the RS code based on the generating polynomial:

$$G(x) = \prod_{i=0}^{15} (x - \alpha^i) \tag{17.3}$$

where α is a root of the binary primitive polynomial $x^8 + x^4 + x^3 + x^2 + 1$. The codeword carries 239 information symbols and 16 redundant symbols. ITU-T G.975 specifies that one payload section is taken for framing and signaling. Therefore, the data rate increases to 255/238 times higher than the uncoded original. The RS(255,239) algorithm operates on symbols instead of bits. It can correct up to eight errored symbols in a single codeword of length 255. For the RS codes, the error correction capability can be calculated on the assumption that errors occur at random, as can be seen in ITU-T G.975:

$$\begin{cases} P_{\mathrm{UE}} = \sum_{i=9}^{n} \frac{i}{n} \cdot \binom{n}{i} \cdot P_{\mathrm{SE}}^{i} \cdot (1 - P_{\mathrm{SE}})^{n-i} \text{ with } n = 255 \\ \mathrm{BER_{in}} = 1 - (1 - P_{\mathrm{SE}})^{1/8}, \quad \mathrm{BER_{out}} = 1 - (1 - P_{\mathrm{UE}})^{1/8} \end{cases} \tag{17.4}$$

with:

P_{UE} = probability of an uncorrectable error
P_{SE} = probability of a symbol error
n = codeword length (255)

According to Eq. (17.4) an input BER of 1.4×10^{-4} can be corrected to 1×10^{-13}. The correction capability is generally evaluated in terms of an NCG as illustrated in Fig. 17.5. This is characterized by the code rate R (the ratio of bit rate without FEC to bit rate with FEC) and the maximum allowable input BER ($\mathrm{BER_{in}}$) of the signal input to the FEC decoder which can be reduced to a reference output BER ($\mathrm{BER_{out}}$) by applying the FEC algorithm. The NCG refers to a BSC with AWGN.

In the case of RS(255,239), a Q_{in} of 11.2 dB can be corrected to 17.3 dB. The gross coding gain is 6.1 dB (= 17.3–11.2). The NCG is recalculated as 5.8 dB taking the redundancy of $10\log_{10}(255/238)$ at the output BER of 1×10^{-13} into account. Note that the NCG increases as the defined output BER decreases. Performance of G. 975 FEC are summarized in Table 17.2.

The FEC format is assembled according to the given frame structure as shown in Fig. 17.6. The parameters in the case of STM-16 are listed in Table 17.3:

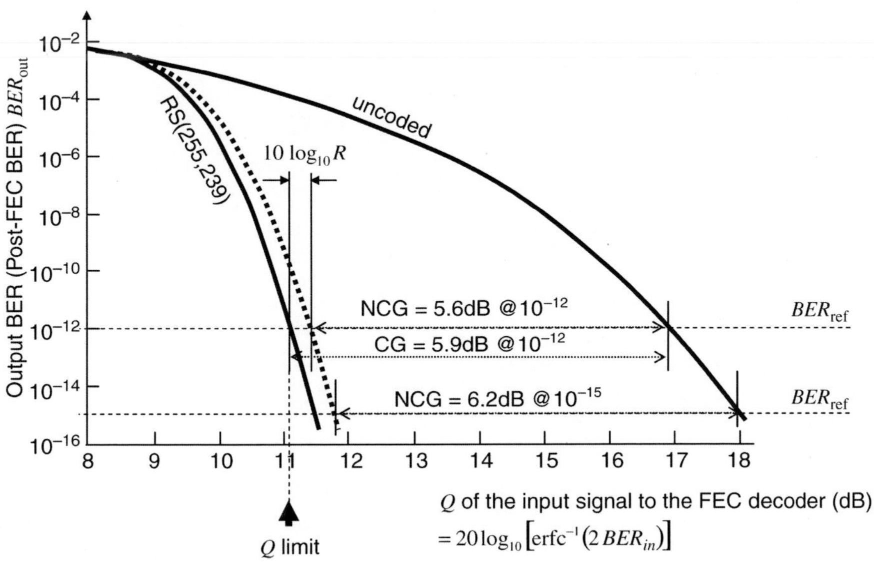

Fig. 17.5 Net coding gain

Table 17.2 Performance of G.975 FEC

Term	Value
$BER_{in}(BER_{out} = 10^{-13})$	1.4×10^{-4}
Coding gain ($BER_{out} = 10^{-13}$) in dB	6.1
Net coding gain ($BER_{out} = 10^{-13}$) in dB	5.8
Code rate	0.937 (= 239/255)
Redundancy	6.69%

Table 17.3 Parameters of G.975 FEC

Term	Value
Signal rate	$2.48832 \times 255/238 = 2.6660571$ Gb/s
Frame length	32,640 bits
Frame speed	12.24279835 μs
Frame cycle rate	81.6807 kHz
Overhead bits	128 bits/frame
Number of sub-frames	128

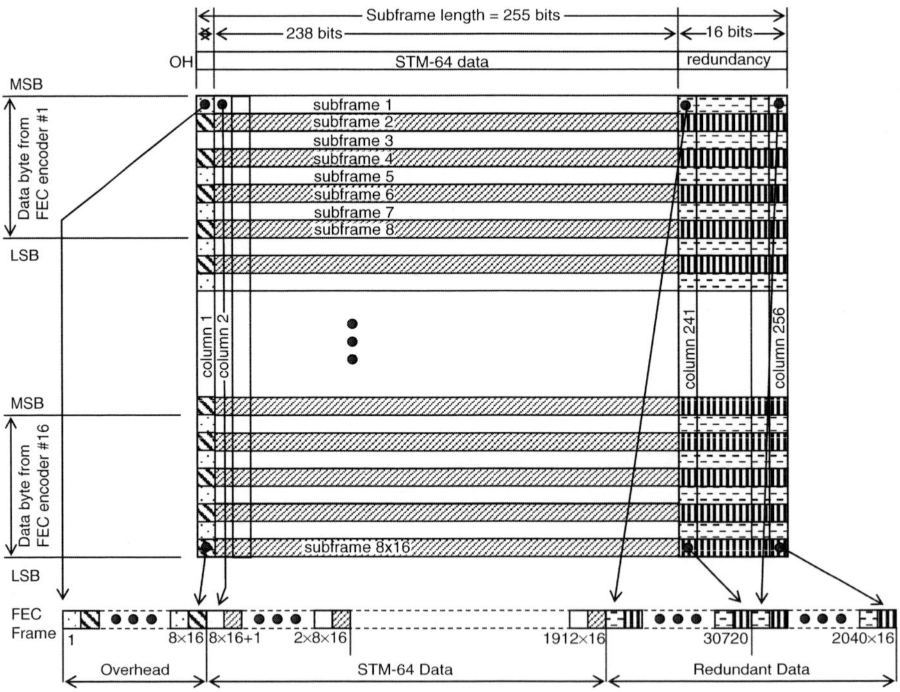

Fig. 17.6 G.975 FEC frame structure

One byte of overhead is placed at the beginning of each sub-frame, and the FEC error correction bits are placed at the end. The overhead consists of a tributary ID, order wire function, frame synchronization, and other IDs specifying each system parameter.

Another important correcting capability issue is its tolerance to burst errors. This intrinsic property is enhanced by interleaving the elementary RS(255,239). ITU-T G.957 puts the error correcting capacity to bursts at 1017 bits maximal length, for 16 interleaves. This is called the *burst error correction length* (BECL).

Based on the capabilities of the first-generation FEC as explained above, and in particular of RS(255,239), a trans-pacific and a trans-atlantic cable were successfully deployed in 1996, carrying 5 Gb/s data over 7000 km [24, 25]. RS(255,239) has also been standardized in IEEE802.3ah for use with gigabit Ethernet PON (passive optical network).

17.3 Second-Generation FEC

As WDM matured in the mid 1990s, system designers started to seek more powerful FEC codes. Because the required signal-to-noise ratio increases in proportion to the number of multiplexed wavelengths, several types of FEC stronger than the first-generation FEC have been developed based on concatenated codes [26] and iterative hard-decision decoding. We categorize this type of stronger FEC as second-generation FEC. The second-generation FEC is sometimes referred to as either *super FEC* or *enhanced FEC* (*EFEC*).

Figure 17.7 depicts the encoder/decoder configuration of a second-generation FEC. The incoming client data to be mapped into the FEC frame is interleaved then encoded by the outer encoder, and the output from the encoder is then de-interleaved and encoded with the inner code. Conversely, the received data is first decoded by the inner decoder, and the output from this is interleaved and then decoded by the outer decoder. The decoding is iterated to improve the correction capability without increasing the line rate.

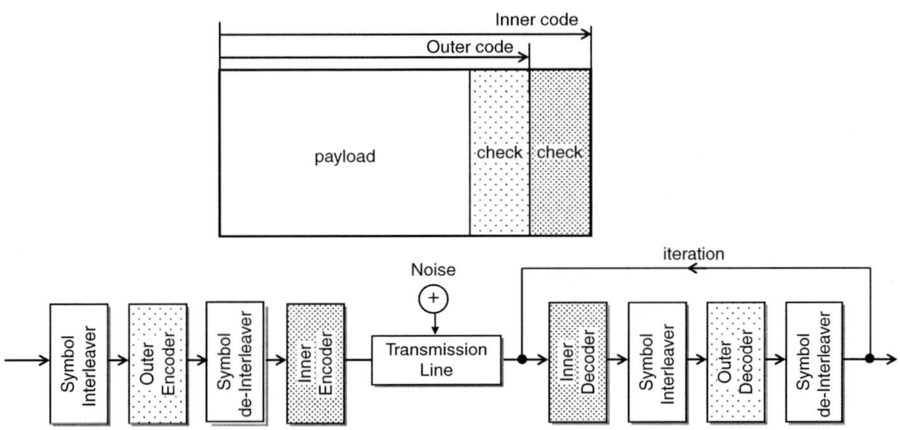

Fig. 17.7 Block diagram for concatenated codes

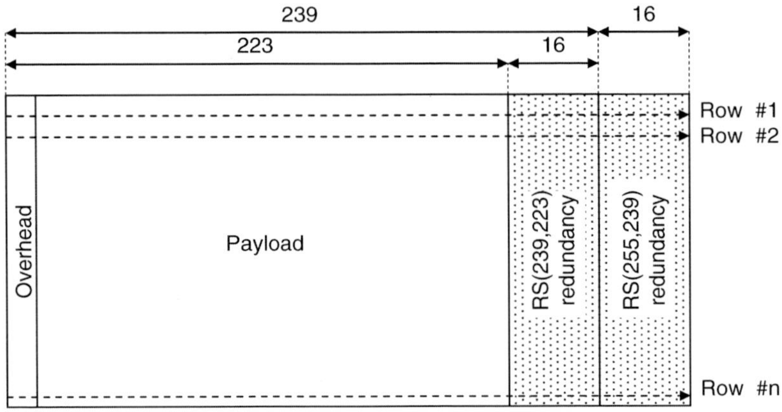

Fig. 17.8 Frame format of RS(239,223) + RS(255,239)

The first proposal for second-generation FEC for optical communications was concatenated Reed–Solomon codes with iterative decoding. Ait Sab et al. proposed RS(255,239) + RS(255,223) with 22% redundancy and two iterations [27] Its NCG of 8.4 dB was demonstrated by Monte-Carlo simulation. Yoshimura et al. proposed RS(248,232) + RS(144,128) having 20.8% redundancy [28]. The NCG was 7.8 dB. They implemented these concatenated codes in submarine line terminal equipment. Seki et al. implemented a RS(255,239) + convolutional self-orthogonal code (CSOC) [29]. It provides 8 dB NCG at 1×10^{-12} output BER with 25% redundancy, using three times iterative CSOC decoding. Thanks to the advent of second-generation FEC, the attainable transmission capacity per fiber has been drastically increased [30–32].

Several types of second-generation FEC have been standardized in ITU-T G.975.1: RS(255,239) + CSOC ($n_0/k_0 = 7/6, \mathrm{J} = 8$), BCH (3860,3824) + BCH(2040,1930), RS(1023,1007) + BCH(2047,1952), RS(1901,1855) + extended Hamming product code (512,502) × (510,500), two orthogonally concatenated BCH codes, and two interleaved extended BCH (1020,988) codes are discussed in this recommendation. One of the best performing second-generation FECs shows an NCG of 9.4 dB with a code rate R of 0.8. It appears in ITU-T G.975.1 as two orthogonally interleaved BCH codes. Several types of second-generation FEC chip have been commercialized for 10 Gb/s systems. They have NCGs of greater than 8 dB. Optical transport network (OTN) compatible second-generation FEC LSI has been developed with a Q limit of 9.1 dB at 43 Gb/s [33]. Second-generation FECs may introduce transmission delay due to the FEC encoding, decoding, interleaving, and de-interleaving operations. This physical time delay is known as latency. We have to pay attention to latency for the sake of network operation.

Let us examine a typical example of second-generation FEC using concatenated RS codes as illustrated in Fig. 17.8 [34]. The frame structure comprises sixteen RS(255,239) codes and sixteen RS(239,223) codes, where 259 and 239 are the code length and information length, respectively. The first column is for the FEC overhead, the following 222 columns are for the information, the next 16 columns are for the redundancy for RS(239,223), and the last 16 columns are for the redundancy for RS(255,239). RS(239,223) + RS(255,239) needs 14.2% redundancy for its RS codes. The error correction performance varies with the depth of interleaving and the number of iterations. Greater depth of interleaving and more iterations enable higher coding gain.

Figure 17.9 shows the measured and calculated error correction performance of RS(239,223) + RS(255,239). The NCG at an output BER of 1×10^{-11} is 7.5 dB when the number of iterations is 16. Transmission experiments using RS(239,223) + RS(255,239) have been demonstrated at 10 Gb/s [35] and 20 Gb/s [36].

Figure 17.10 shows a typical block diagram of a second-generation-based FEC LSI for a 10 Gb/s system. The 130–65 nm CMOS technology enabled full

Fig. 17.9 Error correction performance of RS(239,223) + RS(255,239)

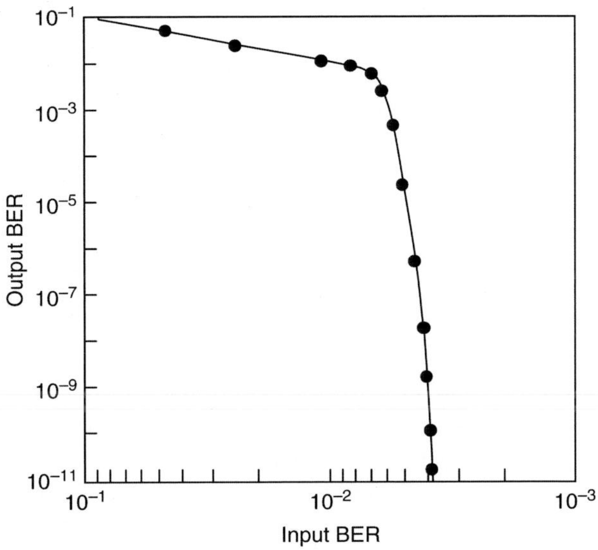

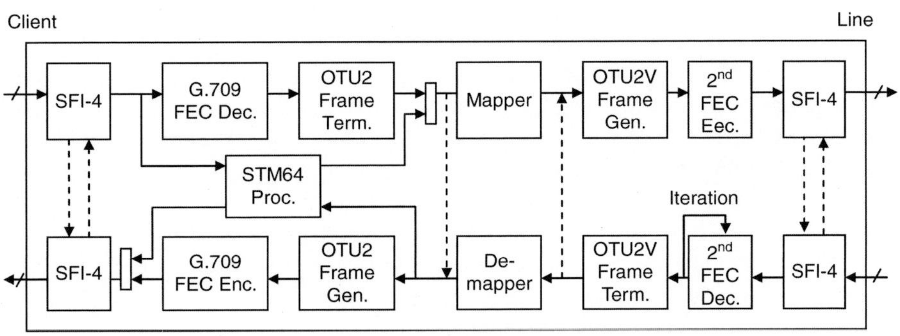

Fig. 17.10 Typical block diagram of OTN compatible FEC LSI

integration into a ball grid array package. The circuit volume is approximately tens or sub tens of M-gates with Mbit class RAM. The LSI chip usually has an SFI-4 (SERDES Framer Interface-4) compliant interface configured to support a single 10 Gb/s stream. The LSI provides a SONET/SDH compliant section and line processing for OC-192/STM-64. The input client signal is mapped in two ways. The G.709 standardized FEC in the incoming OTN signal is decoded first, then the OTU2 overhead is extracted. In the case of an STM-64 client signal, the overhead is terminated and its performance is monitored. The client signal is then mapped on to the second-generation FEC framer with an OPU2/ODU2/OTU2V overhead. The encoded signal is output to the line side. At the receiver side, the incoming errored signal is iteratively decoded by the second-generation FEC decoder. The OTU2V/ODU2/OPU2 overhead is terminated and demapped by the de-framer. The OTN overhead is generated before launching to the client side. The LSI usually accommodates three loop-back functions: at the client interface, line interface, and before/after client signal termination. The final loop-back acts as a regenerator. The LSI also supports tandem connection monitoring (TCM).

17.4 Third-Generation FEC

In this section, we discuss more powerful FEC having an NCG of over 10 dB. The motivation was not only more capacity, but also reducing the initial expense of the system as a whole. Because terabit systems need expensive optical technologies, i.e., ultra-wide band optical amplifiers, complex optical channel equalizers, and special grade premium fibers, an important role of stronger FEC is to relax the requirements on such expensive optical building blocks. In this context, FECs stronger than the second generation, based on soft-decision decoding, will be classified as third-generation FEC.

Figure 17.11 shows examples of the clear and degraded eye patterns at 10 Gb/s of a return-to-zero optical signal. We see a clear eye opening in (a), with an input BER and corresponding Q factor of $<1 \times 10^{-13}$ and >17.3 dB, respectively. The degraded eye opening (b), having an input BER of 2×10^{-2}, and Q of only 6.3 dB, is suitable for the third-generation FEC discussed in this section. We can no longer see where the eye opening is. To correct such a degraded signal by using any previous technologies would be a momentous matter indeed.

The first- and second-generation FECs are based on hard-decision decoding. Soft-decision decoding is one of the most promising technologies for correcting an input Q of 6.3 dB to error free. The soft decision is performed by establishing $2^{\mathrm{M}}-1$ decision thresholds, where M is the number of quantization bits. In the case of $M = 3$ as shown in Fig. 17.12, soft decision between the two possible received signals, "one" and "zero," lies in regions represented by a binary vector ranging from [011] to [111]. The left most bit is the hard-decision digit, and the other two digits are information bits indicating the probability of "one" or "zero."

In the soft-decision decoder, the k_{th} channel value r_k is input to the decoder. The soft-in-soft-out (SISO) decoder generates candidate constituent codewords $\{C_{\mathrm{j}}\}$

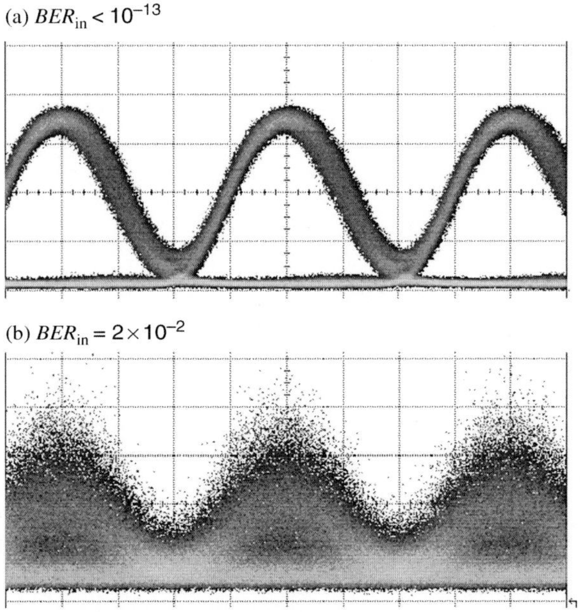

Fig. 17.11 Examples of eye patterns through additive noise channels that should be corrected by each generation of FEC: (**a**) $Q \geq 17.3\,\text{dB}$, $\text{BER}_{\text{in}} < 1 \times 10^{-13}$, (**b**) $Q = 6.3\,\text{dB}$, $\text{BER}_{\text{in}} = 2 \times 10^{-2}$

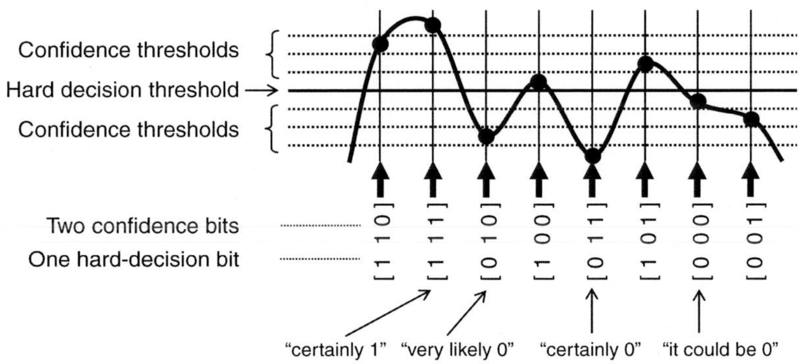

Fig. 17.12 Soft decision

from the soft inputs $\mathbf{Y} = \{y_k\}$ and computes the log-likelihood ratio (LLR) L_{k}, which is defined as follows:

$$L_k = \ln \frac{\sum\limits_{C_j^1} p\left(\mathrm{Y}|C_j^1\right)}{\sum\limits_{C_j^0} p\left(\mathrm{Y}|C_j^0\right)} \tag{17.5}$$

where C_1^{j} is a candidate codeword such that the bit at its k_{th} position is 1 and C_0^{j} is such that the bit at its k_{th} position is 0, and where $p(\mathbf{Y}|C)$ is the probability

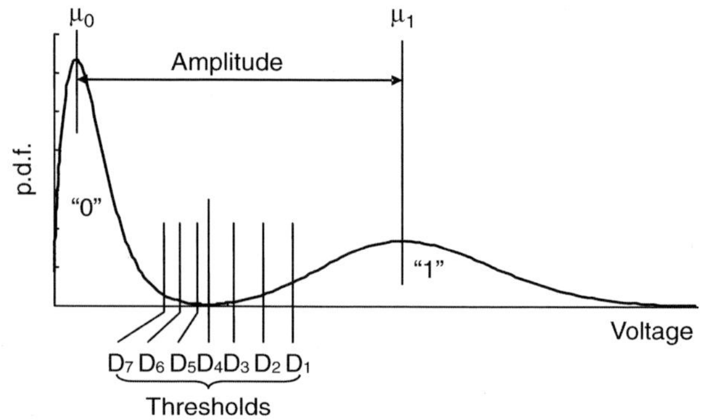

Fig. 17.13 Example of 3-bit soft-decision threshold arrangement for asymmetric pdf

density function of **Y** conditioned on C. When the probability of "one" or "zero" of the received code vector is 0.5, the LLR takes 0. When the probability of "one" approaches 1, LLR goes to $+$.

In optically amplified OOK systems, the probability density function (pdf.) of the received signals is represented by a non-central chi-square distribution as shown in Fig. 17.13. The chi-square distribution, the sum of squares of independent Gaussian random variables with identical variances, is applicable here because the optical noise is additive white Gaussian noise (AWGN). However, because the numerical calculation of this pdf is not straightforward, it is often assumed that the electrical currents in the marks and spaces are asymmetric Gaussian distributions having different variances for mark and space. On making a soft decision, the LLR of the received signals is calculated to obtain their confidence bits by:

$$\mathrm{LLR} = \ln \frac{p(x|1)}{p(x|0)} = \frac{1}{2}\left(\frac{1}{\sigma_0{}^2} - \frac{1}{\sigma_1{}^2}\right)x^2 + \frac{x}{\sigma_1{}^2} + \ln \frac{\sigma_0}{\sigma_1} - \frac{1}{2\sigma_1{}^2} \tag{17.6}$$

where $p(x|1)$ and $p(x|0)$ are a posteriori probabilities, and σ and σ_1 are the variances of the received spaces and marks. An example of LLR under Gaussian distribution is shown in Fig. 17.14. We have previously derived the best arrangement of threshold settings for 3-bit soft decision, where a wider spacing for mark and a narrower spacing for space are preferable [37].

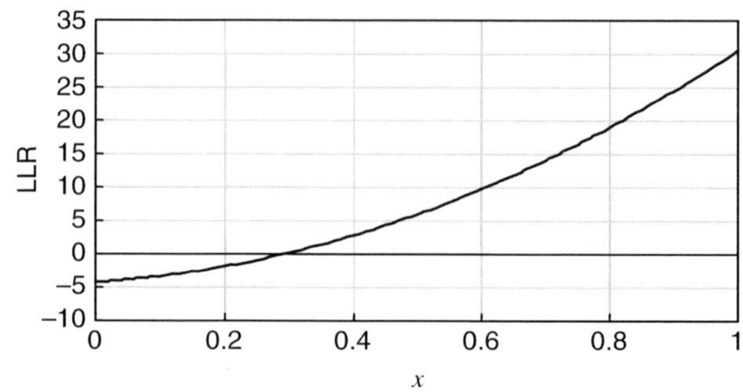

Fig. 17.14 LLR under Gaussian distribution ($Q = 5.5$ dB)

Soft decision for optical communications was first demonstrated by Puc et al. using an RS code concatenated with a Viterbi convolutional code. They demonstrated an NCG of 10.3 dB at 2.5 Gb/s [38], but the redundant overhead was all of 113%. For practical application to optical fiber communications at 10 Gb/s or more, a redundant overhead of no more than 25% is highly desirable.

One of the most powerful FECs with soft-decision decoding is a turbo code, which is a promising candidate for the third-generation FEC as it takes us close to the Shannon limit. Turbo codes were introduced by Berrou, Glavieux, and Thitimajshima (from ENST, Bretagne, France) in 1993 [39–41]. Then, a comparable performance was demonstrated using a block code with a SISO decoder.

Turbo codes are classified into two categories: Turbo convolutional codes (TCC) and block turbo codes (BTC). Both codes show significant improvement over concatenated codes. In general, TCC shows greater correction ability than BTC at low code rates. Conversely, BTC shows better performance at high code rates, and no puncturing is required. It also has lower complexity. This results in lower power consumption and cost. A BTC shows no error floor because of its large minimum distance (16 or greater is typical). It also shows only minor flaring starting at very small output BER. All of these advantages seem to make it suitable for optical communication systems. A BTC is built on a 2- or 3-dimensional array of codes, in which the encoding is done in a single iteration. The minimum distance of a 2-dimensional product code is the square of the constituent codes.

Ait Sab was the first to discuss this kind of code in the context of optical fiber communication systems [42]. Several numerical simulations have followed, with the potential to achieve around 10 dB NCG with an overhead of less than 25% using BTC [43]. A hardware demonstration has been reported using field programmable gate arrays (FPGAs) having a throughput of 194 Mb/s for a prescaled FEC encoder/decoder [44]. The demonstration showed an NCG of 10.1 dB for a BTC at 12.4 Gb/s with a 3 bit soft-decision LSI operating at 12.4 Gb/s serial [45]. The BTC was implemented in the LSI pictured in Fig. 17.15 [46]. The 0.13 μm CMOS technology enabled full integration into a 1024 pin ball grid array package 40 × 40 mm square. The circuit volume is approximately 16 M-gates with 4 Mbits RAM. The power consumption is less than 10 W. The LSI chip features an SFI-4 (SERDES

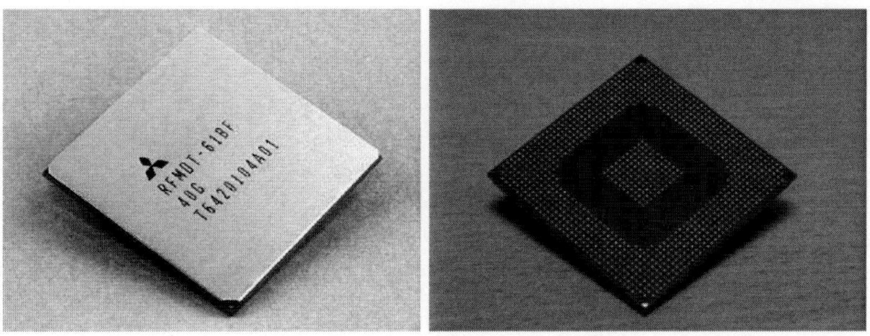

Fig. 17.15 Block Turbo Code LSI for 10 Gb/s optical communications

Framer Interface-4) compliant interface configured to support a single 10 Gb/s stream. The resulting bit rate for STM-64 is 12.4 Gb/s with a redundancy of 23.6%.

Let us consider the BTC in detail taking the LSI shown above as an example. A product of two constituent Bose, Chaudhuri, and Hocquenghem (BCH) codes is a desirable candidate for the BTC. BCH codes are considered to among the best understood, encoding k information bits into n code bits by adding n–k parity-checking bits for the purpose of detecting and correcting the errors. Figure 17.16 is an example of a frame format of a BCH(144,128) × BCH(256,239) product code having a minimum distance of 30.

Figure 17.17 shows the block diagram of a BTC decoder. Two elementary SISO decoders are interconnected with each other. The kth channel value r_{k} is input to the delay line and added to the mth extrinsic information $w^{(m)}{}_k$ to generate the soft input $y_{\text{k}} = r_{\text{k}} + w^{(\text{m})}{}_{\text{k}}$, which is input to the SISO decoder for the rows. Then the interleaved vector is input into the SISO decoder for the columns. The SISO decoders generate candidate constituent codewords $\{C_{\text{j}}\}$ from the soft inputs $Y = \{y_k\}$ and compute the LLR. The channel value r_k is subtracted from the LLR L_k and the result is multiplied by the optimum weighting factor $\alpha^{(m+1)}$ to generate the next stage extrinsic information $w^{(m+1)}{}_k$. Finally, the transmitted data value is estimated from the sign of the LLR. The whole process is then iterated again. This process has given rise to the term "turbo code," since the original inventors likened the process

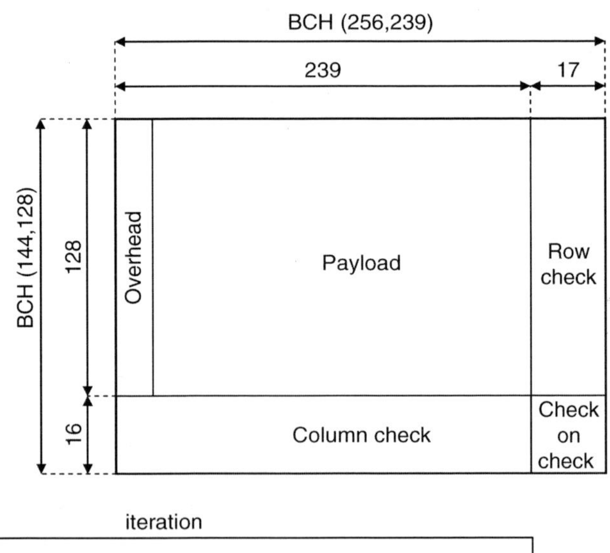

Fig. 17.16 Frame format of BCH product code for the BTC

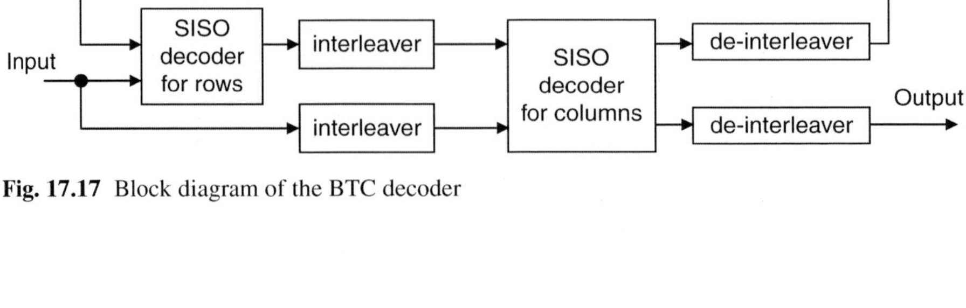

Fig. 17.17 Block diagram of the BTC decoder

to a turbo-charged engine, in which part of the power at the output is fed back to the input to boost the performance of the whole system. Thus the term "turbo" should really be applied to the decoder structure rather than to the codes themselves.

The number of iterations is an important factor for the error correction ability. The coding gain improves with the number of iterations, and at a BER of 10^{-6} the performance of the BTC after four iterations is about 0.2 dB better than after three iterations. However, five or more iterations do not bring any significant coding gain. A total of four iterations is thought to be optimum, taking circuit volume practicalities and latency into account.

The number of soft-decision bits, in other words the resolution of the soft-decision process, is also an important parameter for soft-decision decoding. As expected, better quantization improves the performance of the BTC, but soft decision is difficult to implement at high speeds such as tens of gigabits per second. The simulation predicts that for 10^{-6} BER, the performance of a BTC with 3-bit soft decision is superior by about 0.6 dB to that of a BTC with 2-bit soft decision and is superior by about 0.4 dB to that of a BTC with hard decision. From the viewpoint of hardware implementation, it is concluded that 3-bit soft decision is the optimum solution in terms of quantization.

A typical block diagram of a BTC receiver is depicted in Fig. 17.18. The receiver consists of an optical front end, a 3-bit (in this case) soft-decision circuit, a turbo decoder, a D/A converter, and a control processor. A soft-decision circuit is essential for the BTC receiver. The 3-bit soft-decision circuit includes sevens decision units, a 3-bit encoder, a clock recovery circuit, three sets of 1:16 demultiplexers, and their control circuitry. Decisions are made on the received signal by seven decision units with separate thresholds, obtaining a 3-bit output with two confidence bits. An error filter was incorporated in the IC, which prevents the degradation of FEC performance caused by signal noise or the yield of the semiconductor manufacturing

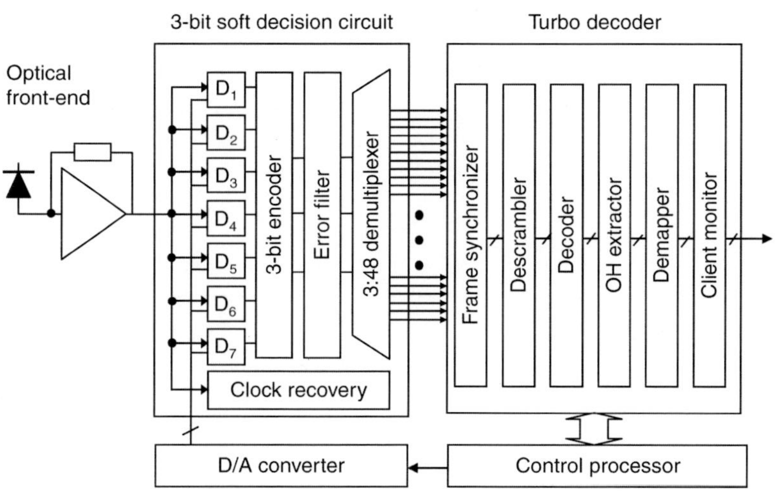

Fig. 17.18 Schematic diagram of BTC receiver

process. Although there should only be eight types of 3-bit output information, impossible outputs in fact occur due to sensitivity variations. In the event that an impossible output is generated, the error filter provides the most probable pattern.

The error correction capability of the BTC is shown in Fig. 17.19. In order to emulate a severely degraded received optical signal, amplified spontaneous emission (ASE) noise generated from an optical amplifier is added to the transmitter signal. Zero errors are observed over 1×10^{14} bit counts at an input BER of 1.95×10^{-2}. No indication of bit-error flare is observed. An uncorrected BER of 1.98×10^{-2} could thus be corrected to 1×10^{-13}. This corresponds to an input Q of 6.26 dB being corrected to 17.32 dB. Taking the bit rate increase from 9.95 to 12.4 Gb/s into consideration, the NCG is 10.1 dB at the same output BER. In the hard-decision case, only an input BER of 9.13×10^{-3} can be corrected to 1×10^{-13}. The NCG here is 8.9 dB. This is 1.2 dB worse than that for 3-bit soft-decision, which agrees well with the numerical simulation. The typical parameters of the BTC are summarized in Table 17.4.

Optimization of the multiple decision threshold arrangement is important to maintain superior error correction performance. In optically amplified systems, the probability density function of the dominant signal-ASE beat noise is asymmetric

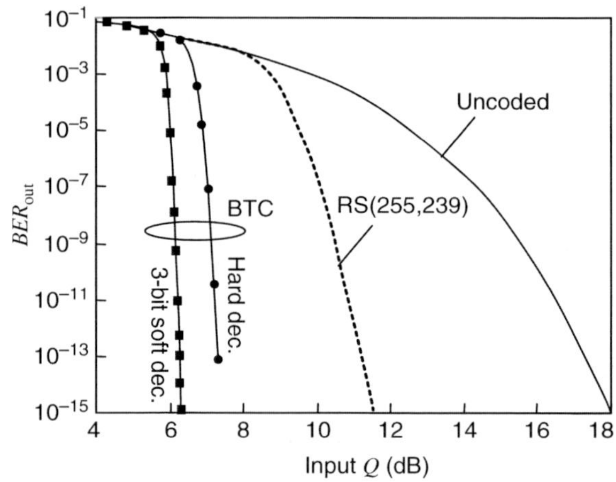

Fig. 17.19 Error correction capability of BTC

Table 17.4 Typical BTC parameters

Term	Parameter
Code	BCH(144,128) × BCH(256,239)
Redundancy	23.6%
Bit rate	12.409 Gb/s @ STM64
Q limit	6.26 dB @ 10^{-13}
BER_{in}/BER_{out}	$1.9 \times 10^{-2}/1.1 \times 10^{-15}$
Gross coding gain @ $BER_{out} = 10^{-13}$	11.1 dB
Net coding gain @ $BER_{out} = 10^{-13}$	10.1 dB
Number of soft-decision bit	3

additive white Gaussian noise. The best performance was obtained with wider spacing on the "1" side and narrower spacing on the "0" side, which is consistent with expectations from theory [37].

Another potential solution for achieving 10 dB NCG is a low-density parity-check code (LDPC). The LDPC code is a linear code, obtained from the sparse parity-check matrix invented by Gallager [47]. This technique lay unrecognized for a long time, then was re-discovered by MacKay [48], leading to further intensive investigation for wireless communication systems. We also categorize LDPC coding as a third-generation FEC, since it is based on soft decision and has the potential to achieve turbo code-like performance.

Studies into applying LDPC to optical communications were instituted by Vasic and Djordjevic. The irregular LDPC(3367,2821) code with 19% redundancy was studied, and with the iterative soft-decision min-sum algorithm yielded an NCG of 8.1 dB at a post-FEC BER of 10^{-9} [49]. This performance was improved on by the generalized LDPC(3639,3213) code with 23.6% redundancy, with which a record NCG of 10.9 dB at a post-FEC BER of 10^{-13} was demonstrated in a Monte-Carlo simulation [50]. The calculations showed that the LDPC has the potential to be improved further and, for example, the project geometry PG(2,26)-LDPC(4161,3431) code with 21.2% redundancy was reported to achieve an NCG of 11.27 dB at a post-FEC BER of 10^{-13} [51].

The *shuffled belief propagation* (Shuffled BP) algorithm [52] is well known to give remarkable performance, closing in on the Shannon limit. However, the hardware implementation for optical communications can be quite complex, requiring six or more soft-decision bits and an enormously complex circuit. Circuit complexity is a severe problem when we try to implement very high-speed FEC in a single LSI chip. The latest LSIs for 40 Gb/s made with the 90 nm CMOS process have equivalent gate counts of 30∼40 million. For 100 Gb/s, at least three times higher throughput and a higher speed I/O are essential. Therefore use of a 45 nm process and a high-level system design would be necessary. If we can realize the same error correction performance with 1/3 the circuit complexity, more mature and more economical CMOS processes, which for practical reasons are mandatory, become available. Therefore, reduced circuit complexity is a very important issue for high-speed FEC implementation.

In order to reduce the circuit complexity, Miyata et al. proposed a cyclic-approximated δ-minimum algorithm [53], which only requires real additions and comparisons, together with fewer soft-decision bits. A hardware configuration for this algorithm is shown in Fig. 17.20, where the cubes and squares represent memories and operators. The operators and paths combine to calculate the LLR of a bit which corresponds to a column having a weight of three in the parity-check matrix. The most significant LLRs of bits sent from parity are calculated cyclically using the input LLR λ_n, the sign of $S_{m,n} = \mathrm{sgn}(\bullet)$, and the previously stored LLR_m, and stored in memory.

Now, calculating the LLR of a bit sent from row parity is an essential operation for an LDPC code. However, such a calculation requires a huge amount of memory. Using an approximation to calculate the minimum function is a useful way of

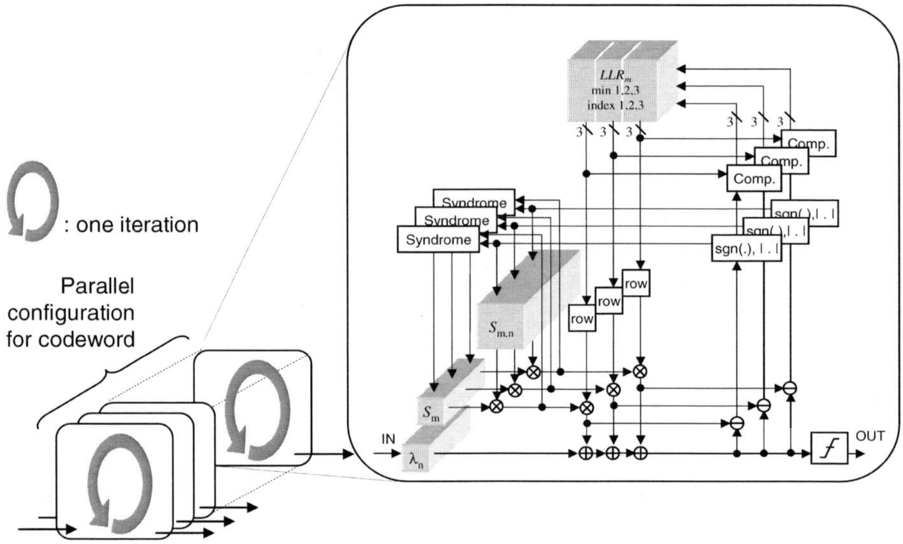

Fig. 17.20 Circuit configuration for the cyclic-approximated δ-minimum algorithm

significantly reducing the required memory. In this calculation, only the smallest top three LLRs, i.e., the most significant terms, are derived. The fourth candidate is compared cyclically during the row calculation. This procedure can drastically reduce the required memory.

In addition, a second essential operation for LDPC codes is the calculation of the expression $a \oplus b = 2\tanh^{-1}[\tanh(a/2) \cdot \tanh(b/2)]$ (a and b are absolute values) for BP decoding. However, a huge number of operations are necessary to calculate such an equation precisely. To achieve a significant reduction in the required number of operations, an approximation using $\delta = |a-b|$ as $\max[\min(a,b) - \max(0.9-\delta/2, 0), 0]$ is effective.

Figure 17.21 shows the calculated error corrected BER vs. input Q factor for the irregular Quasicyclic LDPC code having codeword and information lengths of 8148

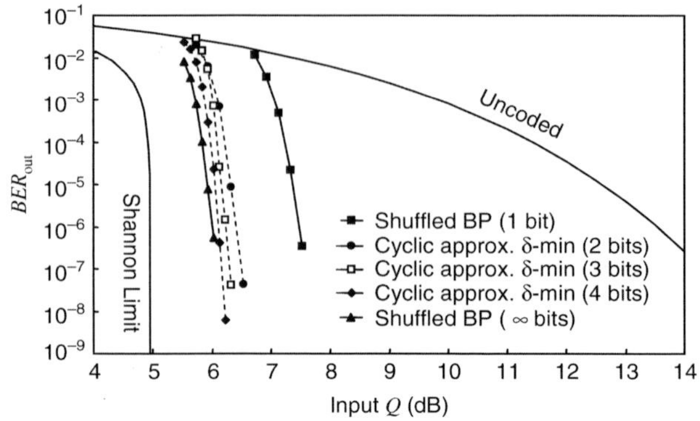

Fig. 17.21 Simulated error correction performance for various decoding algorithms

and 6984, respectively. The redundancy is 16.67%. Additive white Gaussian noise (AWGN) is assumed for the communication channel. The number of iterations is set to only 16 (including initialization), which is sufficiently small to hold down the latency.

The best performance is exhibited by the shuffled BP with infinite quantization bits, with which an input Q of 6.0 dB can be corrected to 10^{-6} BER. With only one quantization bit for the shuffled BP, the required Q rises to 7.5 dB. Three different cyclic-approximately δ-minimum algorithms with different numbers of quantization bits were calculated, and as seen in Fig. 17.21, as the number of quantization bits is increased, the error correction capability starts to approach that of the infinite-bit shuffled BP case.

It is useful to estimate the required number of operations per iteration when comparing the shuffled BP and the cyclic-approximately δ-minimun algorithms. During the decoding procedure, each algorithm involves the following operations: add, compare, EXOR, and table lookup. The required number of operations for cyclic-approximately δ-minimum is 10 times smaller than that for the shuffled BP algorithm. Using an approximate calculation contributes to a drastic reduction in operations. The required memory size is also crucial in implementing a decoder LSI. The required memory for cyclic-approximately δ-minimum is 5 times less than that for the shuffled BP algorithm.

Even with the superior codes and superior algorithm proposed, we have to pay great attention to error floor, frequently observed in measured post-FEC BER as shown in Fig. 17.22. Even though an FEC may show superior performance at

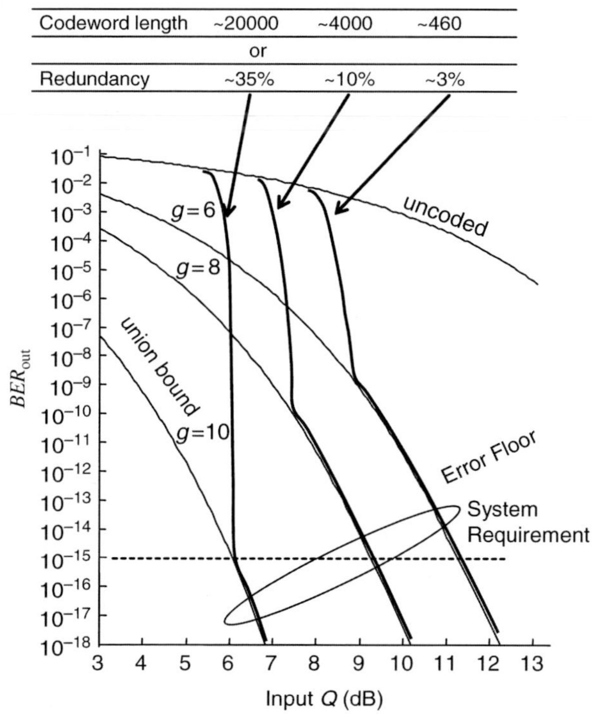

Fig. 17.22 LDPC Error floor

a post-FEC BER of 10^{-7}, if there is an error floor at 10^{-8}, such an FEC is unusable.

In order to reduce the error floor of an LDPC code, a large girth (∼10) is necessary. There are two ways to increase the girth. One is by expanding the codeword length. According to a rough estimate, girths of 6, 8, and 10 are obtained from codeword lengths of 460, 4000, and 20,000, respectively. By increasing the codeword length, a possible parity-check matrix pattern which has only large loops can be discovered. However, longer codewords require increased circuit size. For 100 Gb/s class LSI, a codeword length of 20,000 is very difficult to implement. Another way to increase girth is to increase the redundancy. Regular LDPC codes having redundancies of 3, 10, and 35% enable their girths to be 6, 8, and 10, respectively. However, increased redundancy raises the bit rate. In the case of 100 GbE, 35% redundancy is impractical because its bit rate would approach 140 Gb/s. In slower speed systems, e.g., mobile telephones and other wireless systems, we can use either of the above methodologies. However, especially for 100 GbE, very long codewords and/or large redundancy are simply impossible.

As an alternative approach, the concatenation of another code can be effective in suppressing the unwanted error floor, and concatenated LDPC and RS codes have been proposed for this purpose [54]. LDPC(9252,7967) having a redundancy of only 16.1% is selected as the inner code, balancing error correction capability, and circuit complexity due to codeword length. For the outer code, RS(992,956) can correct 18 symbols with only 3.8% redundancy. The additional hardware volume due to the RS code is very small compared to that of LDPC codes, and the total redundancy of the concatenated LDPC + RS is 20.5%. The error correction performance of LDPC(9252,7967) alone and the concatenated LDPC(9252,7967) + RS(992,956) are compared by Monte-Carlo simulation. Figure 17.23 shows the simulated pre-FEC Q vs. post-FEC BER for the LDPC code under test. The unwanted error floor which appeared in the case of LDPC alone at a post-FEC BER of 10^{-9} is eliminated by concatenation with the RS code, at least down to a post-FEC BER of 10^{-11}.

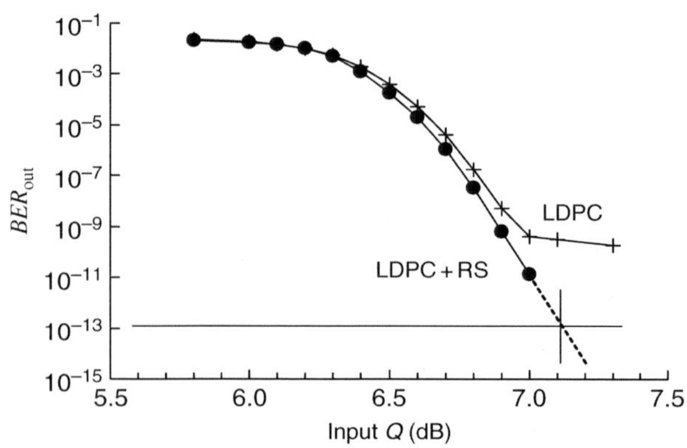

Fig. 17.23 Simulated BER performance w and w/o concatenated RS

17.5 Comparison with the Shannon limit

Figure 17.24 summarizes the typical error correction performances of the three generations of FEC. The features of each generation are listed in Table 17.5. The horizontal axis of Fig. 17.24 indicates the degraded Q factor when received through an additive noise channel. The first-generation FEC uses RS(255,239), having a code rate of 0.93 (7% redundancy). It corrects an input BER of 1.4×10^{-4} up to an output BER of better than 1×10^{-13}. The second-generation FEC is based on concatenated RS and/or BCH codes. Various code rates ranging from 0.93 to 0.79 have been reported and released on the market to date. The third-generation FEC is based on soft-decision decoding.

Figure 17.25 plots the progress in FECs for optical communication systems over the past 20 years. The vertical axis shows the product of (linear) net coding gain defined in terms of an output (post-FEC) BER of 10^{-13} and bit rate in Gb/s. The three sets of points show the different FEC schemes: first-generation FEC using linear block codes represented by RS(255,239); second-generation concatenated codes; and third-generation soft-decision and iterative decoding-based FEC, e.g., block turbo codes. A clear trend can be seen in that an improvement of 1.4 times has been achieved every year. This improvement has been achieved not by FEC algorithm improvements, but by LSI technology evolution. The dashed lines at the right indicate the NCG bit rate product for RS(255,239) at each bit rate. For example, RS(255,239) operating at 10 Gb/s with an NCG of 3.8 (= 5.8 dB) has an NCG

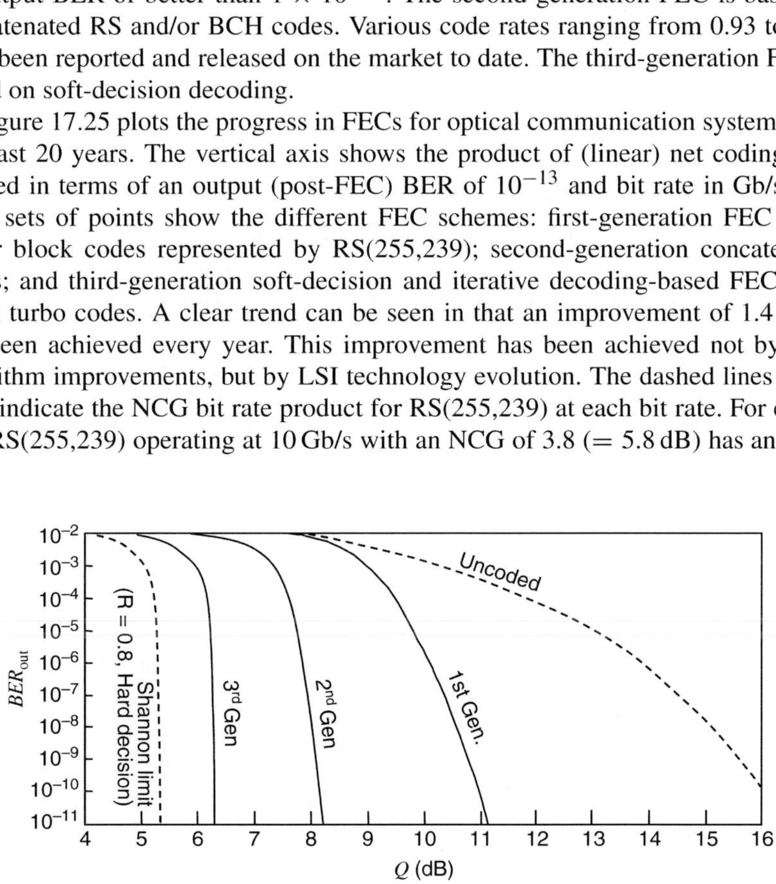

Fig. 17.24 Typical error correction performance of the three generations of FEC

Table 17.5 Comparison of the three generations of FEC

Generations	Decision	Codes	Code rate	NCG (dB)
First	Hard	Linear codes e.g., RS(255,239)	0.93	5.8
Second	Hard	Concatenated codes	0.93–0.79	<10
Third	Soft	Block turbo code Viterbi convolutional codes LDPC	0.80–0.47	>10

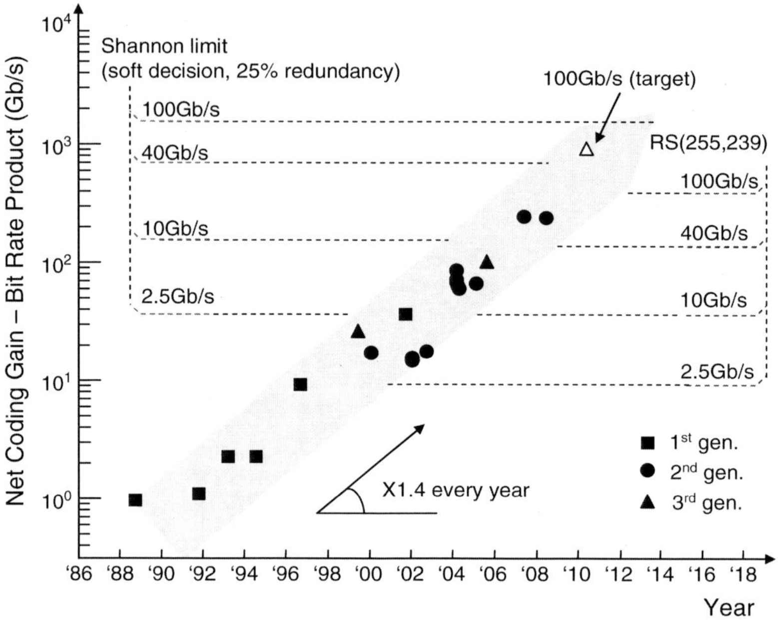

Fig. 17.25 Progress of FEC in optical communication systems

bit rate product of 38. The dashed lines at the left indicate the Shannon limit in the case of a redundancy of 25%. Regardless of how excellent its error correction capability may be, no FEC system can ever surpass that limit. The open triangle at the top right shows the research target for strong FEC for 100 Gb/s.

Let us consider the maximum attainable NCG for each generation of FEC. The ultimate NCG, the so-called Shannon limit, is derived from Shannon's capacity theorem [2]. Given an input Q per symbol for an AWGN channel, the $\mathrm{BER_{in}}$, is derived from Eq. (17.2). The channel capacity C of the binary symmetric channel is given by

$$C = 1 + \mathrm{BER_{in}} \log_2(\mathrm{BER_{in}}) + (1 - \mathrm{BER_{in}}) \log_2(1 - \mathrm{BER_{in}}). \tag{17.7}$$

Now we set $C = R$ (code rate) in Eq. (17.7), and then determine the Q per symbol bits that satisfies this equation. The NCG is simply derived from 17.3 dB – $Q_{\mathrm{in}} + 10 \log_{10} R$, where 17.3 dB denotes the output Q corresponding to an output BER of 1×10^{-13}. The result is shown as a graph of NCG vs. R at an output BER of 1×10^{-13} (Fig. 17.26). For $R = 0.8$, the ultimate NCG for hard decision is 11.0 dB.

As for soft-decision decoding, the capacity of the binary input AWGN channel can be calculated as indicated in references [55]. The result of this calculation is also shown in Fig. 17.26. The difference in NCG between hard- and soft-decision decoding with infinite quantization bits when R approaches zero is $\pi/2$. This is an approximately 2 dB difference. In the case of $R = 0.8$, the difference is about 1.5 dB. Note that all of the calculations above are based on a binary symmetric channel (BSC). In the case of a phase-shift keying system, the probabil-

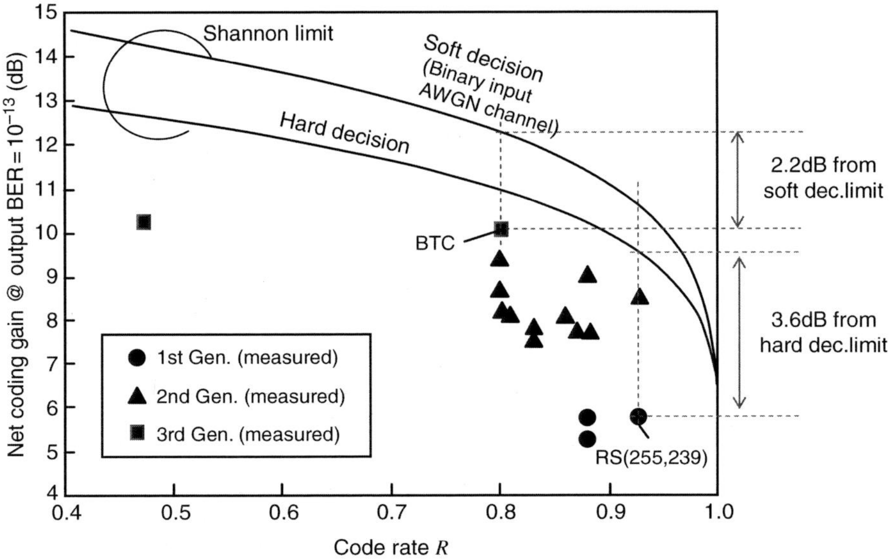

Fig. 17.26 Several FECs reported to date and the Shannon limits

ity density function (pdf) does correspond to a BSC. However, in the case of an on–off keying system, the pdf of the chi-square function is a binary asymmetric channel (BAC). The Shannon limit for a BAC is slightly different from that for a BSC [56].

Various NCGs reported to date for optical fiber communications are also depicted in Fig. 17.26. Note that the code rate R of the horizontal axis includes not only the FEC redundancy but also the functional overhead. The NCG for the first-generation FEC RS(255,239) is 3.6 dB away from the hard-decision Shannon limit. The best performing second-generation FEC has an NCG of greater than 9 dB, which is still well away from the hard-decision Shannon limit. The block turbo code indicated by the filled circles having an NCG of 10.1 dB is 0.9 dB, closing in on the hard-decision limit, but still 2.2 dB away from the soft-decision limit.

In order to keep the BER quality at 112 Gb/s the same as that of a standard RS(255,239) FEC-based 43 Gb/s system having a Q limit of 11.2 dB, at least 4.2 dB ($= 10\ \log_{10}(112/43)$) greater net coding gain (NCG $>$ 10.0 dB at post-FEC BER of 10^{-13}) is necessary. Figure 17.27 shows the various NCGs reported to date, with the hard- and soft-decision Shannon limits also depicted. The best performing enhanced FEC LSI for 40 Gb/s with a redundancy of 7% is still far from the hard-decision limit. The BTC indicated by the filled square at an NCG of 10.1 dB does exceed the required NCG, but it has only been implemented in 10 Gb/s LSI. In order to obtain that NCG, at least 12% redundancy is essential even for the hard-decision limit. However, no FEC which achieves the hard limit has been seen yet. In terms of practical circuit implementation, 20∼25% redundancy would be necessary to obtain an NCG of 10 dB.

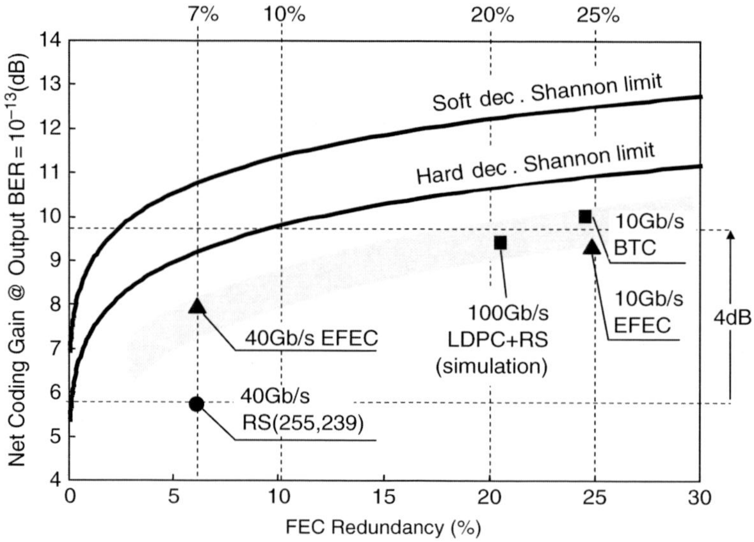

Fig. 17.27 FEC Redundancy and NCG

17.6 FEC Error Count

Most FECs provide an error count function. This is so the FEC system can know how many errored bits the FEC has corrected. A parameter related to the transmission impairment can be optimized if it is subjected to control by an appropriate method in which the number of corrected errors is automatically minimized. A wide range of applications of FEC error count has been started to be put into practice. In this section, some remarkable functional applications of this are reviewed, i.e., automatic decision threshold adjustment, adaptive chromatic dispersion compensation, and adaptive polarization mode dispersion (PMD) mitigation.

The most useful application of the error count is decision threshold optimization at the receiver [57]. The optimum decision threshold is a time-varying characteristic with various sources of variation: e.g., transmission power fluctuations in the fiber, received power changes, the state of polarization rotation, and aging of the optical link. As illustrated in Fig. 17.28, the probability density function (pdf) of an optically amplified system follows the chi-square distribution. Only when the decision threshold is set to the optimum value can the minimum BER be obtained. Figure 17.29 shows a simplified block diagram for automatic decision threshold adjustment. The incoming signal is compared with a decision threshold V_{th}. The error count is extracted from the FEC decoder. The decision threshold is then varied so as to minimize the error count using a "hill-climbing search". Using heuristics, this finds which direction will take it closest to the goal. Some FEC chips output flag information indicating whether a "1" or a "0" was corrected. Using such a flag,

Fig. 17.28 Adjustment of decision threshold

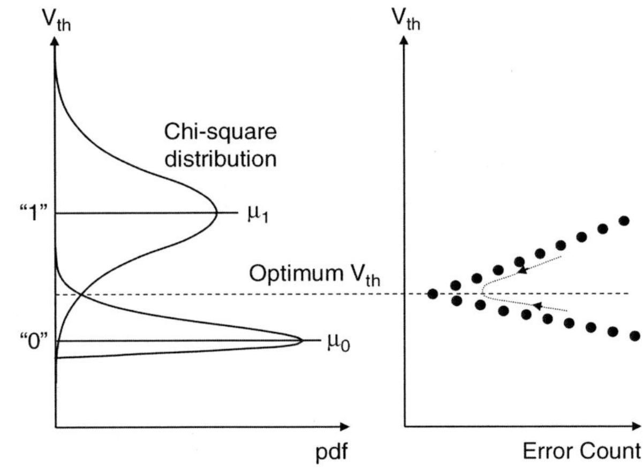

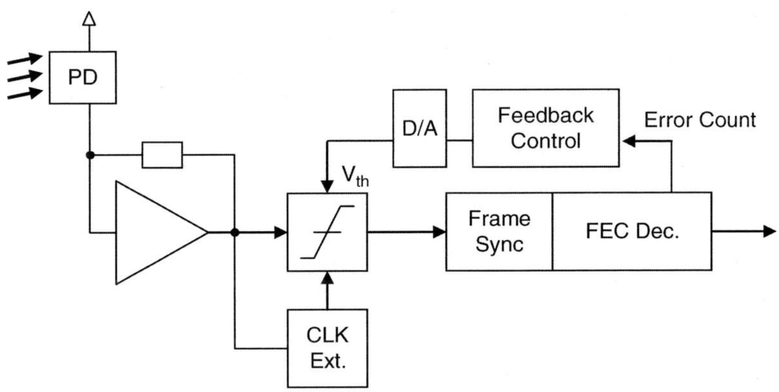

Fig. 17.29 Schematic diagram for decision threshold adjustment using FEC error count

the decision threshold can be optimized more quickly without using a hill-climbing search.

Compared with a hard-decision receiver, it would appear to be more difficult to automatically track each soft-decision threshold. However, once the amplitude and the optimum hard-decision threshold are known, the other thresholds can be established using pre-calculated values. Automatic threshold tracking has been demonstrated using a block turbo code LSI with 3-bit soft-decision at 12.4 Gb/s [58].

The dispersion of the installed fiber varies with changes in temperature and/or strain, and, in a 40 Gb/s system, dispersion fluctuation results in severe BER degradation. To avoid this, it is necessary to continuously control a tunable dispersion equalizer so as to minimize the number of errors. In order to realize automatic tracking of a tunable dispersion equalizer, the FEC error count is again useful, since no additional high-speed circuitry is necessary. An automatic dispersion equalization using FEC error count on a 40 Gb/s test bed has been demonstrated [59].

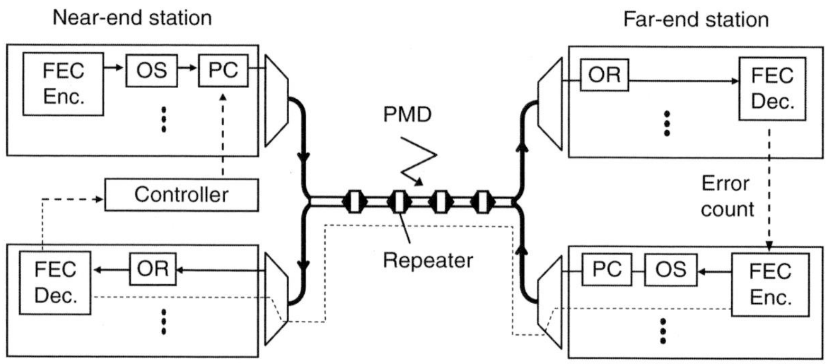

Fig. 17.30 Functional block diagram for PMD mitigation using error count

The FEC error count can also be applied to another approach to PMD mitigation. If the input state of polarization (SOP) is aligned to correspond to the principal state of polarization (PSP) of the transmission line, the BER degradation due to PMD can be reduced. Figure 17.30 depicts the functional block diagram for this method. At the transmitter side, each optical sender (OS) at the near end station is followed by a polarization controller (PC) so as to independently align its input SOP with the transmission line. The received signal, which suffers severe waveform distortion from PMD in the course of the long-haul transmission, is detected by the optical receiver (OR) at the far end station. The received bit errors are corrected by an FEC decoder, which counts the corrected errors over time. The far end error count is placed into the FEC frame overhead of the transmitted signal by the FEC encoder at that station and is returned immediately to the near end station. Each PC is adjusted continuously to minimize the number of errors under an endless (reset-free) SOP control methodology. Although the feedback delay is large due to the round-trip time, e.g., 45 ms for 9000 km, the simple methodology gives this method a large advantage over other schemes.

An experiment in input SOP control using a 42-channel 20-Gb/s WDM transmission test bed has been conducted [60]. The time-varying Q was measured after transmission over 3540 km. The Q was drastically improved by the application of input SOP control. Without control, loss of frame (LOF) occurred frequently. The probability of outages, defined as BER $> 3.9 \times 10^{-3}$, was reduced from 2.1×10^{-2} to zero with input SOP control.

References

1. R.W. Hamming, Error detecting and error correcting codes. Bell Syst. Tech. J. **29**(2) 147–160, April (1950).
2. C.E. Shannon, A mathematical theory of communication. Bell Syst. Tech. J. **27**, 379–423, 623–656, July, Oct. (1948)..
3. E. Prange, Cyclic error-correcting codes in two symbols. *Air Force Cambridge Research Center*, AFCRC-TN-57–103, Cambridge, MA, Sept. (1957).

4. A. Hocquenghem, Codes correcteurs d'Erreurs. Chiffres, **2**, 147–156, (1959).
5. R.C. Bose, D.K. Ray–Chaudhuri, On a class of error correcting binary group codes. Inf Control **3**, 78–79, March (1960).
6. I.S. Reed, G. Solomon, Polynomial codes over certain finite fields. SIAM J. Appl. Math., **8**, 300–304, June (1960).
7. D.E. Müller, Application of Boolean algebra to switching circuit design and to error detection. IRE Trans. Electron. Computers, **EC-3**(3) 6–12, Sept. (1954).
8. P. Elias, Coding for noisy channels. IRE Convention Record, **3**(part 4) 37–46 (1955).
9. E.R. Berlekamp, Binary BCH codes for correcting multiple errors. Algebraic Coding Theory, (McGraw-Hill, New York, 1968).
10. J.L. Massey, Shift-register synthesis and BCH decoding. IEEE Trans. Inform. Theory, **IT-15**(1) 122–127, Jan. (1969).
11. Y. Sugiyama, M. Kasahara, S. Hirasawa, T. Namekawa, A method for solving key equation for decoding Goppa codes. Inform. Control **27**(1) 87–99, Jan. (1975).
12. A.J. Viterbi, Error bounds for convolutional codes and an asymptotically optimum decoding algorithm. IEEE Trans. Inform. Theory **IT-13**(2) 260–269, Apr. (1967).
13. L.H. Charles Lee, *Error-control block codes for communications engineers*, (Artech House Publishers, New York, 2000).
14. P.A. Humblet, M. Azizoglu, On the bit error rate of lightwave systems with optical amplifiers. IEEE J. Lightwave Technol. **9**(11) 1576–1582, Nov. (1991).
15. J. Zweck, I.T. Lima Jr., Y. Sun, A.O. Lima, C.R. Menyuk, G. Carter, Modeling receivers in optical communication systems with polarization effects. OSA Opt. Photon. News, **14**(11) 31–35, Nov. (2003).
16. W.D. Grover, Forward error correction in dispersion–limited lightwave systems. IEEE J. Lightwave Technol. **6**(5) 643–645, May (1988).
17. P. Moro, D. Candiani, 565 Mb/s optical transmission system for repeaterless sections up to 200 km. in Proceedings of IEEE Int. Conf. on Commun. (ICC'91), Chicago, IL, June (1991) pp. 1217–1221,.
18. P.M. Gabla, J.L. Pamart, R. Uhel, E. Leclerc, J.O. Frorud, F.X. Ollivier, S. Borderieux, 401 km, 622 Mb/s and 357 km, 2.488 Gb/s IM/DD repeaterless transmission experiments using erbium-doped fiber amplifiers and error correcting code. IEEE Photon. Technol. Lett. **4**(10), 1148–1150, Oct. (1992).
19. S. Yamamoto, H. Taga, N. Edagawa, H. Wakabayashi, Observation of BER degradation due to fading in long distance optical amplifier system. IEE Electron. Lett. **29**(2), 209–210, Jan. (1993).
20. S. Yamamoto, H. Takahira, M. Tanaka, 5 Gbit/s optical transmission terminal equipment using forward error correcting code and optical amplifier. IEE Electron. Lett. **30**(3), 254–255, Feb. (1994).
21. J.L. Pamart, E. Lefranc, S. Morin, G. Balland, Y.C. Chen, T.M. Kissell, J.L. Miller, Forward error correction in a 5 Gbit/s 6400 km EDFA based system. IEE Electron. Lett. **30**(4) 342–343, Feb. (1994).
22. M. Tomizawa, Y. Yamabayshi, K. Murata, T. Ono, K. Kobayashi, K. Hagimoto, Forward error correcting codes in synchronous fiber optic transmission systems. IEEE J. Lightwave Technol. **15**(1) 43–52, Jan. (1997).
23. Y. Sugiyama, M. Kasahara, S. Hirasawa, T. Namekawa, *A method for solving key equation for decoding Goppa codes*. ed. by M. Eden. Information and Control, vol 27. (Academic Press, New York and London, 1975), pp. 87–99.
24. W.C. Barnett, H. Takahira, J.C. Baroni, Y. Ogi, The TPC-5 cable network. IEEE Commun. Mag. **34**(2) 36–40, Feb. (1996).
25. P. Trischitta, M. Colas, M. Green, G. Wuzniak, J. Arena, The TAT-12/13 cable network. IEEE Commun. Mag. **34**(2) 24–28, Feb. (1996).
26. G.D. Forney, *Concatenated Codes*, (The MIT Press, Cambridge, MA, 1966).
27. O. Ait Sab, J. Fang, Concatenated forward error correction schemes for long-haul DWDM optical transmission systems. in Proceedings of ECOC'99, II-290, Nice, France, Sept. (1999).

28. J. Yoshimura, K. Suzuki, T. Ohsumi, 10 Gb/s × 105 DWDM SLTE with super FEC transponder. in Proceedings of SubOptic 2001, T.4.4.4, Kyoto, Japan, May (2001).
29. K. Seki, K. Mikami, M. Baba, A. Katayama, H. Tanaka, Y. Hara, M. Kobayashi, N. Okada, Single-chip FEC codec LSI using iterative CSOC decoder for 10 Gb/s long-haul optical transmission systems. in Proceedings of IEEE Custom Integrated Circuits Conf. (CICC2002), 9–3, 155–158, Orlando, FL, May (2002).
30. G. Vareille, O. Ait Sab, G. Bassier, J.P. Collet, B. Julien, D. Dufournet, F. Pitel, J.F. Marcerou, 1.5 terabit/s submarine 4000 km system validation over a deployed line with industrial margins using 25 GHz channel spacing and NRZ format over NZDSF. in Proceedings of OFC2002, WP5, Anaheim, CA, March (2002).
31. C. Rasmussen, S. Dey, F. Liu, J. Bennike, B. Mikkelsen, P. Mamyshev, M. Kimmitt, K. Springer, D. Gapontsev, V. Ivshin, "Transmission of 40 × 42.7 Gbit/s over 5200 km UltraWave® fiber with terrestrial 100 km spans using turn-key ETDM transmitter and receiver. in Proceedings of ECOC2002, PD4.3, Copenhagen, Denmark, Sept. (2002).
32. J.-X. Cai, M. Nissov, A.N. Pilipetskii, A.J. Lucero, C.R. Davidson, D. Foursa, H. Kidorf, M.A. Mills, R. Menges, P.C. Corbett, D. Sutton, N.S. Bergano, 2.4 Tb/s (120 × 20 Gb/s) transmission over transoceanic distance using optimum FEC overhead and 48% spectral efficiency. in Proceedings of OFC2001, PD20, Anaheim, CA, March (2001).
33. Y. Kisaka, S. Aisawa, M. Tomizawa, Y. Miyamoto, K. Terada, N. Iwasaki, A. Sano, H. Masuda, M. Koga, Fully transparent multiplexing and transport of 10 GbE-LANPHY signals in 44.6-Gbit/s-based RZ-DQPSK WDM transmission. in Proceedings of OFC/NFOEC 2007, OThL1, Anaheim, CA, (2007).
34. T. Mizuochi, K. Kubo, M. Akita, M. Imai, S. Kurahashi, N. Takemura, A. Hamaoka, Y. Osaki, K. Shimokasa, K. Kuno, H. Yoshida, T. Kitayama, Transparent multiplexer featuring super FEC for optical transport networking. in Proceedings of SubOptic2001, P.4.2.3, Kyoto, June, (2001).
35. Y. Yamada, S. Nakagawa, T. Kawazawa, H. Taga, K. Goto, 2 Tbit/s (200 × 10 Gbit/s) over 9200 km transmission experiments using C-band EDFA and VSB format with 53% spectral efficiency. in Proceedings of SubOptic2001, PDP-1, Kyoto, Japan, May (2001).
36. T. Mizuochi, K. Ishida, K. Kinjo, T. Kobayashi, S. Kajiya, K. Shimizu, T. Tokura, K. Motoshima, K. Kasahara, 1.7 Tb/s (85 × 22.8 Gb/s) transmission over 9,180 km using symmetrically collided transmission methodology. IEE Electron. Lett., **38**(21) 1264–1265, (2002).
37. T. Mizuochi, Y. Miyata, T. Kobayashi, K. Ouchi, K. Kuno, K. Kubo, K. Shimizu, H. Tagami, H. Yoshida, H. Fujita, M. Akita, K. Motoshima, Forward error correction based on block turbo code with 3–bit soft decision for 10 Gb/s optical communication systems. IEEE J. Sel. Topics Quant. Electron. Opt. Commun. **10**(2) 376–386, March/April (2004).
38. A. Puc, F. Kerfoot, A. Simons, D.L. Wilson, Concatenated FEC experiment over 5000 km long straight line WDM test bed. in Proceedings of OFC'99, ThQ6, San Diego, CA, Feb. (1999).
39. C. Berrou, A. Glavieux, P. Thitimajshima, Near Shannon limit error-correcting coding and decoding: Turbo-codes (1). in Proceedings of Int. Conf. on Commun. (ICC'93), 1064–1071, Geneva, May (1993).
40. R. Pyndiah, A. Glavieux, A. Picart, S. Jacq, Near optimum decoding of products codes. in Proceedings of IEEE GLOBCOM'94, 339–343, San Francisco, CA, Dec. (1994).
41. R. Pyndiah, Near-optimum decoding of product codes: block turbo codes. IEEE Trans. Commun. **46**(8) 1003–1010, Aug. (1998).
42. O. Ait Sab, V. Lemaire, Block turbo code performances for long-haul DWDM optical transmission systems. in Proceedings of OFC2000, ThS5, Baltimore, MD, March (2000).
43. M. Akita, H. Fujita, T. Mizuochi, K. Kubo, H. Yoshida, K. Kuno, S. Kurahashi, Third generation FEC employing Turbo Product Code for long-haul DWDM transmission systems. in Proceedings of OFC2002, WP2, Anaheim, CA, March (2002).
44. T. Mizuochi, K. Ouchi, T. Kobayashi, Y. Miyata, K. Kuno, H. Tagami, K. Kubo, H. Yoshida, M. Akita, K. Motoshima, Experimental demonstration of net coding gain of 10.1 dB using 12.4 Gb/s block Turbo code with 3-bit soft decision. in Proceedings of OFC2003, PD21, Atlanta, GA, March (2003).

45. H. Tagami, T. Kobayashi, Y. Miyata, K. Ouchi, K. Sawada, K. Kubo, K. Kuno, H. Yoshida, K. Shimizu, T. Mizuochi, K. Motoshima. A 3–bit soft–decision IC for powerful forward error correction in 10–Gb/s optical communication systems," IEEE J. Solid–State Circuits, **40**, 8, 1695–1705, Aug. (2005).
46. K. Ouchi, K. Kubo, T. Mizuochi, Y. Miyata, H. Yoshida, H. Tagami, K. Shimizu, T. Kobayashi, K. Shimomura, K. Onohara, K. Motoshima, A fully integrated block turbo code FEC for 10 Gb/s optical communication systems. in Proceedings of OFC/NFOEC2006, OTuK4, Anaheim, March (2006).
47. R.G. Gallager, Low density parity check codes. IRE Trans. Inform. Theory, **IT-8**(21–28), Jan. (1962).
48. D. MacKay, Good error correcting codes based on very sparse matrices. IEEE Trans. Inform. Theory **45**(2) 399–431, March (1999).
49. I.B. Djordjevic, S. Sankaranarayanan, B. Vasic, Irregular low-density parity-check codes for long-haul optical communications. IEEE Photon. Technol. Lett, **16**(1), 338–400 (2004).
50. I.B. Djordjevic, O. Milenkovic, B. Vasic, Generalized low-density parity-check codes for optical communication systems. IEEE J. Lightwave Technol. **23**(5) 1939–1946 (2005).
51. I.B. Djordjevic, B. Vasic, Iteratively decodable codes from orthogonal arrays for optical communication systems. IEEE Commun. Lett. **9**(10) 924–926 (2005).
52. J. Zhang, M.P.C. Fossorier, Shuffled iterative decoding. IEEE Trans. Commun. **53**(2), 209–213, Feb. (2005).
53. Y. Miyata, R. Sakai, W Matsumoto, H. Yoshida, T. Mizuochi, Reduced-complexity decoding algorithm for LDPC codes for practical circuit implementation in optical communications. in Proceedings of OFC/NFOEC2007, OWE5, Anaheim, CA, March (2007).
54. Y. Miyata, W. Matsumoto, H. Yoshida, T. Mizuochi, Efficient FEC for optical communications using concatenated codes to combat error-floor. in Proceedings of OFC/NFOEC2008, OTuE4, San Diego, CA, Feb. (2008).
55. J.G. Proakis, *Digital communications*, 4th edn. (McGraw-Hill, Electrical Engineering Series, 2000).
56. Y. Cai, N. Ramanujam, J.M. Morris, T. Adali, G. Lenner, A.B. Puc, A. Pilipetskii, Performance limit of forward error correction codes in optical fiber communications. in Proceedings of OFC2001, paper TuF2, Anaheim, CA, March (2001).
57. T. Yasuda, E. Shibano, T. Inoue, M. Ichikawa, A. Okayasu, N. Yoshikane, H. Taga, K. Goto, T. Mizuochi, Y. Osaki, K. Shimokasa, T. Kitayama, DWDM terminal equipment for OSW submarine cable system. in Proceedings of SubOptic2001, T.4.4.2, Kyoto, Japan, May (2001).
58. K. Onohara, K. Ouchi, T. Mizuochi, Automatic tracking of 3-bit soft decision thresholds in Block Turbo Code based forward error correction. in Proceedings of ECOC2007, We1.5.6, Cannes, Sept. (2006).
59. T. Sugihara, K. Shimomura, K. Shimizu, Y. Kobayashi, K. Matsuoka, M. Hashimoto, T. Hashimoto, T. Hirai, S. Matsumoto, T. Ohira, M. Takabayashi, K. Yoshiara, T. Mizuochi, Automatically tracked dispersion compensation with penalty–free tunable dispersion equalizer for 40 Gbit/s systems. in Proceedings of OFC2002, ThAA2, Anaheim, CA, March (2002).
60. K. Ishida, T. Mizuochi, T. Sugihara, Demonstration of PMD mitigation in long–haul WDM transmission using automatic control of input state of polarization. in Proceedings of ECOC2002, 11.1.1, Copenhagen, Denmark, Sept. (2002).

Index

M. Nakazawa et al. (eds.), *High Spectral Density Optical Communication Technologies*, Optical and Fiber Communications Reports 6, DOI 10.1007/978-3-642-10419-0,

MIX
Papier aus verantwortungsvollen Quellen
Paper from responsible sources
FSC® C105338

Printed by Books on Demand, Germany